Hypericum

Medicinal and Aromatic Plants – Industrial Profiles

Individual volumes in this series provide both industry and academia with in-depth coverage of one major genus of industrial importance.

Edited by Dr Roland Hardman

Hypericum

The genus *Hypericum*

Edited by
Edzard Ernst

Taylor & Francis
Taylor & Francis Group

LONDON AND NEW YORK

First published 2003
by Taylor & Francis
11 New Fetter Lane, London EC4P 4EE

Simultaneously published in the USA and Canada
by Taylor & Francis Inc,
29 West 35th Street, New York, NY 10001

Taylor and Francis is an imprint of the Taylor & Francis Group

© 2003 Taylor & Francis

Typeset in Garamond by
Newgen Imaging Systems (P) Ltd, Chennai, India
Printed and bound in Great Britain by
T.J. International Ltd, Padstow, Cornwall

Every effort has been made to ensure that the advice and information in this book
is true and accurate at the time of going to press. However, neither the publisher
nor the authors can accept any legal responsibility or liability for any errors or
omissions that may be made. In the case of drug administration, any medical
procedure or the use of technical equipment mentioned within this book,
you are strongly advised to consult the manufacturer's guidelines.

British Library Cataloguing in Publication Data
A catalogue record for this book is available from the British Library

Library of Congress Cataloging in Publication Data
A catalog record for this book has been requested

ISBN 0–415–36954–1

Contents

List of contributors

S.K. Bhattacharya
Department of Pharmaceutics
Institute of Technology
Banaras Hindu University
Varanasi 221 004
India

Eva Čellárová
PJ Safárik University
Department of Experimental Botany
  and Genetics
Faculty of Science
Mánesova 23
04167 Kosice
Slovakia

William E. Court
1 Maes Yr Haul
Mold
Clwyd CH7 1NS

N. Debrunner
Valplantes, CP18
CH-1933 Sembrancher
Switzerland

Alberto C.P. Dias
Department of Biology
University of Minho
Campus de Gualtar
4710-057 Braga Codex
Portugal

E. Ernst
Department of Complementary Medicine
School of Postgraduate Medicine and Health
  Sciences

University of Exeter
25 Victoria Park Road
Exeter
Devon EX2 4NT, UK

Frauke Gaedcke
H. Finzelberg's Nachf. GmbH & Co KG
Koblenzer Str. 48–56
D 56626 Andernach
Germany

M. Gaudin
Médiplant
Station Fédérale de Recherches en
  Production Végétale de Changins
CH-1964 Conthey
Switzerland

Josef Hölzl
Philipps-Universität Marburg
Fachbereich Pharmazie und
  Lebensmittelchemie
Institut für Pharmazeutische Biologie
Deutschheusstrasse 17 1/2
D 35032 Marburg
Germany

A.A. Izzo
Department of Experimental
  Pharmacology
University of Naples Federico II
Via D. Montesano 49
80131 Naples
Italy

Hartmut Kegler
Bäckerstieg 11
D-06449 Aschersleben
Germany

Vikas Kumar
Division of Pharmacology and Toxicology
Indian Herbs Research & Supply Co. Ltd.
Saharanpur 247 001, U. P.
India

Werner Likussar
Institute of Pharmaceutical Chemistry
University of Graz
Schubertstrasse 1
A-8010 Graz
Austria

Beat Meier
Research and Development
Max Zeller und Söhne AG
Seeblickstraße 4
CH-8590 Romanshorn 1
Switzerland

Astrid Michelitsch
Institute of Pharmaceutical Chemistry
University of Graz
Schubertstrasse 1
A-8010 Graz
Austria

Maike Petersen
Philipps-Universität Marburg
Fachbereich Pharmazie und
 Lebensmittelchemie
Institut für Pharmazeutische Biologie
Deutschheusstrasse 17 1/2
D 35032 Marburg
Germany

Norman K.B. Robson
The Natural History Museum
Cromwell Road
London, SW7 5BD, UK

Manfred Schubert-Zsilavecz
Institute of Pharmaceutical Chemistry
University of Frankfurt
Germany

Katarzyna Seidler-Łożykowska
Instytut Roslin i Przetworow Zielarskich
ul. Libelta 27
61-707 Poznań
Poland

X. Simonnet
Médiplant
Station Fédérale de Recherches en
 Production Végétale de Changins
CH-1964 Conthey
Switzerland

P.N. Singh
Department of Pharmaceutics
Institute of Technology
Banaras Hindu University
Varanasi 221 004
India

C. Stevinson
Department of Complementary Medicine
School of Postgraduate Medicine and
 Health Sciences
University of Exeter
25 Victoria Park Road
Exeter
Devon EX2 4NT, UK

Mario Wurglics
Institute of Pharmaceutical Chemistry
University of Frankfurt
Germany

Preface to the series

There is increasing interest in industry, academia and the health sciences in medicinal and aromatic plants. In passing from plant production to the eventual product used by the public, many sciences are involved. This series brings together information which is currently scattered through an ever increasing number of journals. Each volume gives an in-depth look at one plant genus, about which an area specialist has assembled information ranging from the production of the plant to market trends and quality control.

Many industries are involved such as forestry, agriculture, chemical, food, flavour, beverage, pharmaceutical, cosmetic and fragrance. The plant raw materials are roots, rhizomes, bulbs, leaves, stems, barks, wood, flowers, fruits and seeds. These yield gums, resins, essential (volatile) oils, fixed oils, waxes, juices, extracts and spices for medicinal and aromatic purposes. All these commodities are traded worldwide. A dealer's market report for an item may say 'Drought in the country of origin has forced up prices'.

Natural products do not mean safe products and account of this has to be taken by the above industries, which are subject to regulation. For example, a number of plants which are approved for use in medicine must not be used in cosmetic products.

The assessment of safe to use starts with the harvested plant material which has to comply with an official monograph. This may require absence of, or prescribed limits of, radioactive material, heavy metals, aflatoxin, pesticide residue, as well as the required level of active principle. This analytical control is costly and tends to exclude small batches of plant material. Large scale contracted mechanised cultivation with designated seed or plantlets is now preferable.

Today, plant selection is not only for the yield of active principle, but for the plant's ability to overcome disease, climatic stress and the hazards caused by mankind. Such methods as *in vitro* fertilization, meristem cultures and somatic embryogenesis are used. The transfer of sections of DNA is giving rise to controversy in the case of some end-uses of the plant material.

Some suppliers of plant raw material are now able to certify that they are supplying organically-farmed medicinal plants, herbs and spices. The European Union directive (CVO/EU No. 2092/91) details the specifications for the *obligatory* quality controls to be carried out at all stages of production and processing of organic products.

Fascinating plant folklore and ethnopharmacology leads to medicinal potential. Examples are the muscle relaxants based on the arrow poison, curare, from species of *Chondrodendron*, and the anti-malarials derived from species of *Cinchona* and *Artemisia*. The methods of detection of pharmacological activity have become increasingly reliable and specific, frequently involving enzymes in bioassays and avoiding the use of laboratory animals. By using bioassay linked fractionation of crude plant juices or extracts, compounds can be specifically targeted which, for example, inhibit blood platelet aggregation, or have anti-tumour, or anti-viral, or any other required activity. With the assistance of robotic devices, all the members of a genus may be readily screened. However, the plant material must *be fully* authenticated by a specialist.

The medicinal traditions of ancient civilisations such as those of China and India have a large armamentaria of plants in their pharmacopoeias which are used throughout South-East Asia. A similar situation exists in Africa and South America. Thus, a very high percentage of the World's population relies on medicinal and aromatic plants for their medicine. Western medicine is also responding. Already in Germany all medical practitioners have to pass an examination in phytotherapy before being allowed to practise. It is noticeable that throughout Europe and the USA, medical, pharmacy and health related schools are increasingly offering training in phytotherapy.

Multinational pharmaceutical companies have become less enamoured of the single compound magic bullet cure. The high costs of such ventures and the endless competition from 'me too' compounds from rival companies often discourage the attempt. Independent phytomedicine companies have been very strong in Germany. However, by the end of 1995, eleven (almost all) had been acquired by the multinational pharmaceutical firms, acknowledging the lay public's growing demand for phytomedicines in the Western World.

The business of dietary supplements in the Western World has expanded from the health store to the pharmacy. Alternative medicine includes plant-based products. Appropriate measures to ensure the quality, safety and efficacy of these either already exist or are being answered by greater legislative control by such bodies as the Food and Drug Administration of the USA and the recently created European Agency for the Evaluation of Medicinal Products, based in London.

In the USA, the Dietary Supplement and Health Education Act of 1994 recognised the class of phytotherapeutic agents derived from medicinal and aromatic plants. Furthermore, under public pressure, the US Congress set up an Office of Alternative Medicine and this office in 1994 assisted the filing of several Investigational New Drug (IND) applications, required for clinical trials of some Chinese herbal preparations. The significance of these applications was that each Chinese preparation involved several plants and yet was handled as a *single* IND. A demonstration of the contribution to efficacy, of *each* ingredient of *each* plant, was not required. This was a major step forward towards more sensible regulations in regard to phytomedicines.

My thanks are due to the staffs of Harwood Academic Publishers and Taylor & Francis who have made this series possible and especially to the volume editors and their chapter contributors for the authoritative information.

Roland Hardman, 1997

Preface

St John's wort (*Hypericum perforatum*) is one of the best-selling herbal medicines worldwide. In the US alone, the annual sales figure is around US$200 million. It is, therefore, understandable that research into all aspects of St John's wort continues to be intense. This book provides a summary of our current knowledge on a wide range of issues. It covers botany and includes plant infections, cultivation, manufacturing, standardisation, quality control, biochemistry, pharmacology and clinical application.

Within this wide spectrum of topics, numerous significant advances have been made. Yet many questions remain insufficiently answered. Which are the pharmacologically active compounds? What is the best way for standardisation of *Hypericum* products? How relevant are herb–drug interactions? Is St John's wort effective for severe depression or for any other conditions for which trial data are scarce? What are its mechanisms of action? What are the long-term effects and risks? Should St John's wort products be marketed as dietary supplements or as drugs? How does it compare to synthetic drugs for the same indication? And these are just the questions that go through my mind as a physician! Obviously biochemists, botanists, etc. would want to add many more.

The recent history of St John's wort is revealing in more than one way. It tells us, for instance, that investments into research can be well worth it. The unprecedented boom in sales must be viewed in direct relation to the fact that the evidence for clinical effectiveness had reached a level where even sceptics could no longer remain in denial. Those who had invested in research received high rewards (and many who had not, did very nicely too). Even more, recent history shows how fragile the markets still are for herbal medicinal products. As soon as the news about the possibility of herb–drug interactions appeared in the mass media, the sales figures fell sharply. No doubt they will recover once we can show convincing data demonstrating that such risks are controllable and minor compared to those of competing drugs.

Despite a number of problems, the 'St John's wort story' is by and large a success story. It shows that the best way to success is via rigorous research. This volume is a significant landmark on the way this research has taken us. We have come far but we still have a long way to go until the majority of answers to important open questions have been found. I hope that this book provides a good basis for the work to come.

Edzard Ernst

1 *Hypericum* botany

Norman K.B. Robson

Introduction

The genus Hypericum

Hypericum is a genus of about 450 species of trees, shrubs and herbs that occurs in all temperate parts of the world but has only one species in southern South America and two species in Australia and New Zealand. It is absent from habitats that are extremely dry, hot or cold, is very rarely found in water that is other than very shallow, and in the tropics is almost always confined to high elevations. Most species of *Hypericum* can be recognised by (i) opposite simple entire exstipulate leaves containing translucent and often black or red glandular secretions, (ii) flowers with a 5-merous perianth comprising green (sometimes red-tinged) sepals and free yellow (often red-tinged) petals, stamens in 3–5 bundles or fascicles and an ovary with 3–5 slender styles (free or ± united) and (iii) a capsular fruit containing many small cylindrical seeds. As will become apparent, there are exceptions to all those character generalisations.

Hypericum belongs to the family Clusiaceae (alternative name Guttiferae) and shares with all or most of the other genera of that family opposite simple entire exstipulate leaves, the presence of glandular secretions, free petals, fascicles of stamens and seeds lacking endosperm.

In the following account, evidence for unsupported statements and references will be found in the author's as yet incomplete monograph of *Hypericum* (Robson 1977, 1981, 1985, 1987, 1990, 1996, 2001).

The name Hypericum

Despite the attempts of some lexicographers to derive it from *hypo-* or *hyper-ericum* (beneath or above the heath), the meaning and derivation of *Hypericum* is quite clear. The name ὑπέρεικον was given by the ancient Greeks to a plant or plants that they hung above their religious figures to ward off evil spirits (ὑπέρ – above, εἰκών – image). Exactly which species was so used is not known with certainty, although various authors have suggested that it was *H. empetrifolium* Willd. or *H. triquetrifolium* Turra (*H. crispum* L.). The earliest use of the name that has been traced so far is in the second century BC by Nikander (Alexipharmaca V, line 603):

τῷ δ᾽ ὁτὲ μὲν σμύρνης ὁδελοῦ πόρε διπλόον ἄχθος,
   ἄλλοτε δ᾽ ὁρμίνοιο νέην χύσιν, ἄλλοτε κόψαις
   οὐρείην ὑπερέικον, ὅθ᾽ ὑσσώπον ὁροδάμνους.

'And take a double 12-grain dose of myrrh, or a fresh draught of horminium, or pounded mountain hypericum or branches of hyssop'.

The qualification 'mountain' here suggests that more than one species of *Hypericum* was recognised at that time.

The name was mentioned by Dioscorides (Mat. Med., i. 3 cap. 171), Galen (12, 148) and Pliny (26. s. 53) and the illustration of *Uperikon* in the *Codex Aniciae Iulianae nunc Vindobonensis*: 357r, representing the traditional use of the name, clearly portrays *H. empetrifolium*. Dioscorides also described *Akuron* (M. M. 3, 172) – possibly *H. triquetrifolium* or *H. perforatum* L. – and *Androsaimon* (M. M. 3, 173). The last name appears to have been given to any species that had red sap or glandular secretion that stained the fingers like blood (ἀνήρ, ἀνδρός – man, 'αἷμα – blood), but it has become associated particularly with one species, *H. androsaemum* L., which has black fleshy capsules with red sap.

The power to ward off evil spirits was especially important at times when such spirits were believed to be most abundant, for example, on Halloween (31 October) and Midsummer's Eve (23 June), and *Hypericum* was picked on the latter day to decorate religious images. The pagan feast celebrated on Midsummer's Day was eventually Christianised and dedicated to St John the Baptist, whose birthday was the 24th of June; and the plant used on that day became St John's wort (*Johanniskraut, Erba di San Giovanni*, etc.). In regions away from the Aegean, that is, where *H. empetrifolium* did not occur as a native, other species of *Hypericum* were used for decoration, particularly *H. perforatum*, the commonest species elsewhere in Europe. Hence a medieval name for this plant was *fuga daemonum* ('flight of the demons' or, more loosely, 'make the demons flee'). It is now frequently known merely as St John's wort but is more correctly Common St John's wort.

Hypericum has thus been associated with pharmacy and folklore for many centuries; so its recent 'discovery' by western medicine is not surprising, though it may be regarded as rather belated.

Hypericum – variation and classification

Characters and trends

Morphology

The above initial short description of typical characters gives little impression of the range of variation in *Hypericum*, which is shortly elaborated below (see Colour Plate I–VIII). For sections of the genus, see Tables 1.1–1.3.

Habit   This ranges from the arboreal to the ephemeral. At one extreme, the east African *H. bequaertii* and relatives (sect. *Campylosporus*) can form trees up to 12 m high, with a true trunk. At the other, one finds, for example, *H. gentianoides* (sect. *Brathys*), an annual herb from eastern North America with leaves reduced to scales. True trees, however, are rare, most of the woody species being many-stemmed. The lower stems of these species may arise from below ground, but, with the conspicuous exception of *H. calycinum*, they do not normally root. Many of the perennial herbs, however, spread by means of runners that root before becoming erect or ascending shoots; and at least two of them, *H. perforatum* and *H. pulchrum*, regularly produce buds from the roots. On the other hand, the annual species do not root at lower nodes.

Vestiture   Simple uniseriate hairs are found on stems, leaves and sepals in various parts of the genus, especially in sects *Hirtella* and *Adenosepalum*.

Stems   The young stems of woody species and mature stems of herbs often have prominent lines decurrent from the nodes (Figure 1.9). Those decurrent from the midrib have been termed 'principal lines' and those from between the leaves 'subsidiary lines'. Both pairs of lines are present in the more primitive species, and there are various evolutionary trends in the genus

Figure 1.1 *H. addingtonii* (sect. 3 *Ascyreia*) – ex China, cult. England; note five stamen fascicles (photograph by E. Robson). (See Colour Plate I.)

Figure 1.2 *H. orientale* (sect. 16 *Crossophyllum*) – ex N. Turkey, cult. England; note gland-fringed leaves (photograph by E. Robson). (See Colour Plate III.)

towards elimination of one and then the other pair (i.e. from 4-lined to 2-lined then terete). The subsidiary lines are the usual ones to be eliminated first; but in one section (sect. *Androsaemum*) it is the principal ones that are absent. In some species of the N. American sect. *Myriandra*, the subsidiary lines split into two, making the stem 6-lined; and where the leaves are in whorls of three or four (e.g. in *H. coris* and other species of sect. *Coridium*), the two lines on stems of related species are replaced by three or four.

Leaves and glands Although the vast majority of species have opposite decussate pairs of entire leaves, (i) three- to four-leaved whorls occur sporadically as aberrations and have become

Figure 1.3 H. capitatum (sect. 17 *Hirtellum*) – ex S. Turkey, cult. New Zealand; note red flowers (photograph by A.R. Mitchell). (See Colour Plate IV.)

Figure 1.4 H. frondosum (sect. 20 *Myriandra*) – SE United States, cult. England; note stamens in a wide ring (photograph by E. Robson). (See Colour Plate V.)

'fixed' in sect. *Coridium* and in parts of two S. American species of sect. *Trigynobrathys* and (ii) the leaves in the two species of sect. *Crossophyllum* (*H. orientale, H. adenotrichum*) and the related *H. thasium* (sect. *Thasia*) have paired basal auriculate structures. In the former two species the leaf margin is gland-fringed, an extreme case of prominence of the marginal row of glands (see below) that also occurs in sect. *Hypericum* and less markedly in sect. *Adenosepalum*. Prominent glands are frequent on the margin of bracts and sepals, less so on the margin of petals. In sects *Hypericum* and *Adenosepalum* the bracts of some species have paired basal clusters of stalked glands (glandular auricles).

Trends in leaf venation in *Hypericum* run from parallel with the occasional dichotomy, in *H. bequaertii* (sect. *Campylosporus*), to increasingly dense reticulation; and there is a parallel variation in the leaf glands from lines to dots. These are of two types: pale (pellucid or translucent) and dark (black or red). The pale glands are schizogenous, that is, intercellular spaces lined with

Figure 1.5 *H. montanum* (sect. 27 *Adenosepalum*) – England; note gland-fringed sepals (photograph by E. Robson). (See Colour Plate VI.)

Figure 1.6 *H. annulatum* (sect. 27 *Adenosepalum*) – ex Balkan Peninsula, cult. England (photograph by E. Robson). (See Colour Plate II.)

cells that secrete an essential oil, whereas the dark glands are solid groups of cells that contain wax impregnated with dark red hypericin (a naphtho-dianthrone related to emodin) or occasionally the chemically almost identical pseudohypericin. It is hypericin that stains the fingers with 'man's blood' and photosensitises the unpigmented areas of the skin of animals that ingest

Figure 1.7 H. elodes (sect. 28 *Elodes*) – England (photograph by E. Robson). (See Colour Plate VII.)

Figure 1.8 H. irazuenuse (sect. 29 *Brathys*) – Costa Rica (photograph by C.J. Humphries). (See Colour Plate VIII.)

Hypericum (hypericism) (see e.g. Roth 1990). Its presence, distribution within the plant and strength (black or red), as revealed by the presence and distribution of dark glands, has proved to be very useful in classifying the genus.

Pale glands are also of two types, which have had quite different evolutionary origins:

(i) In *H. bequaertii* the parallel veins are interspersed with weaker dichotomising vein-like glands that may well have originated from previously existing veins. As cross veins and then increasingly denser tertiary reticulate veins evolved, these vein-like glands were apparently cut into smaller and smaller pieces, eventually becoming dots in the areolae of the vein reticulum (Figure 1.10).

(ii) Small dot glands occur between the veins and glands of type (i) in the primitive species of the African sect. *Campylosporus* and on the ventral leaf surface of some species in the closely related Asian sect. *Ascyreia*. They are found elsewhere in the genus (immersed in the lamina) only in sects *Brathys* and *Trigynobrathys*, the sections in which glands of type (i) are absent, and appear to be homologous with dot glands in related genera, for example, *Santomasia* and *Vismia*.

Dark glands are mostly black, but where the amount of hypericin is small they are red (e.g. in the sepals of *H. elodes*). Where hypericin is completely absent, the wax in the cell mass gives the gland an amber colour. In addition, dark glands often occur in positions in the plant (in the stem, leaf, sepal, petal or anther) that are occupied in closely related species by pale glands. The glands at or near the margin of leaves, sepals and petals (i.e. marginal, submarginal and inframarginal) seem to be of the first type (black, red or amber) and those elsewhere in the blade (i.e. laminar) of the second (pale evolving into black).

Table 1.1 Distribution of hypericin and pseudo-hypericin in sections of *Hypericum*

Section		No. of species	Stem	Leaf	Sepal	Petal	Anther	Capsule
1	Campylosporus	11	−/+	−/+ +−	−/+	−/+	−	−
2	Psorophytum	1	−	−	−	−	−	−
3	Ascyreia	44	−	−	−	−	−	−
4	Takasagoya	5	−	−	−	−	−	−
5	Androsaemum[a]	4	−	−	−	−	−	−
6	Inodora	1	−	−	−	−	−	−
6a	Umbraculoides	1	−	−	−	−	−	−
7	Roscyna	2	−	−	−	−	−	−
8	Bupleuroides	1	−	−	−/+	+	−	−
9	Hypericum	52	−/+	+	−/+	+	−	
9a	Concinna	1	−	+	−/+	−/+		
10	Olympia	2	−	−/+	−/+	−/+	−	−
11	Campylopus	1	−	−/+	−/+	+	−	−
12	Origanifolia	10	+(−)	+(−)	+(−)	+(−)	+	−
13	Drosocarpium	11	−(+)	+	+	+	+	−(+)
14	Oligostema	5	−(+)	+	+	+/−	+	
15	Thasia	2	+	+	−/+	−/+	+	
16	Crossophyllum[b]	2	−	+/−	+/−	+/−	−	−
17	Hirtella	24	−(+)	−(+)	(−)+	+	−	−
18	Taeniocarpium	24	−	−(+)	(−)+	+/−	−	−
19	Coridium	6	−	−	+(−)	−(+)	−	−
20	Myriandra	29	−	−	−	−	−	−
21	Webbia	1	−	−	−	−	−	−
22	Arthrophyllum	5	−	−	−/+	−	−	−
23	Triadenioides	5	−(+)	−(+)	−(+)	−(+)	−(+)	−
24	Heterophylla	1	−	−	−	−	−	−
25	Adenotrias	3	−	−	−	−	−	−
26	Humifusoideum	10	−(+)	−/+	−/+	−/+	−/+	−
27	Adenosepalum	24	−/+	+(−)	+(−)	+(−)	+(−)	−
28	Elodes	1	−	−	+	−	−	−
29	Brathys	88	−	−	−	−	−	−
30	Trigynobrathys	52	−	−	−	−	−	−

Notes

a The fruits of *H. androsaemum* are black when ripe and have red juice.

b The dark glands in *H. adenotrichum* are all black, whereas in *H. orientale* they are all amber.

 −/+ = pale or dark; −(+) = pale or rarely dark; etc.

Table 1.2 Classification of *Hypericum* (Robson 1977 onward); some characters, excluding hypericin distribution

Section		Habit	Leaves united	Petal no./dec	Fascs no./dec	Fascs united	Styles no./union	Capsule vittae	Chromosome no./ploidy
1	Campylosporus	T/S	−	5/−	5/−	−	5/± a	−	12/2
2	Psorophytum	S	−	5/+	5/+	−	5(4)/f	+	12/2
3	Ascyreia	S(DS)	−	5/+	5/+	−	5/f(± u)	−	12 − 9/2,4,6
4	Takasagoya	S	−	5/+	5/+	−	5/u	−	?
5	Androsaemum	S	−	5/+	5/+	−	3/f	−	10/4
6	Inodora	S	−	5/−	5/−	−	3/f	−	10/4
6a	Umbraculoides	S	−	5/−	5/−	−	3/f	?	?
7	Roscyna	H	−	5/−	5/−	−	5/± u − f	+	9 − 8/2
8	Bupleuroides	H	+(−)	5/−	5/−	'3'('4−')	3/fa	+	?
9	Hypericum	H	−(+)	5(4)/−	5(4)/−	'3'	3/f	+ − (ves)	8 − 7/2 − 6
10	Olympia	DS	−	5/−	5/−	'3'	3/f	−(+)	9/2
11	Campylopus	H	−	5/−	5/−	'3'(O)	3(4)/f	+	8/2
12	Origanifolia	H	−	5/−	5/−	'3'	3(5)/f	+(ves)	9 − 8/2
13	Drosocarpium	H	−	5/−	5/−	'3'	3/f	int/ves	8 − 7/2
14	Oligostema	H	−	5(4)/−	5(4)/−	'3'	3/f	+	9 − 8/2
15	Thasia	H	−	5/−	5/−	−/'3'	5 − 3/f	+	8/2
16	Crossophyllum	H	−(+)	5/−	5/−	'3'	3/f	+	8/2
17	Hirtella	DS/H	−	5/−	5/−	'3'	3/f	+	10 − 12(14)/2
18	Taeniocarpium	DS/H	−	5/−	5/−	'3'	3/f	+	9/2
19	Coridium	DS/H	−	5/−(+)	5/−(+)	'3'	3/f	+(ves)	9/2
20	Myriandra	S/H	−	5 − 4/+	5 − 4/+ −	O	3 − 5/a	−	9/2(3 − 4)
21	Webbia	S	−	5/−	5/−	'3'	3/f	(+)	10/4
22	Arthrophyllum	DS	−	5/−	5/−	'3'	3/f	+	?
23	Triadenioides	S/DS	−	5/−	5/−	'3'	3/fa	+(ves)	8/2
24	Heterophylla	DS	−	5/−	5/−	'3'	3/fa	+	9/2
25	Adenotrias	S/DS	−	5/−/+	5/−	'3'+stf	3/fa	+	10/2
26	Humifusoideum	S/H	−	5/−	5/−	± irr O	3(−5)f	+	12,9 − 8/2
27	Adenosepalum	S/H	−/+	5/−	5/−	'3'	3/f	+	10 − 8/2,4
28	Elodes	H	−	5/−	5/−	'3' + stf	3/f	+	(10?)8/2,4?
29	Brathys	T/S/H	−/+ −	5/−	5/−	O(−)	5/3fa	(+)(+ ves)	12/2
30	Trigynobrathys	S/H	−/+	5(4)/−	5/−	O(irr)	5 − 3/fa	(+)−(+ ves)	12,9 − 8/2,4

Notes
Habit: T − tree; S − shrub; DS − dwarf shrub; H − herb. *Leaves*: − − free; + − basally united to perfoliate. *Petals* and [*Stamen*] *fascicles*: + − dec[iduous]; − − persistent. [*Stamen*] *fascicles united*: '3' − 2 + 2 + 1; '4' − 2 + 1 + 1 + 1; stf − sterile fascicle; irr − irregular; O − all united in a ring. *Style union*: f − free; fa − basally appressed; a − appressed; u − united. *Capsule vittae*: (+) − obsolete; + − present, narrow; int − interrupted; ves − vesicles present. Character states in brackets are occasional or rare.

Inflorescence   Flowers are terminal on the shoot and primitively single. Elaboration is basically cymose and is of two types:

(i)   The commonest is by formation of repeated dichasia (flower formation from the uppermost pair of axillary buds) and sometimes monochasia (flowers from one of each pair of uppermost axillary buds). Similar elaboration frequently occurs at adjacent lower stem nodes; and at even lower nodes these may be succeeded by (several-noded) inflorescence branches.

(ii)   In the woody species of a few sections (e.g. *Brathys* and *Trigynobrathys*), short several-noded flower-bearing branches, not single flowers, develop from one or both terminal leaf axils, producing a 'pseudo-dichotomy'. Elaboration is by repetition of such branches.

In sect. *Humifusoideum*, as well as sects *Brathys* and *Trigynobrathys*, mixed inflorescences with both types of branching sometimes occur (Robson 1993).

Flowers These are radially symmetrical, bisexual and homostylous or very rarely heterostylous (sect. *Adenotrias*), with free petals and a superior ovary.

Sepals The calyx is usually 5-merous and quincuncial (imbricate with two outer sepals, one intermediate and two inner), the sepals being free or partly united; but in sects *Myriandra* and

Table 1.3 Hypericum – sections, types, selected species and distributions

Section	Type species	Another species	Distribution
1 Campylosporus	*H. lanceolatum* Lam.	*H. revolutum* Vahl	Africa, Madag., SW. Asia
2 Psorophytum	*H. balearicum* L.	—	Balearic Is
3 Ascyreia	*H. calycinum* L.	*H. patulum* Thunb.	S. & E. Asia, N. Turkey
4 Takasagoya	*H. formosanum* Maxim.	*H. geminiflorum* Hemsley	Taiwan, Luzon
5 Androsaemum	*H. androsaemum* L.	*H. hircinum* L.	Mediterr., W. Eur., Atl. Is
6 Inodora	*H. xylosteifolium* (Spach) N. Robson	—	NE. Turkey, SW. Georgia
6a Umbraculoides	*H. umbraculoides* N. Robson		Mexico (Oaxaca)
7 Roscyna	*H. ascyron* L.	*H. przewalskii* Maxim.	NE. Asia, E. N. America
8 Bupleuroides	*H. bupleuroides* Griseb.	—	NE. Turkey, SW. Georgia
9 Hypericum	*H. perforatum* L.	*H. erectum* Thunb.	N. temperate regions
9a Concinna	*H. concinnum* Benth.	—	California
10 Olympia	*H. olympicum* L.	*H. polyphyllum* Boiss. & Balansa	S. Balkan Peninsula, W. Turkey
11 Campylopus	*H. cerastioides* (Spach) N. Robson	—	NE. Aegean region
12 Origanifolia	*H. origanifolium* Willd..	*H. aviculariifolium* Jaub. & Spach	Cyprus, Turkey, Georgia
13 Drosocarpium	*H. barbatum* Jacquin	*H. perfoliatum* L.	Mediterr., Balkan Penin., SW. Asia
14 Oligostema	*H. humifusum* L.	*H. linariifolium* Vahl	Eur., NW. Africa, Atl. Is
15 Thasia	*H. thasium* Griseb.	*H. aucheri* Jaub. & Spach	NE. Aegean region
16 Crossophyllum	*H. orientale* L.	*H. adenotrichum* Spach	N. & W. Turkey, Caucas.
17 Hirtella	*H. hirtellum* (Spach) Boiss.	*H. hyssopifolium* Vill.	W. Mediterr. to Altai reg.
18 Taeniocarpium	*H. linarioides* Bosse	*H. pulchrum* L.	Europe to Altai & Iran
19 Coridium	*H. coris* L.	*H. empetrifolium* Willd.	W. Mediterr. to Caucasus
20 Myriandra	*H. prolificum* L.	*H. fasciculatum* Lam.	E. N. America, Gr. Antil., Bahamas, Bermuda
21 Webbia	*H. canariense* L.	—	Atlantic Is
22 Arthrophyllum	*H. rupestre* Jaub. & Spach	*H. nanum* Poiret	S. Turkey, Levant
23 Triadenioides	*H. pallens* Banks & Solander	*H. scopulorum* Balf. f.	Socotra, Levant, S. Turkey
24 Heterophylla	*H. heterophyllum* Vent.	—	NW. Turkey
25 Adenotrias	*H. russeggeri* Fenzl	*H. aegypticum* L.	S. Morocco to Levant
26 Humifusoideum	*H. peplidifolium* A. Rich.	*H. papuanum* Ridley	New Guinea, SE Asian Is, trop. & S. Africa, Madag.
27 Adenosepalum	*H. montanum* L.	*H. tomentosum* L.	Atl. Is, Afr., Medit., Eur.
28 Elodes	*H. elodes* L.	—	W. Europe, Azores
29 Brathys	*H. juniperinum* Kunth	*H. caracasanum* Willd.	N. & S. America
30 Trigynobrathys	*H. myrianthum* Cham. & Schlecht.	*H. mutilum* L.	N. & S. America, Africa, E. Asia, Austral., N.Z.

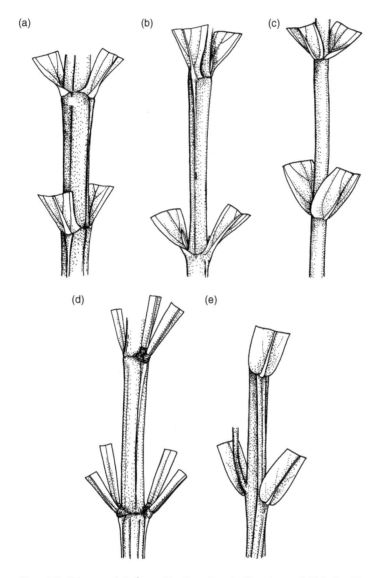

Figure 1.9 [Monograph 2, figure 8] – Stem lines in *Hypericum* and *Ploiarium* (Bonnetiaceae, sometimes included in the Clusiaceae): (a) *H. revolutum* (sect. 1); (b) *H. perforatum* (sect. 9); (c) *H. bithynicum* (sect. 13); (d) *H. fasciculatum* (sect. 20); (e) *Ploiarium alternifolium*. (Drawn by Margaret Tebbs.)

Hypericum some species have only four sepals, which are then opposite and decussate. The sepal margin varies from entire to glandular- or eglandular-fimbriate (i.e. fringed) (Figure 1.11).

Petals   Like the calyx the corolla is usually 5-merous, but the aestivation is contorted. As a consequence the petals are nearly always asymmetrical. The line of the outer margin in bud is often interrupted by a more or less evident projection (apiculus) marking the end of the midrib, while the inner part in bud is usually thinner and larger (Figure 1.12). The petal colour is usually a shade of yellow (lemon to deep orange-yellow); but rarely the yellow pigment (a xanthone)

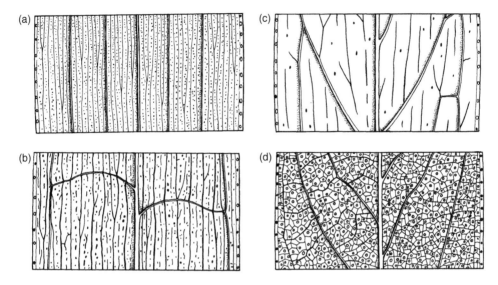

Figure 1.10 [Monograph 2, figure 11] – One origin of pale glands in leaves of *Hypericum* (sect. 1. *Campylosporus*): (a) *H. bequaertii* (×6); (b) *H. revolutum* subsp. *keniense* (×12); (c) *H. revolutum* subsp. *revolutum* (×12); *H. roeperianum* (×4). (Drawn by Margaret Tebbs.)

is absent, so that the petal is white. The part outside in bud is frequently tinged or streaked red, making the unopened bud red-tipped and thus less attractive to foraging insects than the open flowers. Very rarely this red colour diffuses throughout the petal making it deep red or, if the base colour is white, pink. Marginal petal glands are frequently slightly protruding but, unlike those of the sepals, not very often on cilia or fimbriae. The behaviour of the petals after flowering (i.e. whether they are deciduous and, if not, how they fade) is of importance in the classification of *Hypericum*.

Stamens   In the flower of members of the Clusiaceae each perianth member basically has a group of stamens opposite it; but in *Hypericum* those opposite the sepals are absent and there are usually only five groups (fascicles), one opposite each petal. The stamens in each fascicle are nearly always united near the base only, leaving most of the filament free. This pattern has been modified in different ways in different parts of the genus:

(i)   Most frequently two pairs of fascicles have merged, resulting in a 3-fascicled androecium with one single antipetalous fascicle and two antisepalous ones. This pattern is found, for example, in sect. *Hypericum* (which includes *H. perforatum*). Merging of only one pair of fascicles to produce a 4-fascicled androecium is much less common (Figure 1.13).

(ii) In four relatively large sections, however, all five fascicles have merged to form a ring, wide in sect. *Myriandra*, narrower in sects *Brathys*, *Trigynobrathys* and *Humifusoideum*. Where the perianth in sect. *Myriandra* has become 4-merous, the androecium comprises four united fascicles. Where the ring of stamens is narrow, reduction in the number of stamens in each fascicle (which occurs in all parts of the genus) has resulted in an irregular androecium. The extreme case is *H. gentianoides* (sect. *Brathys*), where each 'fascicle' often consists of only one stamen. In contrast, the monospecific sect. *Campylopus* (*H. cerastioides*) sometimes has a ring of stamens formed secondarily by fusion of '3' fascicles.

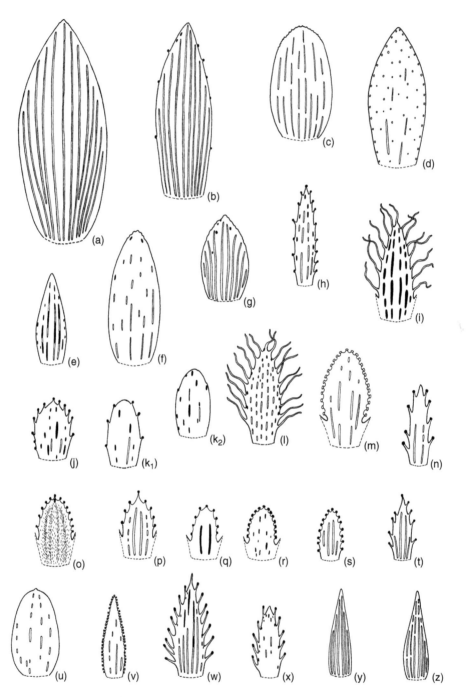

Figure 1.11 {Monograph 2, figure 15} – Variation in marginal contour and in glandularity of sepals in *Hypericum* (numbers indicate sections): (a) *H. bequaertii* (1); (b) *H. quartinianum* (1); (c) *H. forrestii* (3); (d) *H. formosanum* (4); (e) *H. erectum* (9); (f) *H. maculatum* subsp *obtusiusculum* (9); (g) *H. polyphyllum* subsp. *polyphyllum* (10); (h) *H. montbretii* (13); (i) *H. barbatum* (13); (j) *H. linariifolium* (14); (k) *H. humifusum* (14); (l) *H. thasium* (15); (m) *H. orientale* (16); (n) *H. hirtellum* (17); (o) *H. salsolifolium* (17); (p) *H. uniglandulosum* (17); (q) *H. retusum* (17); (r) *H. asperulum* (17); (s) *H. pumilio* (18); (t) *H. fragile* (18); (u) *H. prolificum* (20); (v) *H. canariense* (21); (w) *H. elodeoides* (9); (x) *H. wightianum* (9); (y) *H. strictum* (29); (z) *H. brasiliense* (30) (all × 5). (Drawn by Margaret Tebbs.)

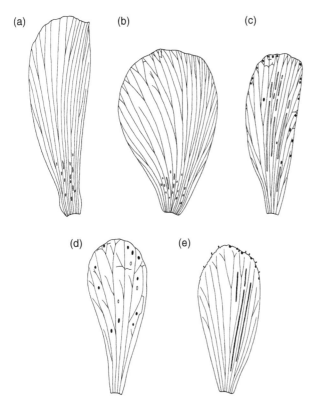

Figure 1.12 [Monograph 2, figure 18] – Variation in venation and glandularity of petals in *Hypericum* (numbers indicate sections): (a) *H. styphelioides* (29); (b) *H. forrestii* (3); (c) *H. perforatum* (9); (d) *H. asperulum* (17); (e) *H. elodeoides* (9) (a, b × 2.5; c × 3.5; d × 4; e × 5). (Drawn by Margaret Tebbs.)

In *Hypericum* the filament is always relatively slender and the anther relatively small and dorsifixed. The connective terminates in a 'dark' gland, amber or black.

Ovary   The gynoecial elements ('carpels') vary from five to two and, in isomerous flowers, alternate with the stamen fascicles. The reduction in number is real, unlike that in the androecium; and it begins evolutionarily before the androecial one, that is, species with five free stamen fascicles and three carpels exist, but none with '3' fascicles and five carpels (Figure 1.13). The number of carpels is reflected in the number of styles, which are slender and vary from free to completely united, the latter state occurring only in sects *Campylosporus* and *Takasagoya*. The stigma is relatively small and may be wider than the style (capitate) or not (punctate). It is often red. Although the trend towards reduction in number of floral parts is general in *Hypericum*, it has been reversed in a few unrelated advanced species. For example, *H. thasium* (sect. *Thasia*), *H. peplidifolium* (sect. *Humifusoideum*) and *H. pleiostylum* (sect. *Trigynobrathys*) all have five styles although their immediate relatives have three or four.

The ovary placentation in *Hypericum* is primitively loosely axile, that is, the placentae are in contact at the centre but not united (e.g. sect. *Campylosporus*), and becomes either truly axile with firmly united placentae (e.g. in *H. empetrifolium*, sect. *Coridium*) or parietal (with the placentae increasingly widely separate, e.g. in *H. elodes*, sect. *Elodes*). Each placenta in the primitive (woody) species bears more or less numerous ovules, which gradually become fewer in various

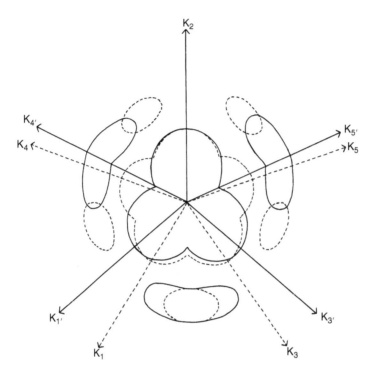

Figure 1.13 [Monograph 2, figure 21] – Transition from 5 s to 3 s in the inner floral whorls of
Hypericum (after Breindl 1934). K_{1-5} – sepal radii when whorls are in 5 s; $K_{1'-5'}$ when
whorls are in 3 s. (Drawn by Margaret Tebbs.)

parts of the genus until in *H. russeggeri* (sect. *Adenotrias*) they are reduced to two. In *H. olivieri*
(sect. *Hirtellum*) only one ovule ripens in each loculus (Figure 1.14).

Fruit   The fruit of *Hypericum* is almost always a dry capsule that dehisces septicidally, that is,
along the septae bearing the placentae. In axile ovaries the septae split, leaving a central seed-
bearing column distinct from the diverging capsule valves; but in parietal ones the placentae
part completely as the valves separate. In three species in distantly related parts of the genus,
however, the capsule has become more or less fleshy and berry-like, notably in sect.
Androsaemum, in which there is a trend to incomplete dehiscence (*H. hircinum*) or slight initial
fleshiness and tardy dehiscence (*H. foliosum*) and then a fleshy 'berry' (*H. androsaemum*) that
ripens from red to shiny black. That it is not a true berry can be seen if slight lateral pressure is
exerted on the fruit, which then easily splits into three valves, especially when older. It would
nevertheless seem to be dispersed at least sometimes by birds, like truly baccate fruits.

The ovary wall contains glands, elongate and narrow (vittae) or short and broad (vesicles),
which contain essential oils or possibly resins. In some woody sections (e.g. *Campylosporus* and
Brathys) they are inconspicuous in fruit, that is, the capsule valves are 'not vittate'. Mostly, however, they become more or less prominent as the fruit develops. Evolutionarily the trends are
from numerous narrow vertical vittae to increasingly interrupted, divergent and enlarged vittae
and eventually, in separate parts of the genus, to short protuberent vesicles that in one or two
species of sect. *Drosocarpium* are black, that is, they contain hypericin. In *H. elongatum* (sect.
Hirtella) scattered capsular vesicles have resulted from the enlargement of small parts of the
numerous vertical vittae (Figure 1.15).

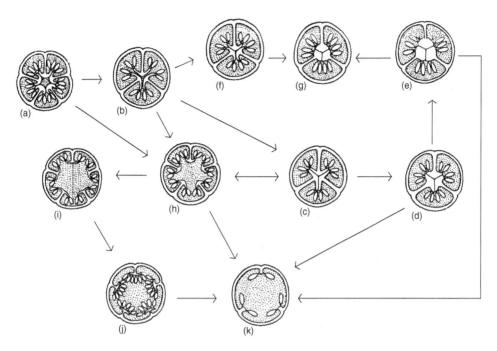

Figure 1.14 [Monograph 2, figure 24] – Ovaries of *Hypericum* species in L. S. section (diagrammatic) (numbers indicate sections): (a) *H. revolutum* (1); (b) *H. orientale* (16); (c) *H. prolificum* (20); (d) *H. elodes* (28) (a × 5; b × 7.5; c × 6.5; d × 8). (Drawn by Margaret Tebbs.)

Seeds   In *Hypericum* these are small (*c*.0.5–2 mm long), cylindrical or the smaller often ovoid, and dark purplish-brown to yellow-brown. The more primitive species have a unilateral narrow papery wing and sometimes also a similar terminal wing. In more advanced parts of the genus the wing is reduced to a narrow keel (carina) or disappears altogether. In one section (*Adenotrias*) the terminal wing has become a fleshy caruncle, a modification that in other families is associated with ant dispersal, and in the above-mentioned *H. olivieri* the single seed is shed in the deciduous capsule valve.

The cells of the testa (seed coat) have thickened inner and radial walls that resist collapse on drying, producing a varying cellular pattern that is useful in classification. This is primitively a prominent reticulum (reticulate). In some evolutionary lines the cells first become aligned (linear-reticulate), then their cross walls become parallel (scalariform) and may almost disappear (ribbed). In other evolutionary lines the angles of the cells are 'filled in' (foveolate). Sects *Hirtella*, *Taeniocarpium* and *Coridium* are characterised by testa cells in which the outer walls do not collapse initially but expand, forming a rugulose to papillose pattern.

The seeds, as in the rest of the Clusiaceae, are non-endospermous or almost so, and the embryo is slender and straight with cotyledons somewhat shorter than the hypocotyl. At germination the emergence of the radicle is almost immediately followed by the development of a ring of strong root hairs that appear to serve as an anchor for the developing seedling (Roth 1990: figure 27).

Anatomy

It is not necessary to give a detailed account of the vegetative anatomy of *Hypericum* here. Comparative accounts and discussions will be found in Vestal (1938), Metcalfe and Chalk (1950),

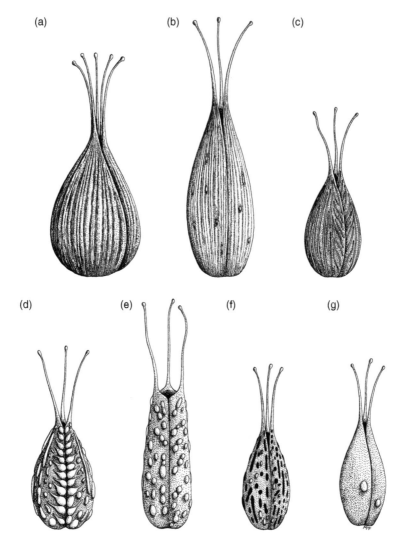

Figure 1.15 [Monograph 2, figure 26] – Capsules of *Hypericum* species, showing patterns of vittae and vesicles (numbers indicate sections): (a) *H. revolutum* (1); (b) *H. elongatum* (17); (c) *H. maculatum* subsp. *obtusiusculum* (9); (d) *H. perfoliatum* (13); (e) *H. montbretii* (13); (f) *H. richeri* (13); (g) *H. paucifolium* (30) (all × 4). (Drawn by Margaret Tebbs.)

Schofield (1968) (nodal anatomy), Baretta-Kuipers (1976) (wood anatomy) and Gibson (1980) (wood anatomy). What is desirable is to give an account of the secretory system.

Distribution of secretory cavities and canals   Secretory canals containing essential oils are present in *Hypericum* in the stem and root. They also penetrate, as has already been explained, into the leaves, sepals and petals, where (i) they are frequently dissected into streaks and dots and (ii) their contents are often denser and darker in colour. Lastly, they occur in the ovary wall, where they may also be dissected and enlarged. The external stem glands are always dots or short streaks, and their contents are resinous or waxy and in colour pale, amber, reddish or black.

In the vegetative parts the glandular canals are distributed internally as follows: root – phloem, pericycle; stem – medulla (occasionally one central canal), phloem, pericycle, cortex (rare); leaf – mesophyll, phloem. A detailed chemo-taxonomic study of the secretions of *Hypericum* and their distribution was made by Mathis and Ourisson (1963, 1964a–d).

Distribution of hypericin and pseudohypericin   The general distribution of these secretions has been outlined above. Their occurrence in the various sections of *Hypericum* is summarised in Table 1.1. For other characters of the sections adopted in this classification, see the section *Classification* below.

Except where hypericin and pseudohypericin have been detected by chemical methods, for example, by Mathis (1963), their presence or absence has been deduced by the occurrence of black or red glands, a practice that, from the reported chemical evidence, seems to be quite valid.

Chromosome numbers

The primitive basic chromosome number (x) in *Hypericum* is 12, and there are decreasing series of basic numbers in different evolutionary lines from 11 (very rare) and 10 to 8 and rarely 7. Tetraploidy occurs on all these numbers except 7, and higher degrees of polyploidy are found in hybrids (including *H. perforatum*, see p. 20). There is a secondary basic number (21) in sect. 3. *Ascyreia* and an apparently secondary ascending series from 10 to 14 in sect. 17. *Hirtella*.

Classification

Following early surveys of *Hypericum* and its relatives by Choisy (1821, 1824) and Spach (1836a,b), a complete classification of the genus was first attempted by Keller (1893, 1925). His scheme, however, was unsatisfactory in several ways, particularly his treatment of the numerous species with persistent petals and stamens, '3' stamen fascicles and three styles as sect. *Eu-hypericum*. The classification adopted here (Robson 1977 onward) has resulted from an attempt to sort out the natural evolutionary lines (clades) in *Hypericum*, although cladistic methods of classification have not been used (Tables 1.2 and 1.3). Since Tables 1.1–1.3 were compiled, sect. 9. *Hypericum* has been subdivided into six sections, one of which is sect. 9a. *Concinna* (Robson 2001) (Figure 1.16); but the other additional sections have not been included in the tables, as the variation in their characters could not be accommodated in the format of the tables.

Hypericum perforatum

Introduction

Hypericum perforatum L. is a perennial herb, belonging to sect. *Hypericum*, with a wide natural distribution from the Azores via Siberia to China (south to Yunnan) and via the Mediterranean region to the western Himalayas. It has been introduced into many other parts of the world, and is the commonest species of *Hypericum* over a considerable part of its range. Being, therefore, the most familiar species to many people in folklore and medicine, it was selected as the type species of the genus. It may be recognised by the following summary description (Figure 1.17).

Perennial herb, usually erect, much branched, especially distally. *Stem* internodes 2-lined, with black glands on the raised lines. *Leaves* sessile or very shortly petiolate; blade oblong to elliptic or linear, apex rounded to apiculate or acute, base cordate-amplexicaul to cuneate; reticulate venation lax or scarcely visible; laminar glands pale and sometimes black, few, dots; intramarginal glands mostly black. *Inflorescence* from one to three nodes, with flowering branches curved-ascending from up to 15 or more nodes below, the whole cylindrical to broadly pyramidal or

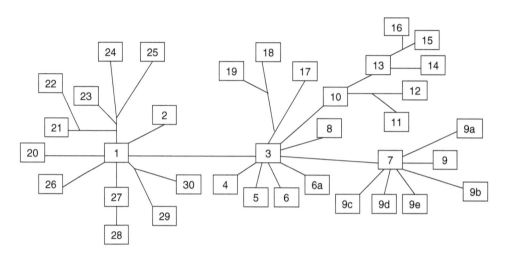

Figure 1.16 [Monograph 2, figure 2, modified] – Relationships of sections of *Hypericum*.

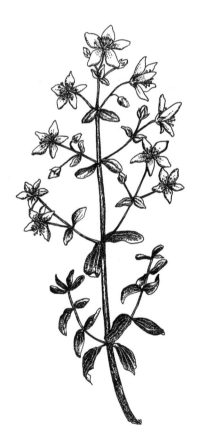

Figure 1.17 *Hypericum perforatum* – drawn by Victoria Brown. Please note that this drawing is a diagrammatic representation of the species.

flat-topped. *Flowers* 15–35 mm in diameter; buds acute. *Sepals* equal, narrowly oblong or lanceolate to linear, acute to finely acuminate with acumen sometimes glandular, entire; laminar glands pale and often a few black, basal streaks to all dots; intramarginal glands few, black or none. *Petals* golden yellow, not tinged red, three to four × sepals, asymmetric, distally ± crenate; laminar glands all pale to mostly black, all lines or partly streaks or dots; intramarginal glands black, in sinuses of crenations. *Stamens* '3'-fascicled; anther gland black. *Ovary* 3-locular; styles three, free, *c*.1.5–2 × ovary, spreading. *Capsule* narrowly ovoid to ovoid-conic; valves with dorsal narrow vittae and lateral narrow vittae or yellowish elongate to round vesicles. *Seeds* dark brown, *c*.1 mm long, not carinate or appendiculate; testa pattern finely linear-foveolate. *Chromosomes*: $2n = 32$ (48).

Origin, variation and classification

Origin

There is good morphological, geographical and cytological evidence indicating that *H. perforatum* originated through hybridisation between two closely related species with subsequent doubling of the chromosomes, that is, it is an allopolyploid. One of these species is *H. maculatum* Crantz (more specifically one of its diploid subspecies, $2n = 16$). Subsp. *maculatum* is distributed from western Europe to central Siberia and hybridises easily with *H. perforatum*; but subsp. *immaculatum* (Murb.) Fröhl. is morphologically more similar to *H. perforatum*. The other parent of the cross would appear to be *H. attenuatum* Choisy ($2n = 16$), which ranges from western Siberia to Korea and eastern China and has the characters of *H. perforatum* that are missing from *H. maculatum*. Thus,

H. maculatum subsp. *immaculatum* [16] × *H. attenuatum* (16) × 2 = *H. perforatum* (32).

The location of this hybridisation was almost certainly somewhere in Siberia, although *H. maculatum* subsp. *immaculatum* is now confined to south-eastern Europe.

Variation, distribution and classification

Although *Hypericum perforatum* varies considerably throughout its natural range, the variation is apparently continuous and, therefore, theoretically impossible to classify. For practical purposes, however, one can recognise four main variants as varieties or preferably subspecies.

The two most variable organs of *H. perforatum* are the leaves and the capsules. The leaves vary (i) from sessile to very shortly petiolate and (ii) from broadly oblong or elliptic in mesophytic habitats to linear and/or small in dry habitats. The capsules vary in shape from ovoid to pyramidal; but, more importantly, the valve glands vary from (i) several dorsal vittae and numerous oblique lateral vittae (all narrow) to (ii) fewer dorsal vittae and fewer oblique vesicles interrupted and enlarged especially towards the base, and then to (iii) one or no dorsal vittae and scattered ovoid to round vesicles. Like trend (ii) in the leaves, the capsule trends (i)–(iii) accompany increasing habitat dryness.

Only two of the four subspecies have valid names at that rank. The other two (Subspp. 2 and 4) have valid names at the rank of species, and these have been cited. They will be named as subspecies in the forthcoming Part 4(2) of the *Hypericum* monograph (Robson 2002).

1   Subsp. *perforatum*   The primitive form of *H. perforatum* may thus be envisaged as having broad leaves, sessile like those of its parents and with a rounded base and a wholly vittate capsule. Some plants of subsp. *perforatum* from Russia and Scandinavia agree with this description; but this variety typically has petiolate leaves narrowed at the base. Its range

     extends from western Siberia (Altai region) to western Europe (except the Mediterranean region).

2   *H. songaricum*   Ledeb. ex Rchb. To the south-west of the Altai mountains, in north-eastern Kazakhstan, Kirghizstan and adjacent China, the leaves become cordate and stem-clasping and the capsule glands are (rarely) first interrupted then vesicular. The range of this variety extends around the north of Kazakhstan into southern Russia, the Ukraine and the Crimea, the leaves becoming thicker, glaucous beneath and (in the drier areas) with margins recurved.

3   Subsp. *veronense* (Schrank) H. Lindb.   From the Caucasus region southwards and westwards the leaves become increasingly narrow, often with revolute margins ('var. *angustifolium*'), and then small as the habitats become drier ('var. *microphyllum*'); and there is a parallel trend from interrupted capsule vittae to scattered vesicles. The leaves, however, usually remain sessile. The natural range of this subspecies extends from north of the Caucasus region eastward to Tajikistan and NW India, westward to Turkey, the Mediterranean region and Macaronesia as far as the Azores, and southward to western Saudi Arabia (Asir).

4   *H. foliosissimum* Makino.   There is a gap in distribution from subsp. *perforatum* in the Altai region and adjacent Mongolia to NW China (Gansu), where the species reappears sometimes in a scarcely different form. Further east towards the coast and south to Yunnan, however, the form changes to that typical of China. This is similar to subsp. *perforatum* as regards the capsule vittae; but the leaves become gradually narrower and the flowers smaller, the latter being in dense few-flowered clusters at the end of spreading branches.

Cytology and hybrids

Variation in *Hypericum perforatum* is complicated by its cytological behaviour (Noack 1939, Mártonfi *et al.* 1996). The pollen undergoes normal reduction division, but lagging chromosomes lead to sterility that is usually about 30% but can be up to 70% (Nielsen 1924, Hoar and Haertl 1932). Only *c*.3% of the embryo sacs are produced by normal meiosis ($n = 16$), the remaining (unreduced) ones being produced parthenogenetically ($n = 32$). The latter, however, are pseudogamous, that is, they require pollination for seed development. Fertilisation is therefore possible though not necessary. Occasionally the $n = 32$ embryo sacs are fertilised, resulting in hexaploid ($2n = 48$) plants and, very rarely, a reduced embryo sac may develop, resulting in a polyhaploid plant ($2n = 16$). This cytological variation results in considerable morphological variation, which is complicated in the wild by hybridisation, especially with *H. maculatum* ($2n = 16, 32$) and *H. tetrapterum* ($2n = 16$). With tetraploid *H. maculatum*, the hybrids are also tetraploid ($2n = 32$) and fertile, so that a whole series of intermediate forms can be found; but with diploids of both species, two types of hybrid result. Where *H. perforatum* is the pollen parent, the hybrids are triploid ($8 + 16 = 24$); but where it is the ovule parent, the hybrids are pentaploid ($32 + 8 = 40$).

Conclusion

The use of *Hypericum perforatum* in medicine in preference to other species of *Hypericum* would appear to be related to its availability and consequent historical causes, rather than because it contains the most effective substances. It may do so, but the onus is on the medical profession to prove this. Unless its hybrid origin brought together two such substances or groups of substances, which seems unlikely, then at least its near relatives in sect. *Hypericum* (e.g. *H. maculatum* and *H. attenuatum*) would be expected to be similarly effective. As a chemical widespread in the genus (*hypericin*) is apparently involved, then many species in various sections are likely to have pharmaceutical uses. Only future research will reveal if this is so.

References

Baretta-Kuipers, J. (1976) Comparative wood anatomy of Bonnetiaceae, Theaceae and Guttiferae. *Leiden Botanical Series* 3, 76–101.

Breindl, M. (1934) Zur Kenntnis der Baumechanik des Blutenkelches der Dikotylen. *Bot. Arch.* 36, 191–268.

Choisy, J.D. (1821) *Prodromus d' une monographie de la famille des Hypéricinées*. Geneva and Paris.

——(1824) Hypericineae. In *Prodromus systematis naturalis Regni vegetabilis*. A.P. De Candolle (Ed.) 1, 541–56. Paris.

Gibson, A.C. (1980) Wood anatomy of *Thornea*, including some comparisons with other Hypericaceae. *IAWA Bulletin* n.s. 1, 87–92.

Hoar, C.S. and Haertl, E.J. (1932). Meiosis in the genus *Hypericum. Bot. Gaz.* 92, 396–406.

Keller, R. (1893) *Hypericum*. In *Die natürlichen Pflanzenfamilien*. A. Engler and K. Prantl (Ed.) 3(b), 208–15.

——(1925) *Hypericum*. In op. cit., 2nd edn 21, 175–83.

Mártonfi, P., Brutovská, R., Čellárová, E. and Repčák, M. (1996) Apomixis and hybridity in *Hypericum perforatum. Folia Geobot. Phytotax.* 31, 389–96.

Mathis, C. (1963) *Étude chimio-taxonomique du genre* Hypericum *L.* Thesis Fac. Sci. Univ. Strasbourg.

Mathis, C. and Ourisson, G. (1963) Étude chimio-taxonomique du genre *Hypericum*. I. Répartition de l' hypéricine. *Phytochem.* 2, 157–71.á

——(1964a) Étude chimio-taxonomique du genre *Hypericum*. II. Identification de constituants de diverses huiles essentielles d' *Hypericum. Phytochem.* 3, 115–31.

——(1964b) Étude chimio-taxonomique du genre *Hypericum*. III. Répartition des carbures saturés et des monoterpènes dans les huiles essentielles d' *Hypericum. Phytochem.* 3, 133–141.

——(1964c) Étude chimio-taxonomique de genre *Hypericum*. IV. Répartition des sesquiterpènes des alcools monoterpeniques et des aldehydes saturés dans les huiles essentielles d' *Hypericum. Phytochem.* 3, 377–8.

——(1964d) Étude chimio-taxonomique du genre *Hypericum*. V. Identification de quelques constituants non volatils d' *Hypericum. Phytochem.* 3, 379.

Metcalfe, C.R. and Chalk, L. (1950) *Anatomy of the Dicotyledons*. Oxford.

Nielsen, N. (1924) Chromosome numbers in the genus *Hypericum* (A preliminary note). *Hereditas* 5, 378–82.

Noack, K.L. (1939) Über *Hypericum* Kreuzungen. VI. Fortpflanzungsverhältnisse und Bastarde von *H. perforatum* L. *Z. indukt. Abstamm.- u. Vererbungslehre* 76, 569–601.

Robson, N.K.B. (1977) Studies in the genus *Hypericum* L. (Guttiferae). 1. Infrageneric classification. *Bull. Br. Mus. Nat. Hist.* (Bot.) 5, 291–355.

——(1981) Studies in the genus *Hypericum* L. (Guttiferae). 2. Characters of the genus. *Bull. Br. Mus. nat. Hist.* (Bot.) 8, 55–226.

——(1985) Studies in the genus *Hypericum* L. (Guttiferae). 3. Sections 1. *Campylosporus* to 6a. *Umbraculoides. Bull. Br. Mus. nat. Hist.* (Bot.) 12, 163–325.

——(1987). Studies in the genus *Hypericum* L. (Guttiferae). 7. Section 29. *Brathys* (part 1). *Bull. Br. Mus. nat. Hist.* (Bot.) 16, 1–106.

——(1990) Studies in the genus *Hypericum* L. (Guttiferae). 8. Sections 29. *Brathys* (part 2) and 30. *Trigynobrathys. Bull. Br. Mus. nat. Hist.* (Bot.) 20, 1–151.

——(1993) Parallel evolution in tropical montane *Hypericum. Opera Botanica* 121, 263–74.

——(1996) Studies in the genus *Hypericum* L. (Guttiferae). 6. Sections 20. *Myriandra* to 28. *Elodes. Bull. nat. Hist. Mus. Lond.* (Bot.) 26, 75–217.

——(2001) Studies in the genus *Hypericum* L. (Guttiferae). 4(1). Sections 7. *Roscyna* to 9. *Hypericum* sensu lato (part 1). *Bull. Nat. Hist. Mus. Lond.* (Bot.) 31, 37–87.

——(2002) Studies in the genus *Hypericum* L. (Guttiferae). 4(2). Section 9. *Hypericum* sensu lato (part 2). *Bull. Nat. Hist. Mus. Lond.* (Bot.) 32, 61–123.

Roth, L. (1990) *Hypericum – Hypericin. Botanik – Inhaltsstoffe – Wirkung*, Landsberg, Lech.

Schofield, E.K. (1968) Petiole anatomy of the Guttiferae and related families. *Mem. N. Y. Bot. Gdn* 18, 1–55.

Spach, E. (1836a) Hypericacearum monographiae fragmenta. *Annls Sci. nat.* (Bot.) II, 5, 157–76.

——(1836b) Conspectus monographiae Hypericacearum. *Annls Sci. nat.* (Bot.) II, 5, 349–69, t. 6.

Vestal, P.A. (1938) The significance of comparative anatomy in establishing the relationship of the Hypericaceae to the Guttiferae and their allies. *Philipp. J. Sci.* 64, 199–256.

2 *Colletotrichum gloeosporioides* as the cause of St John's wort (*Hypericum perforatum*) dieback in Switzerland and breeding for a tolerant variety

M. Gaudin, X. Simonnet and N. Debrunner

Introduction

St John's wort (*Hypericum perforatum*) has been used for its medical properties throughout the ages (Czygan 1993). It is currently recommended in plant therapy for its antiviral, vulnerary and antidepressive properties (Schauenberg and Paris 1977, Hobbs 1989, Schaffner *et al*. 1992, Hostettmann 1997). *H. perforatum* flower based formulations are used to cure light or mild depression (Hölzl 1993). This ability to fight depression naturally attracts the pharmaceutical industry's interest.

Hypericum perforatum has been the subject of many pharmacological studies and is one of the most thoroughly investigated medicinal plants in western Europe (Roth 1990, Debrunner and Simonnet 1998). However, pharmaceutically relevant compounds in *H. perforatum* flower extracts are unknown. Flavonoids, naphtodianthrones (hypericin and pseudo-hypericin), phloroglucinols (hyperforin) and xanthones are concentrated in flowers. The antidepressive effect was for a long time attributed to hypericin (Bruneton 1993), this is the standardised analytical reference for extracts preparation. Recent research also emphasises the probable significance of hyperforin (Chatterjee *et al*. 1998, Erdelmeier 1998, Laakmann *et al*. 1998).

Antidepressants represent a huge market thus providing the impetus for *H. perforatum* development. Although limited a few years ago, the acreage now covers several hundred hectares in Europe. A few selected varieties, for example, Hyperimed, Elixir and Topas are commercially available. Topas, a Polish variety registered in 1982, is probably the most extensively available today (Kartnig *et al*. 1997).

In Switzerland, an unidentified fungal disease has infected *H. perforatum* fields since 1995. The majority of the 20 hectares of St John's wort planted in Switzerland are grown organically. The normal harvest has been thwarted by disease. The dieback can destroy this perennial crop in the first year of cultivation, otherwise productivity is over a three year period. Since organic farming does not allow the use of fungicides, cultures are irretrievably lost. Typical symptoms observed in the field were brown, sunken stem-girdling and the reddish colour of infected plants. A later stage shows aerial plant parts which had dried completely, thus, killing the plants (Figure 2.1). Literature (Hildebrand and Jensen 1991, Shepherd 1995, Schwarczinger and Vajna 1998, Debrunner *et al*. 2000) mentions that the worldwide increase in St John's wort production is accompanied by the appearance of anthracnose caused by *Colletotrichum gloeosporioides* (Penz.). *C. gloeosporioides* symptoms seem to accord with those described in our country.

Pathogen identification

Isolates of the pathogen were obtained from multiple lesions on the stems of St John's wort plants growing near Sion (Switzerland) on the south side of the Alps (Figure 2.2). Acervuli

Figure 2.1 Dieback symptoms observed in Swiss St John's wort fields as from 1995. (See Colour Plate IX.)

containing conidia and dark setea emerged from the lesions. Segments with lesions were placed on potato dextrose agar (4%) amended with 25 μg/ml aureomycin and incubated at room temperature. Spore masses, growing from the plant tissue were subcultured two to three times.

Colletotrichum gloeosporioides grew rapidly on the nutrient medium and sporulated abundantly. Two multi-spored strains were sent to CABI Bioscience Identification Services (Egham, UK), who confirmed the identification of *C. gloeosporioides*. Isolates were stored on PDA (4%) at 4°C and periodically subcultured.

Breeding for a St John's wort variety tolerant to *C. gloeosporioides*

In response to the serious anthracnose damages observed in the *H. perforatum* fields, principally in Topas crops, a protocol was set up in 1997 to select a new variety of St John's wort (Gaudin *et al.* 1999). This variety was to be both *C. gloeosporioides* tolerant and productive. We bred accordingly with high flower top yield and easy harvest in mind. The plant must be rich in secondary metabolites, its chemical profile has to be similar to that of Topas since the antidepressive molecules are unknown. Cost effectiveness, of course, is the goal in perspective.

Plant material and agronomic practices

In early March the commercial varieties Topas (P1), Hyperimed (P2) and Elixir (P3) and 21 wild accessions (P4 to P24) collected in Switzerland, Germany, Italy, Australia and Canada were

Figure 2.1 Dieback symptoms observed in Swiss St John's wort fields as from 1995. (See Colour Plate IX.)

containing conidia and dark setea emerged from the lesions. Segments with lesions were placed on potato dextrose agar (4%) amended with 25 μg/ml aureomycin and incubated at room temperature. Spore masses, growing from the plant tissue were subcultured two to three times.

Colletotrichum gloeosporioides grew rapidly on the nutrient medium and sporulated abundantly. Two multi-spored strains were sent to CABI Bioscience Identification Services (Egham, UK), who confirmed the identification of *C. gloeosporioides*. Isolates were stored on PDA (4%) at 4°C and periodically subcultured.

Breeding for a St John's wort variety tolerant to *C. gloeosporioides*

In response to the serious anthracnose damages observed in the *H. perforatum* fields, principally in Topas crops, a protocol was set up in 1997 to select a new variety of St John's wort (Gaudin *et al.* 1999). This variety was to be both *C. gloeosporioides* tolerant and productive. We bred accordingly with high flower top yield and easy harvest in mind. The plant must be rich in secondary metabolites, its chemical profile has to be similar to that of Topas since the antidepressive molecules are unknown. Cost effectiveness, of course, is the goal in perspective.

Plant material and agronomic practices

In early March the commercial varieties Topas (P1), Hyperimed (P2) and Elixir (P3) and 21 wild accessions (P4 to P24) collected in Switzerland, Germany, Italy, Australia and Canada were

2 *Colletotrichum gloeosporioides* as the cause of St John's wort (*Hypericum perforatum*) dieback in Switzerland and breeding for a tolerant variety

M. Gaudin, X. Simonnet and N. Debrunner

Introduction

St John's wort (*Hypericum perforatum*) has been used for its medical properties throughout the ages (Czygan 1993). It is currently recommended in plant therapy for its antiviral, vulnerary and antidepressive properties (Schauenberg and Paris 1977, Hobbs 1989, Schaffner *et al*. 1992, Hostettmann 1997). *H. perforatum* flower based formulations are used to cure light or mild depression (Hölzl 1993). This ability to fight depression naturally attracts the pharmaceutical industry's interest.

Hypericum perforatum has been the subject of many pharmacological studies and is one of the most thoroughly investigated medicinal plants in western Europe (Roth 1990, Debrunner and Simonnet 1998). However, pharmaceutically relevant compounds in *H. perforatum* flower extracts are unknown. Flavonoids, naphtodianthrones (hypericin and pseudo-hypericin), phloroglucinols (hyperforin) and xanthones are concentrated in flowers. The antidepressive effect was for a long time attributed to hypericin (Bruneton 1993), this is the standardised analytical reference for extracts preparation. Recent research also emphasises the probable significance of hyperforin (Chatterjee *et al*. 1998, Erdelmeier 1998, Laakmann *et al*. 1998).

Antidepressants represent a huge market thus providing the impetus for *H. perforatum* development. Although limited a few years ago, the acreage now covers several hundred hectares in Europe. A few selected varieties, for example, Hyperimed, Elixir and Topas are commercially available. Topas, a Polish variety registered in 1982, is probably the most extensively available today (Kartnig *et al*. 1997).

In Switzerland, an unidentified fungal disease has infected *H. perforatum* fields since 1995. The majority of the 20 hectares of St John's wort planted in Switzerland are grown organically. The normal harvest has been thwarted by disease. The dieback can destroy this perennial crop in the first year of cultivation, otherwise productivity is over a three year period. Since organic farming does not allow the use of fungicides, cultures are irretrievably lost. Typical symptoms observed in the field were brown, sunken stem-girdling and the reddish colour of infected plants. A later stage shows aerial plant parts which had dried completely, thus, killing the plants (Figure 2.1). Literature (Hildebrand and Jensen 1991, Shepherd 1995, Schwarczinger and Vajna 1998, Debrunner *et al*. 2000) mentions that the worldwide increase in St John's wort production is accompanied by the appearance of anthracnose caused by *Colletotrichum gloeosporioides* (Penz.). *C. gloeosporioides* symptoms seem to accord with those described in our country.

Pathogen identification

Isolates of the pathogen were obtained from multiple lesions on the stems of St John's wort plants growing near Sion (Switzerland) on the south side of the Alps (Figure 2.2). Acervuli

Figure 2.2 Lesion on stem of diseased *H. perforatum.*

seeded in the potting medium Brill 3® and grown in a glasshouse. In early April seedlings with two to three leaves were transplanted individually in compressed root balls (Brill 4®). In mid-May they were transplanted in the trial fields. Harvest was conducted manually while the plants were in full bloom. Flowers were harvested by cutting the stems with shears around 10 cm above the inflorescence. Plants were then cut back approximately 10 cm above the surface of the soil. The experimental surfaces were weeded manually and regularly irrigated as long as the cultures lasted, that is, two years (1997 and 1998).

Experimental design

To acquire as much data as possible regarding the behaviour of the 24 accessions under culture, three differing soil and climatic sites were selected (Table 2.1). The sites Fougères, Epines and Bruson are located near Sion, between 480 and 1060 m. The experimental design, Fisher blocks with three replications at each test station, is shown in Table 2.1. Only 18 accessions out of 24 were cultured in Fougères for lack of available space. Ten plants from the six others (P4, P6, P8, P10, P17 and P23) were grown outside the experimental device on the same site. Plant development and the sanitary conditions of the plots were monitored throughout the season. Harvested flowering segments were dried at 35°C for approximately 10 days and weighed. The yields by weight were expressed per plant on a ten-plant per plot basis. Samples were collected and powdered to analyse secondary metabolites (one analysis per accession and site). Ten flavonoids and

Table 2.1 Culture sites and experimental design

	Culture sites		
	Fougères plot 1	Epines plot 2	Bruson plot 3
Altitude (m)	480	480	1060
Top soil texture (loam : clay : sand)	23 : 43 : 34	5 : 17 : 78	14 : 37 : 49
pH	8.0	7.8	6.8
Available phosphorus	+ +	+	+ + +
Available potassium	+ + +	+	+ +
Previous crop	rye	rye	fallow land
Basic plots			
Number of plants	10	10	10
Surface (m²)	3.2	3.2	2.4
Density (plants/m²)	3.1	3.1	4.2

Notes
− − very low; − low; + medium; + + high; + + + very high.

two hypericins were measured by HPLC (Bioforce AG laboratory, Roggwil / TG). These measurements were carried out for every accession collected in 1997. They were repeated again in 1998 for some interesting accessions only.

Field experimentation's results

Out of the 21 wild *H. perforatum* accessions subjected to tests, P7 was the only one that met the requirements of our selection criteria. The results of the study are therefore focused on P7 which became our target plant. Topas served as the reference variety.

Anthracnose tolerance

In general, severe anthracnose developed on most plants, few were healthy after 2 years (Figure 2.3). Respectively, 94% and 89% of plants growing in the Epines and Fougères plots at 480 m were dead or diseased (Table 2.2). Anthracnose virulence decreased radically with only 51% of plants being affected at 1060 m at the Bruson site. Six of the 24 accessions at Bruson were totally symptom free, Topas and P7 amongst them. An analysis of variance (ANOVA) followed by Newman–Keuls test ($\alpha = 5\%$) indicated that genotype P7 was, on the Fougères plot, significantly more anthracnose tolerant than the *H. perforatum* varieties currently available on the market. On the Epines site it was statistically comparable to Topas, the reference variety, as well as genotype P17 but less *C. gloeosporioides* susceptible than the other accessions grown in this experimental device (Table 2.2).

Harvesting time

The accessions showed considerable differences in the dates of blooming and thus harvesting (Figure 2.4). The Topas variety (P1) had the longest growing period before flowering. P7 was an early genotype in the first cultivation year and an intermediate one in the second. Hyperimed (P2) bloomed early and Elixir (P3) was intermediate to late (Figure 2.5).

The first harvest (1997) stretched over the period from 8 July (day 189; P22; Fougères) to 4 September (day 247; P8, P12, P16 and P20; Bruson). The second one (1998) was conducted

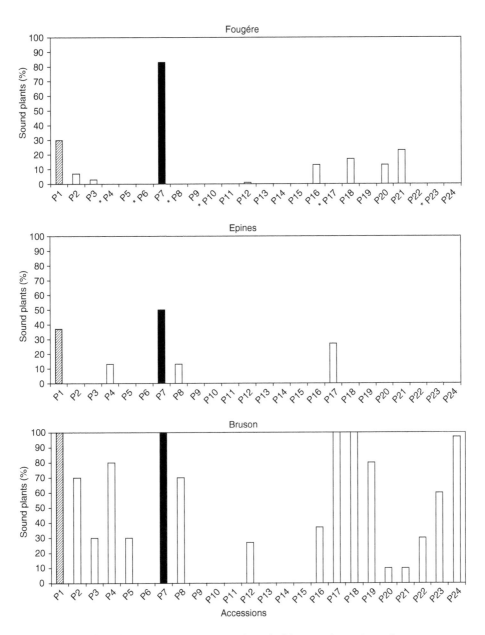

Figure 2.3 Plants free of anthracnose symptoms at the end of the second experimental year.

Note
* Out of the trial scheme.

between 9 June (day 175; P22; Fougères) and 4 August (day 216; Topas; Bruson). During the second year of cultivation, *H. perforatum* was mature about 6 weeks earlier. Cultivation site localisation also had an impact on the date of flowering and the harvest was, at 1060 m, approximately 3 weeks late.

Table 2.2 Anthracnose level after the 1998 harvest at the various sites

Accessions	Plants with symptoms (%)[†]	Diseased plants (%)	Dead plants (%)
Fougères			
P7	17 b	7	10
P1	70 a	67	3
P21	77 a	47	30
P18	83 a	63	20
P20	87 a	27	60
P16	87 a	60	27
P12	90 a	50	40
P2	93 a	70	23
P3	96 a	63	33
P5	100 a	0	100
P9	100 a	0	100
P11	100 a	0	100
P13	100 a	0	100
P14	100 a	0	100
P15	100 a	0	100
P22	100 a	17	83
P19	100 a	50	50
P24	100 a	80	20
Mean	89		
F	8.84***		
Variance ratio	13.3		
Epines			
P7	50 b	50	0
P1	64 ab	57	7
P17	73 ab	73	0
P4	87 a	0	87
P8	87 a	0	87
P5	100 a	0	100
P9	100 a	0	100
P10	100 a	0	100
P11	100 a	0	100
P13	100 a	0	100
P14	100 a	0	100
P15	100 a	0	100
P16	100 a	0	100
P22	100 a	0	100
P23	100 a	0	100
P24	100 a	0	100
P20	100 a	7	93
P6	100 a	17	83
P12	100 a	20	80
P2	100 a	73	27
P21	100 a	73	27
P3	100 a	77	23
P19	100 a	77	23
P18	100 a	83	17
Mean	94		
F	2.64**		
Variance ratio	15.1		
Bruson			
P1	0 c	0	0
P7	0 c	0	0
P17	0 c	0	0

(Continued)

Table 2.2 (Continued)

Accessions	Plants with symptoms (%)[†]	Diseased plants (%)	Dead plants (%)
P18	0 c	0	0
P20	0 c	0	0
P21	0 c	0	0
P24	3 c	0	3
P3	17 bc	10	7
P4	20 bc	3	17
P19	20 bc	20	0
P2	30 bc	20	10
P23	40 b	37	3
P16	64 a	37	27
P22	70 a	43	27
P12	73 a	33	40
P8	93 a	40	53
P5	97 a	7	90
P9	100 a	0	100
P10	100 a	0	100
P11	100 a	0	100
P13	100 a	0	100
P14	100 a	0	100
P15	100 a	0	100
P6	100 a	17	83
Mean	51		
F	31.39***		
Variance ratio	26.2		

Notes
† Different letters indicate statistically significant differences (Newman–Keuls test, $\alpha = 5\%$).
*, **, *** significant at $p = 0.05$, 0.01 and 0.001.

Figure 2.4 Trial field of Epines (1997). Heterogeneity in flowering time is to be noted. (See Colour Plate X.)

Harvest dates of different *H. perforatum* accessions

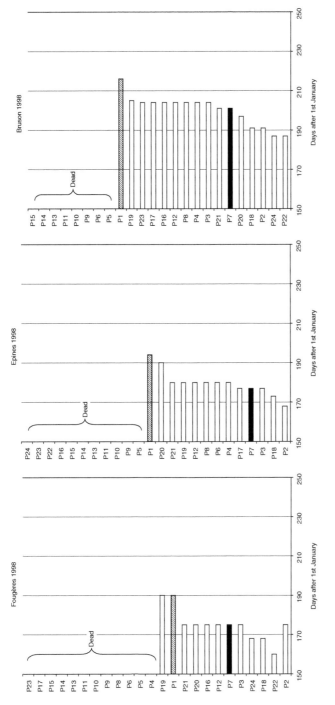

Figure 2.5 Harvest dates of different *H. perforatum* accessions at the various sites.

Note
+ No flowering.

If cultivation in the uplands is intended, an early date of blooming appears to be favourable. Early genotypes like P7 can be harvested at 1060 m from the first year. In the mountains, Topas as five other late genotypes, did not flower during the first year (Figure 2.6).

Harvest quality

Since flowers contain the major part of putative active compounds, the purchasing companies limit stem proportion of the flowering segments processed for pharmaceutical use. The length of the flowering segment is fixed at 15 cm. Highly different growing modes were evident between accessions throughout the first year of culture (Table 2.3). P7's compact flowering horizon, with

Figure 2.6 P7's flowers bloom very early and crop yields are improved in altitude due to this characteristic. On the left: P7 before the second harvest; on the right: Topas, no harvest will be performed (crop production field, Bruson, 1999). (See Colour Plate XI.)

Table 2.3 St John's wort phenotypes throughout the first year of culture

Phenotypes	Plant	Base branching	Flowering horizon
P7, P14, P22	Erect	None	Compact
P9, P18	Erect	Little	Compact
P1, P2, P4, P6, P16	Irregular	Heavy	Large
P3, P8, P12, P17, P19, P20, P23, P24	Irregular	Heavy	Vague
P13, P21		Creeper	
P5, 910, P11		Unclassifiable accessions	

all flowers in the same plane, let cultivators readily calibrate stem cutting. This accession is also easy to harvest thanks to its erect stand (Figure 2.7). The commercial varieties, Topas (P1), Hyperimed (P2) and Elixir (P3) are more difficult to harvest because the corymbs exist at different heights. The morphological differences between accessions were reduced during the second year. Every plant had an erect stand and a varying number of vertical stems. No pronounced morphological variation was observed between plants of the same accession. The specific mode of sexual reproduction in this species, apomixis, is probably responsible for this feature (Martonfi *et al*. 1996, Cellarova *et al*. 1997).

Flower yield

Flower yield is dependent on anthracnose susceptibility and also on soil type, altitude and plant morphology (Table 2.4). Comparatively properly irrigated sand is better than silt soil. Plants developed three times more flowers the first year on the Epines site as compared to the Fougères site.

Figure 2.7 P7's stems are ligneous and relatively rigid, they do not bear secondary stalks and are all of the same length. The flowering horizon is consequently homogeneous and flowers are easy to harvest.

Table 2.4 Dry matter yield of flowering segments at the various sites (g/plant)

	Yield 1997	*Yield 1998*	*Total 1997 + 1998*
Fougères	18	21	39
Epines	52	17	69
Bruson	20	35	55

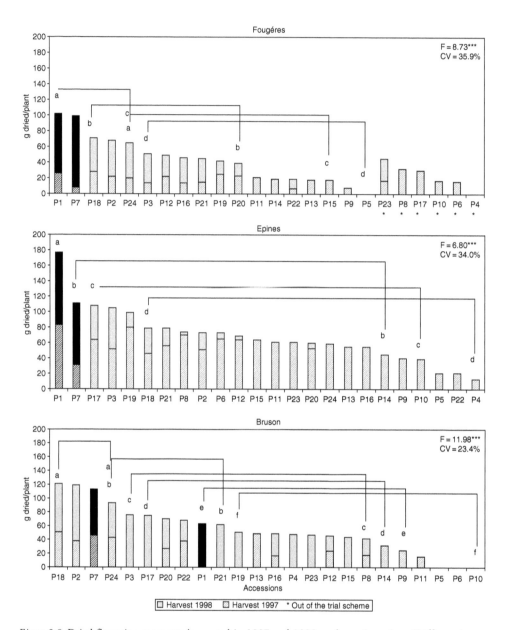

Figure 2.8 Dried flowering segments harvested in 1997 and 1998 at the various sites. (Different letters indicate statistically significant differences; Neuwman–Keuls test, $\alpha = 5\%$.)

Epines' sandy soil was particularly suited for Topas. The 176 g Topas yield over a two year period in sandy soil exceeded by far other accessions similarly tested (21–111 g) (Figure 2.8).

Less flower yield in P7 during the first year at 480 m was attributed to a lack of base branches and early blooming. The yield was 1.5–3 times lower than that of the three commercial varieties. However, the second year yield (91 g in Fougères and 80 g in Epines) was comparable to that of Topas (76 g in Fougères and 93 g in Epines), the best of the three commercial varieties.

This early blooming gives P7 an incomparable advantage at 1060 m. The Bruson yield accumulated over a 2 year period (113 g) exceeded that of Topas (63 g) and Elixir (76 g) which did not bloom until the second season. It provided roughly the same quantity as Hyperimed (119 g). Even if in the plain the accumulated P7 yields were not significantly higher than those of Topas (ANOVA; Newman–Keuls test, $\alpha = 5\%$), this genotype was still one of the most productive plants analysed during our tests (Figure 2.8).

Content of pharmaceutically relevant compounds

In our study, we chose two classes of secondary metabolites which have been discussed as candidates for the various pharmaceutical effects of *H. perforatum* based drugs, flavonoids and hypericins. Presented in Table 2.5 are their 1997 quantified content results for the Epines site. Rutin, hyperosid, isoquercitrin, quercitrin, quercetin and biapigenin were the identified flavonoids, four more remain unidentified. Substance dosages varied greatly between accessions; for example, extreme values of hypericin evolved by a factor of 7.5.

The chemical profile defining quality is dependent on the developmental stage of the plants when harvested (Kartnig *et al.* 1997). It does not seem to be influenced by the soil type, the altitude or the culture's age (Figure 2.9). This principal component analysis demonstrates a very low dispersal for each of the four accessions. These results were consistent at all sites every year.

Hyperimed (P2) and Elixir (P3), and the majority of the wild accessions revealed higher contents of flavonoids and hypericins than our reference variety Topas (P1) (Table 2.5). Genotype P7 is promising for the production of phytopharmaceuticals. It includes the same range of dosed compounds as Topas and also contains 1.26 times more flavonoids and 1.79 times more hypericins as compared to the latter (Table 2.5, Figure 2.10). During the first year in the lowland, the high contents of pharmaceutical compounds in P7 compensate for the low flower yields. So, in terms of yield of secondary metabolites, P7 can be considered as a profitable new variety (Figure 2.11).

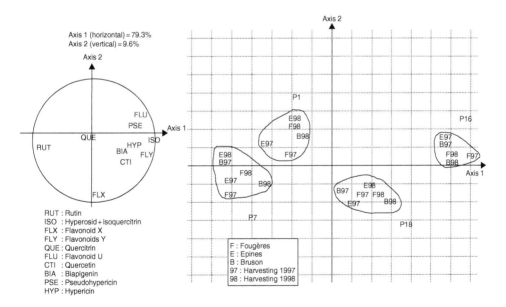

Figure 2.9 Principal component analysis of the chemical composition of the dried flowering segments of four St John's wort accessions grown on various sites over two years.

Table 2.5 Flavonoids and hypericins contents (in mg/100 g of dry flowering segments) for 24 H. perforatum accessions harvested in full bloom during the first year of cultivation (Epines, 1997)

Accessions	Rutin	Hyperosid + isoquercitrin	Flavonoid X	Flavonoids Y[a]	Quercitrin	Flavonoid U	Quercetin	Biapigenin	Pseudohypericin	Hypericin	Total content of flavonoids		Total content of hypericins	
											Content	Index	Content	Index
P1	910	1343	104	39	97	0	119	82	125	34	2693	100	159	100
P2	411	1209	380	408	227	0	169	322	236	108	3125	116	343	216
P3	500	1559	482	408	189	56	200	256	128	58	3649	135	185	116
P4	—	—	—	—	—	—	—	—	—	—	—	—	—	—
P5	504	1388	345	0	118	239	148	165	133	36	2906	108	169	106
P6	493	1434	369	51	120	224	189	159	154	52	3038	113	206	130
P7	1024	1016	453	176	411	0	93	213	204	80	3384	126	284	179
P8	666	940	324	229	153	0	149	123	158	50	2582	96	208	131
P9	0	1841	284	1113	256	132	134	261	112	55	4019	149	167	105
P10	272	847	287	558	301	59	88	94	91	67	2504	93	158	99
P11	206	1430	366	632	126	0	147	80	111	61	2986	111	172	108
P12	1014	1510	176	0	377	182	221	162	138	66	3641	135	203	128
P13	247	1083	270	569	209	0	188	140	243	172	2704	100	415	261
P14	566	1655	569	505	109	336	154	188	140	37	4080	152	177	111
P15	249	1047	248	514	246	0	151	155	213	138	2609	97	350	220
P16	0	2083	254	676	344	264	162	265	327	125	4046	150	452	284
P17	621	1415	510	390	132	0	119	82	81	40	3268	121	121	76
P18	661	1794	573	548	143	0	174	206	182	71	4099	152	253	159
P19	798	1231	144	49	172	249	182	55	293	80	2877	107	373	235
P20	997	1288	464	310	240	0	229	233	252	72	3760	140	324	204
P21	482	1531	180	79	98	256	209	72	144	40	2905	108	183	115
P22	877	1436	249	36	155	0	150	150	113	74	3052	113	187	118
P23	1118	1533	164	51	297	178	166	180	292	97	3685	137	388	244
P24	559	1643	269	50	71	0	114	124	87	23	2830	105	110	69
Mean	573	1402	325	321	200	95	159	164	172	71	3237	—	243	—

Note
a 2 peaks.

Table 2.5 Flavonoids and hypericins contents (in mg/100 g of dry flowering segments) for 24 H. perforatum accessions harvested in full bloom during the first year of cultivation (Epines, 1997)

Accessions	Rutin	Hyperoid + isoquercitrin	Flavonoid X	Flavonoids Y[a]	Quercitrin	Flavonoid U	Quercetin	Biapigenin	Pseudohypericin	Hypericin	Total content of flavonoids		Total content of hypericins	
											Content	Index	Content	Index
P1	910	1343	104	39	97	0	119	82	125	34	2693	100	159	100
P2	411	1209	380	408	227	0	169	322	236	108	3125	116	343	216
P3	500	1559	482	408	189	56	200	256	128	58	3649	135	185	116
P4	—	—	—	—	—	—	—	—	—	—	—	—	—	—
P5	504	1388	345	0	118	239	148	165	133	36	2906	108	169	106
P6	493	1434	369	51	120	224	189	159	154	52	3038	113	206	130
P7	1024	1016	453	176	411	0	93	213	204	80	3384	126	284	179
P8	666	940	324	229	153	0	149	123	158	50	2582	96	208	131
P9	0	1841	284	1113	256	132	134	261	112	55	4019	149	167	105
P10	272	847	287	558	301	59	88	94	91	67	2504	93	158	99
P11	206	1430	366	632	126	0	147	80	111	61	2986	111	172	108
P12	1014	1510	176	0	377	182	221	162	138	66	3641	135	203	128
P13	247	1083	270	569	209	0	188	140	243	172	2704	100	415	261
P14	566	1655	569	505	109	336	154	188	140	37	4080	152	177	111
P15	249	1047	248	514	246	0	151	155	213	138	2609	97	350	220
P16	0	2083	254	676	344	264	162	265	327	125	4046	150	452	284
P17	621	1415	510	390	132	0	119	82	81	40	3268	121	121	76
P18	661	1794	573	548	143	0	174	206	182	71	4099	152	253	159
P19	798	1231	144	49	172	249	182	55	293	80	2877	107	373	235
P20	997	1288	464	310	240	0	229	233	252	72	3760	140	324	204
P21	482	1531	180	79	98	256	209	72	144	40	2905	108	183	115
P22	877	1436	249	36	155	0	150	150	113	74	3052	113	187	118
P23	1118	1533	164	51	297	178	166	180	292	97	3685	137	388	244
P24	559	1643	269	50	71	0	114	124	87	23	2830	105	110	69
Mean	573	1402	325	321	200	95	159	164	172	71	3237	—	243	—

Note
a 2 peaks.

This early blooming gives P7 an incomparable advantage at 1060 m. The Bruson yield accumulated over a 2 year period (113 g) exceeded that of Topas (63 g) and Elixir (76 g) which did not bloom until the second season. It provided roughly the same quantity as Hyperimed (119 g). Even if in the plain the accumulated P7 yields were not significantly higher than those of Topas (ANOVA; Newman–Keuls test, $\alpha = 5\%$), this genotype was still one of the most productive plants analysed during our tests (Figure 2.8).

Content of pharmaceutically relevant compounds

In our study, we chose two classes of secondary metabolites which have been discussed as candidates for the various pharmaceutical effects of *H. perforatum* based drugs, flavonoids and hypericins. Presented in Table 2.5 are their 1997 quantified content results for the Epines site. Rutin, hyperosid, isoquercitrin, quercitrin, quercetin and biapigenin were the identified flavonoids, four more remain unidentified. Substance dosages varied greatly between accessions; for example, extreme values of hypericin evolved by a factor of 7.5.

The chemical profile defining quality is dependent on the developmental stage of the plants when harvested (Kartnig *et al.* 1997). It does not seem to be influenced by the soil type, the altitude or the culture's age (Figure 2.9). This principal component analysis demonstrates a very low dispersal for each of the four accessions. These results were consistent at all sites every year.

Hyperimed (P2) and Elixir (P3), and the majority of the wild accessions revealed higher contents of flavonoids and hypericins than our reference variety Topas (P1) (Table 2.5). Genotype P7 is promising for the production of phytopharmaceuticals. It includes the same range of dosed compounds as Topas and also contains 1.26 times more flavonoids and 1.79 times more hypericins as compared to the latter (Table 2.5, Figure 2.10). During the first year in the lowland, the high contents of pharmaceutical compounds in P7 compensate for the low flower yields. So, in terms of yield of secondary metabolites, P7 can be considered as a profitable new variety (Figure 2.11).

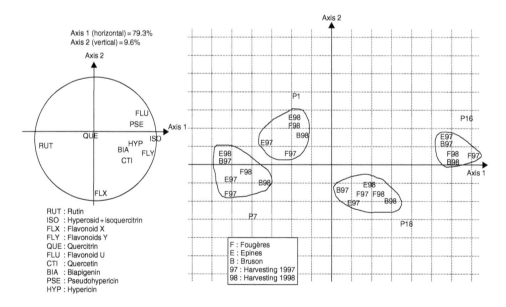

Figure 2.9 Principal component analysis of the chemical composition of the dried flowering segments of four St John's wort accessions grown on various sites over two years.

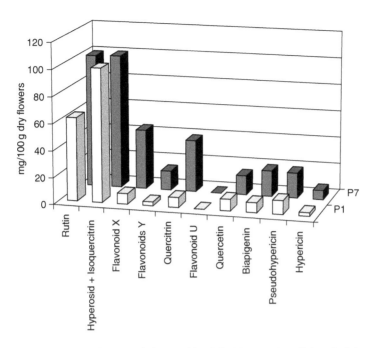

Figure 2.10 Secondary metabolite profile of the Topas variety (P1) and of the selected *H. perforatum* accession (P7). Data are based on means across the three experimental sites (1997).

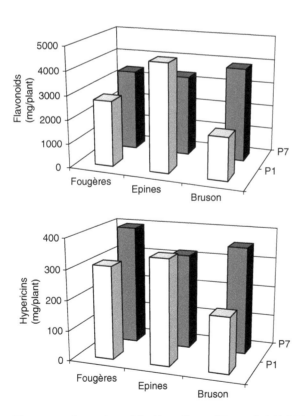

Figure 2.11 Comparison of the Topas (P1) and P7 total yield (1997 + 1998) of compounds.

Laboratory experiment: inoculation of plants

In order to establish unequivocally the P7's tolerance to the pathogenic fungus, we evaluated its response to artificial *C. gloeosporioides* conidial suspension inoculation.

Inoculum production

Our two multi-spored and six mono-spored cultures, provided by the Federal Centre for Breeding Research on Cultivated plants (BAZ)-Quedlinburg-Germany, were used for the test. For production of spores, the fungus was cultured in modified Richard's solution (Daniel *et al.* 1973) on a rotary shaker (100 rpm) at room temperature. Richard's solution contained 50 g of sucrose, 10 g of KNO_3, 5 g of KH_2PO_4, 2.5 g of $MgSO_4 \cdot 7H_2O$, 0.02 g of $FeCl_3$, 150 ml of V_8-juice® (vegetable juice) and distilled water to make the full content 1 l. The solution was adjusted to pH 6 with 50% (w/v) NaOH. Eight cubes (5 mm^3) of PDA with mycelium of the *C. gloeosporioides* isolates were transferred to 250 ml Erlenmeyer flasks containing 100 ml of liquid medium and incubated on the shaker for 4–5 days. Conidia were then harvested by filtration through two layers of cheesecloth and washed twice by centrifugation at 5000 rpm for 10 min with sterile distilled water. The resulting spore pellet was resuspended in sterile distilled water.

Plant material

The commercial varieties Topas (P1), Hyperimed (P2) and Elixir (P3) and 16 of the *H. perforatum* wild accessions (P6–P18, P20, P21 and P24) were sown in trays filled with the potting medium

Figure 2.12 Tray of *H. perforatum* seedlings ready to spore sprays of *C. gloeosporioides* (25 seedlings/row; 8 accessions/tray).

Brill 3® and grown in a greenhouse employing only ambient light. Daytime temperature often exceeded 22°C. After 4–5 weeks seedlings with five to six leaves (Figure 2.12) were infected with *C. gloeosporioides*.

Pathogenicity test

Three replications of 20 seedlings per accession were sprayed till run-off with a spore suspension (10^4 spores/ml), then incubated at high humidity by covering them with a plastic, sprayed internally with water from an atomiser. The plastic was removed 24 h after inoculation. Twenty seedlings per accession sprayed with distilled water served as control plants. Disease symptoms developed on leaves and stems of inoculated St John's wort within 8 days. Foliar lesions were initially <2 mm in diameter but later expanded and coalesced, killing the leaves. Black lesions developed on the stems. With further incubation, green stem tissue between lesions turned brown to give necrotic areas, which girdled the stem. Frequently, lesions affected the entire stem, which became defoliated, and finally killed the seedling (Figure 2.13). Koch's rules were completed by reisolating *C. gloeosporioides* from infected plants.

Three weeks after inoculation, 38.7% of the seedlings were infected and killed by the fungus, 20.6% were affected by anthracnose symptoms and 40.7% were sound (Figure 2.14). As in the case of field results, these findings suggest that the wild accession P7 is less *C. gloeosporioides*

Figure 2.13 Anthracnose-infected *H. perforatum* seedlings inoculated and incubated for 21 days in the greenhouse.

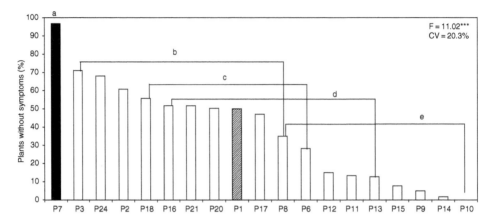

Figure 2.14 Response of the Topas (P1), Hyperimed (P2) and Elixir (P3) commercial varieties and of 16
H. perforatum wild accessions to spore sprays of *C. gloeosporioides* in a greenhouse experiment.
Notations were made 21 days after treatment. (Different letters indicate statistically significant
differences; Neuwman–Keuls test, $\alpha = 5\%$.)

Figure 2.15 Susceptibility difference of two St John's wort accessions to the isolate of *C. gloeosporioides*
(21 days after inoculation).

susceptible than other St John's wort (ANOVA; Neuman–Keuls test, $\alpha = 5\%$). Ninety-seven
per cent of the P7 seedlings were not infected by the organism. Although lesions developed on
some P7 stems (3%), plants did not wilt or die (Figure 2.15). All the water sprayed control
seedlings were healthy at the end of the experiment.

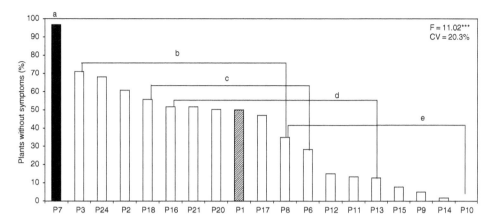

Figure 2.14 Response of the Topas (P1), Hyperimed (P2) and Elixir (P3) commercial varieties and of 16
            H. perforatum wild accessions to spore sprays of *C. gloeosporioides* in a greenhouse experiment.
            Notations were made 21 days after treatment. (Different letters indicate statistically significant
            differences; Neuwman–Keuls test, $\alpha = 5\%$.)

Figure 2.15 Susceptibility difference of two St John's wort accessions to the isolate of *C. gloeosporioides*
            (21 days after inoculation).

susceptible than other St John's wort (ANOVA; Neuman–Keuls test, $\alpha = 5\%$). Ninety-seven
per cent of the P7 seedlings were not infected by the organism. Although lesions developed on
some P7 stems (3%), plants did not wilt or die (Figure 2.15). All the water sprayed control
seedlings were healthy at the end of the experiment.

Brill 3® and grown in a greenhouse employing only ambient light. Daytime temperature often exceeded 22°C. After 4–5 weeks seedlings with five to six leaves (Figure 2.12) were infected with *C. gloeosporioides*.

Pathogenicity test

Three replications of 20 seedlings per accession were sprayed till run-off with a spore suspension (10^4 spores/ml), then incubated at high humidity by covering them with a plastic, sprayed internally with water from an atomiser. The plastic was removed 24 h after inoculation. Twenty seedlings per accession sprayed with distilled water served as control plants. Disease symptoms developed on leaves and stems of inoculated St John's wort within 8 days. Foliar lesions were initially <2 mm in diameter but later expanded and coalesced, killing the leaves. Black lesions developed on the stems. With further incubation, green stem tissue between lesions turned brown to give necrotic areas, which girdled the stem. Frequently, lesions affected the entire stem, which became defoliated, and finally killed the seedling (Figure 2.13). Koch's rules were completed by reisolating *C. gloeosporioides* from infected plants.

Three weeks after inoculation, 38.7% of the seedlings were infected and killed by the fungus, 20.6% were affected by anthracnose symptoms and 40.7% were sound (Figure 2.14). As in the case of field results, these findings suggest that the wild accession P7 is less *C. gloeosporioides*

Figure 2.13 Anthracnose-infected *H. perforatum* seedlings inoculated and incubated for 21 days in the greenhouse.

Conclusion

In the present experiment, *C. gloeosporioides* was identified as the pathogenic agent responsible for *H. perforatum* dieback in Swiss plantations. Susceptible plants which had been sprayed with conidial suspension of the fungus quickly developed typical anthracnose symptoms. Considerable accession effects with regard to anthracnose tolerance, flowering times, plant morphologies, flower and drug yields were observed. The breeding programme consisted of field experimentation and of laboratory inoculations. A new *H. perforatum* variety less anthracnose susceptible than those currently available on the market and well suited for mountain growth was selected. Its morphology guarantees a first rate harvest. Flower and drug yields of this genotype are competitive compared to that of the commercial St John's wort varieties. In addition, its pharmaceutical profile meets industrial requirements. This new cultivar is now being registered. Further research will strengthen agrotechnological methods in order to optimize this superior *Hypericum* variety.

Acknowledgements

This work was funded by Bioforce AG, Postfach 76, CH-9325 Roggwil. We thank André Schwartz and Vincent Michel (Federal Research Station for Plant Production at Changins, CH-1260 Nyon) for their advice and interest in this research. We acknowledge Elke Foltys de Garcia, BAZ, Neuer Weg 22/23, D-06484 Quedlinburg for the supply of *C. gloeosporioides* isolates. We also thank and appreciate Anne-Laure Rauber, Johanna Claude and Yves Dudan for their excellent laboratory work. We gratefully acknowledge Huguette Hausammann and Christian Vergères for their technical assistance.

References

Bruneton, J. (1993) *Pharmacognosie – Phytochimie des plantes médicinales*, Lavoisier TecDoc, London, Paris and New York, 915 pp.

Büter, B., Orlacchio, C., Soldati, A. and Berger, K. (1998) Significiance of genetic and environmental aspects in the field cultivation of *Hypericum perforatum*. *Planta Med.* 64, 431–7.

Cellarova, E., Brutovska, R., Daxnerova, Z., Brunakova, K. and Weigel, R.C. (1997) Correlation between hypericin content and the ploidy of somaclones of *Hypericum perforatum* L. *Acta Biotechnol.* 17, 83–90.

Chatterjee, S.S., Nöldner, M., Koch, E. and Erdelmeier, C. (1998) Antidepressant activity of *Hypericum perforatum* and hyperforin: the neglected possibility. *Pharmacopsychiat.* 31 (supplement), 7–15.

Czygan, F. (1993) Kulturgeschichte und Mystik des Johanniskrautes. *Zeitsch. Phytoth.* 14, 272–8.

Daniel, J.T., Templeton, G.E., Smith, R.J. and Fox, W.T. (1973) Biological control of northern jointvetch in rice with an endemic fungal disease. *Weed Sci.* 21, 303–7.

Debrunner, N. and Simonnet, X. (1998) Le millepertuis: une plante médicinale antidépressive d'un très grand intérêt. *Revue suisse Vitic. Arboric. Horic.* 30, 271–3.

Debrunner, N., Rauber, A.L., Schwarz, A. and Michel, V. (2000) First report of St John's-wort anthracnose caused by *Colletotrichum gloeosporioides* in Switzerland. *Plant Dis.* 84, 203.

Erdelmeier, C.A.J. (1998) Hyperforin, possible the major non-nitrogenous secondary metabolite of *Hypericum perforatum* L. *Pharmacopsychiat.* 31 (supplement), 1–6.

Gaudin, M., Simonnet, X., Debrunner, N. and Ryser, A. (1999) Sélection d'une variété de millepertuis productive et peu sensible au dépérissement – Résultats de deux années de recherche. *Revue suisse Vitic. Arboric. Horic.* 31, 335–3341.

Hildebrand, P.D. and Jensen, K.I.N. (1991) Potential for the biological control of St. John's-Wort (*Hypericum perforatum*) with endemic strain of *Colletotrichum gloeosporioides. Can. J. Plant Path.* 13, 60–70.

Hobbs, C. (1989) St John's Wort (*Hypericum perforatum* L.) – a review. *HerbalGram* 18/19, 24–33.

Hölzl, J. (1993) Inhaltstoffe und Wirkmechanismen des Johanniskrautes. *Zeit. Phytother.* 14, 255–64.

Hostettmann, K. (1997) *Tout savoir sur le pouvoir des plantes sources de médicaments*, Favre, Lausanne, 239 pp.

Kartnig, T., Heydel, B., Lässer, L. and Debrunner, N. (1997) Johanniskraut aus schweizer Arzneipflanzenkultur. *AgraForsch.* 4, 299–302.

Laakmann, G., Schüle, C., Baghai, T. and Kieser, M. (1998) St. John's-Wort in mild to moderate depression: the relevance of hyperforin for the clinical efficacy. *Pharmacopsychiat.* 31 (supplement), 54–9.

Martonfi, P., Brutovska, R. and Cellarova, E. (1996) Apomixis and hybridity in *Hypericum perforatum. Folia Geobot. Phytotax.* 31, 37–44.

Roth, L. (1990) *Hypericum – Hypericin*, Ecomed, Landsberg, 158 pp.

Schaffner, W., Häfelfinger, B. and Ernst B. (1992) *Compendium de phytothérapie*, Arboris-Verlag, Hinterkappelen, Bern, 336 pp.

Schauenberg, P. and Paris, F. (1977) *Guide des plantes médicinales*, Delachaux et Niestlé, Neuchâtel, pp. 386.

Schwarczinger, I. and Vajna L. (1998) First report of St. John's-Wort anthracnose caused by *Colletotrichum gloeosporioides* in Hungary. *Plant Dis.* 82, 711.

Shepherd, R.S.H. (1995) A Canadian isolate of *Colletotrichum gloeosporioides* as a potential biological control agent for St. John's wort (*Hypericum perforatum*) in Australia. *Plant Protec. Quart.* 10, 148–51.

3 A virus causing vein yellowing and necrotic leaf spots of St John's wort (*Hypericum perforatum* L.)

Hartmut Kegler

Introduction

In several European regions, cultivation of St John's wort (*Hypericum perforatum* L.) increased considerably in the course of the last decade. In this connection, occurrence of pests and diseases and their economical consequences were becoming more and more important. However, comparatively little information regarding the occurrence of viruses of *H. perforatum* was found in the literature. Only Achatova *et al.* (1979) found the potato virus Y in Kasachstan and pointed out the danger that St John's wort may be a source of infection for other cultivated plant species.

Recently, Kegler *et al.* (1999) described a virus from St John's wort occurring in a middle German region.

Symptoms

The naturally diseased plants of St John's wort showed severe growth retardation. The young shoots of infected plants were in part smaller than those of healthy plants by approximately one-third to a half. Vein yellowing and pale green as well as brown necrotic spots developed on the leaves. The affected leaves were smaller, narrower and sometimes wavy.

Host range

The experimentally detected host plants of the virus and their reactions after mechanical inoculations are demonstrated in Table 3.1. Altogether 37 plant species have been investigated. The virus could not be transmitted to nine species. These plants did not show any symptoms and re-testings were negative: *Amaranthus cruentus* L., *Asparagus officinalis* L., *Brassica pekinensis* (Lour) Rupr., *Capsicum annuum* L., *Chenopodium capitatum* (L.) Aschers., *Echinochloa frumentaria* Link., *Galega orientalis* Lam., *Lycopersicon esculentum* Mill. 'Moneymaker' and *Zea mays* L. Some more species such as *Celosia argentea* L., *Chenopodium foetidum* Schrad., *Cheonpodium foliosum* Aschers., *Chenopodium murale* L. and *Nicotiana glutinosa* L. could be infected only locally. All other species with the exception of *Lathyrus oderatus* L. were infected systemically and showed distinct symptoms. Some species proved highly sensitive. This specially concerns *Chenopodium quinoa* Willd., *Nicotiana megalosiphon* Heurck et Muell. and *Nicotiana occidentalis* Wheeler which reacted with tip necrosis and frequently died after that. *Nicotiana clevelandii* Grey also developed necrotic lesions on the inoculated leaves whereas the following leaves showed small necrotic spots. The symptoms became milder with progressing age after infection and often there were no symptoms at all. *Lathyrus oderatus* became systemically infected, too, but without developing symptoms.

The virus could be mechanically retransmitted from *N. clevelandii* to the original host plant *H. perforatum* and reisolated from these experimentally infected St John's wort plants to

Table 3.1 Reactions of plant species after mechanical inoculation with the virus isolated from St John's wort (Kegler *et al.* 1999)

Plant species	Symptoms		Biological re-testing	
	Inoculated leaves	Tip leaves	Inoculated leaves	Tip leaves
Antirrhinum majus L.	No symptoms	Small palegreen rings	+	+
Celosia argentea L.	Grey, reddish bordered local lesions, enlarging concentric rings	No symptoms	+	−
Chenopodium amaranticolor Coste et Reyn.	Small pointlike necrotic lesions	Pale green mosaic, leaf distortion	+	+
C. foetidum Schrad.	Yellowish, diffuse or grey bordered lesions	No symptoms	+	−
C. foliosum Aschers.	Grey pointlike necrotic lesions	No symptoms	+	−
C. murale L.	Grey pointlike enlarging lesions	No symptoms	+	−
C. rubrum L.	Grey pointlike necrotic lesions	Yellow spots and vein necroses, leaf deformation	+	+
C. quinoa Willd.	Palegreen, later grey pointlike lesions	Vein clearing leaf distortion tip necroses	+	+
Cucumis sativus L.	Pale green lesions with necrotic centres on cotyledons	Pale green spots and rings	+	+
Cucurbita maxima Duch.	Pale green, later round necrotic spots on cotelydons	Mild pale green mottle	+	+
Datura stramonium L.	Pointlike grey necrotic lesions	No symptoms	+	−
Gomphrena globosa L.	Palegrey necrotic lesions and small rings	Palegreen concentric rings and vein chlorosis	+	+
Hypericum perforatum L. 'Topaz'	No symptoms	Palegreen mosaic, vein yellowing, often latent	+	+
Lathyrus oderatus L.	No symptoms	No symptoms	+	+
Malva sylvestris L.	Palegreen spots	Vein clearing	+	+
Nicotiana benthamiana Domin.	Single necrotic spots	Mild vein clearing	+	+
N. clevelandii Grey	Grey necrotic rings, lines or spots and pointlike grey lesions	Palegreen mosaic, small grey necrotic spots	+	+
N. glutinosa L.	Single grey, brownish bordered lesions	No symptoms	+	+

(*Continued*)

Table 3.1 (Continued)

Plant species	Symptoms		Biological re-testing	
	Inoculated leaves	Tip leaves	Inoculated leaves	Tip leaves
N. megalosiphon Heurck et Muell	Grey necrotic lesions and vein necrosis	Palegreen mosaic, tip necrosis	+	+
N. occidentalis Wheeler	Palegreen and grey necrotic lesions and lines	Palegreen mosaic, small necrotic rings, spots or vein necroses, tip necrosis or crinkling of young leaves	+	+
N. tabacum L. 'Bel'	Large necrotic spots	Mild palegreen mosaic or no symptoms	+	+
N. tabacum L. 'Samsun'	Large necrotic spots and rings	Mild palegreen mottle, rarely necrotic rings	+	+
N. tabacum L. 'Xanthi'	Grey, dark bordered necrotic spots and rings	Grey necrotic rings and lines	+	+
Ocimum basilicum L. ssp. *basilicum* var. *majus*	Dark brown bordered spots	Vein clearing and palegreen mosaic	+	+
O. basilicum L. ssp. *minimum* var. *minimum* f. *minimum*	No symptoms	Palegreen rings and spots	+	+
Petunia hybrida Gaertn.	Grey necrotic lesions	Palegreen mosaic, leaf crinkling	+	+
Phaseolus vulgaris L.	Palegreen spots and single brown lesions	Brown vein necroses	+	+
Spinacia oleracea L. 'Matador'	Yellowish green spots	Palegreen mosaic, leaf crinkling	+	+

Notes
+ = Virus evidence,
− = No virus evidence.

N. clevelandii. Thus, the Koch's postulates for an evidence of a causal connection between disease and isolated pathogene have been fulfilled. However, some plants of the St John's wort cultivar 'Topaz' did not show any symptoms and were proved latent infected. Further investigations will be necessary for clearing up the question of whether these facts are also true for St John's wort plants in the field.

Properties *in vitro* and *in vivo*

The thermal inactivation point of the virus was established at 52°C; thus, it is comparatively low. The dilution end point of 10^{-2} is also not very high and suggests a comparatively low virus concentration. The longevity of the virus *in vitro* in crude leaf extracts of *Nicotiana* species amounts to 3 days but in leaf extracts of St John's wort it was only 1 day. The faster inactivation of the virus in extracts of *H. perforatum* may be caused by antiviral substances which were found in leaves of this species (Schuster and Oschütz 1979).

The virus remains infectious for at least 3 months in frozen leaves of *N. clevelandii* at temperatures of $-15°C$ as well as after air drying of the leaves at temperatures of about $20°C$.

Particle size and morphology

The electron microscopical dip preparations contrasted with uranyl acetate showed isometric particles. The avarage diameter of the particles was 27 nm. But most of the particles (about 80%) appeared as 'empty' virus covers. Intact virions were rarely observed.

Serological properties

The production of antisera suitable for ELISA did not cause many problems. This points to a high immunogenicity of the virus. Titres of 1/2048 could be obtained. The virus was detectable with the produced antisera by a variant of DAS-ELISA in crude saps of *N. clevelandii* and *N. occidentalis* as well as in experimentally infected plants of *H. perforatum*. Furthermore, experiments were carried out concerning the relationship of this virus with other viruses. For this, we used a partially purified and concentrated antigene of the virus isolate from St John's wort which was tested with antisera of arabis mosaic virus, cherry leaf roll virus, cucumber mosaic virus, Havel river virus, *Petunia* asteroid mosaic virus, raspberry bushy dwarf virus, strawberry latent ringspot virus, tobacco necrosis virus (strain D), tomato blackring virus and tomato ringspot virus. Reactions did not appear in any of the cases. Therefore, the virus isolated from St John's wort seems to be neither identical with nor related to the viruses investigated.

Virus transmission

The mechanical transmissibility of the virus was already proved after isolating it from St John's wort to *C. quinoa*. All further experiments on the virus evidence and on the host range confirmed this fact. The virus, therefore, can be considered as easily transmissible mechanically.

The experiments concerning transmissibility of the virus by aphid *Myzus persicae* Sulz. proved negative. Whether other aphid species are able to transmit the virus is unknown but not very probable.

Contact of leaf surfaces as well as injuries by cutting were investigated as further possibilities of natural virus transmission. Different variants of experiments were conducted on contact transmission. It was demonstrated that the virus can be transmitted by slight touching of leaf surfaces of diseased and healthy plants in certain combinations. Contact transmissions succeeded from St John's wort plants to *N. occidentalis* plants as well as from *N. occidentalis* plants to *N. occidentalis* plants and from *N. clevelandii* plants to *N. occidentalis* plants. However, the rates of successful transmission were different. The infection of healthy St John's wort plants succeeded only in a few cases (about 10%). Healthy plants of *N. occidentalis*, however, were infected to almost a hundred per cent. Often there appeared 3–5 lesions per leaf on the leaves in contact. However, no infections were detected at *N. clevelandii*. The results of the transmission experiments were checked by biological re-testings. Probably, the different structures of the leaf surfaces, the kinds of leaf hairs and the sensitivity of the leaf epidermis, respectively, may be responsible for the different results of contact transmissions. In any case, spread of the virus by contact with neighbouring plants should be expected.

Furthermore, the virus could be transmitted from and to St John's wort plants as well as from and to *N. clevelandii* and *N. occidentalis* plants, respectively, by cutting wounds. Presently it

Table 3.2 Contents of compounds of healthy and virus diseased plants of St John's wort (location of the experiments: Artern 1997; second year of vegetation; mean values of each five individual plants) (Kegler *et al.* 1999)

Contents	'Topaz'		Cultivars 'Hyperiflor'		'Hyperimed'	
	Healthy	Infected	Healthy	Infected	Healthy	Infected
Amount of methanolic extract (%)	34.6	27.5*	25.8	26.3	25.9	29.0
Total contents of hypericine after DAC (%)	0.111	0.085*	0.087	0.089	0.080	0.106*
Relation pseudohypericine/ hypericine (mg/g)	1.48	1.28	1.38	1.58*	1.51	1.82*
Total contents of biflavone (mg/g)	0.23	0.22	0.29	0.17	0.28	0.29
Total contents of flavonoids among them (mg/g)	41.22	40.99	34.29	28.96	22.22	27.78
Rutine (mg/g)	9.73	9.18	9.50	8.99	3.60	5.10*
Quercitine (mg/g)	1.78	2.91	4.06	2.34*	3.61	2.74
Quercitrine (mg/g)	0.20	0.16	0.00	0.00	0.04	0.00
Isoquercitrine (mg/g)	7.65	8.39	6.48	5.31	6.80	8.50
Hyperoside (mg/g)	21.86	20.36	14.25	12.32	8.17	11.44*
Contents of hyperforine (mg/g)	12.81	12.64	12.30	10.02	26.51	27.60

Note
* Compared with the mean values differences are significant with $\alpha = 5\%$.

cannot be assessed, but neither excluded, whether this way of virus transmission may have a practical importance.

Plant contents

Interactions of virus infections and biochemical reactions of the plants were investigated by first experiments. It was demonstrated that in comparison with healthy plants the diseased plants showed a changed spectrum of contents (Table 3.2). For example, the amount of methanolic extract was significantly reduced in infected plants of the cultivar 'Topaz' and quercitine was significantly reduced in 'Hyperiflor'. On the other hand, contents of rutine and hyperoside significantly increased in infected plants of the cultivar 'Hyperimed'. Further investigations will be necessary for a significant evidence of reactions typical for the species and specific for cultivars, after virus infections, which includes the influence of the growing site.

Designation of the virus

Obviously, the virus was hitherto unknown in *H. perforatum*. It was proved that it was not identical to and not related to ten other isometric viruses of different taxonomic groups. However, a final assignation of the virus is not yet possible with the present state of knowledge. Therefore, its designation may follow later.

48   *H. Kegler*

References

Achatova, F.C., Elisejeva, Z.N. and Katin, I.A. (1979) Rezervatori Y-virusa. *Vestnik sjelskochosjaistvennoij Nauki Kasachstana* 4, 36–9.

Kegler, H., Fuchs, E., Plescher, A., Ehrig, F., Schliephake, E. and Grüntzig, M. (1999) Evidence and characterization of a virus of St John's wort (*Hypericum perforatum* L.). *Arch. Phytopathologie u. Pflanzenschutz* 32, 205–21.

Schuster, G. and Oschütz, H. (1979) In-vitro- und in-vivo-Inaktivierung des Tabakmosaikvirus durch Extrakte aus *Physarum nudum* Macbr. und *Hypericum perforatum* L. *Berichte Inst. Tabakforschung* 26, 28–35.

4   St John's wort herb extracts

Manufacturing, standardisation and characterisation

Frauke Gaedcke

Introduction

Preparations of St John's wort (*Hypericum perforatum* L.) nowadays are mainly used in the treatment of mild and moderate depressions. According to the German Drug Prescription Report 1996 (Schwabe *et al.* 1996) German physicians prescribed 131 millions of daily doses of herbal medicinal products prepared from St John's wort herb.

General possibilities for manufacturing St John's wort herb preparations

The 1997 German List of Prescription Drugs (Rote Liste 1997) includes more than fifty preparations from St John's wort herb. Table 4.1 lists forms of St John's wort preparations which are available on the market.

The individual manufacturing methods are illustrated in Figure 4.1 by a simplified flow chart. Consequently, the question arises as to which kind of extract is used for which therapeutic indication.

Constituents and their classification

To answer this question, it is important to be aware of the individual St John's wort herb constituents, their structure and their chemical, physical and pharmacological features. In general, according to their therapeutic potential, constituents of herbal drug preparations can be classified into three categories as listed in Table 4.2.

The application of this classification to St John's wort herb and its preparations leads to results which are shown in Table 4.3.

Table 4.1 Commercially available preparations from St John's wort herb

Herbal drugs/extracts	Dosage-form
St John's wort herb fine cut	Teas, tea filter bags
St John's wort powder	Sugar coated tablets, tablets
St John's wort dry extracts	Sugar coated tablets, tablets and hard/soft gelatine capsules
St John's wort liquid extracts/tinctures	Drops, tonics, ointments, ampoules, tinctures
St John's wort pressed juices	Juices
St John's wort oils/oily macerates	Ointments, drops, oils, soft gelatine capsules
Homeopathic remedies (mother tinctures/dilutions)	Drops, tablets, ampoules, injection solutions

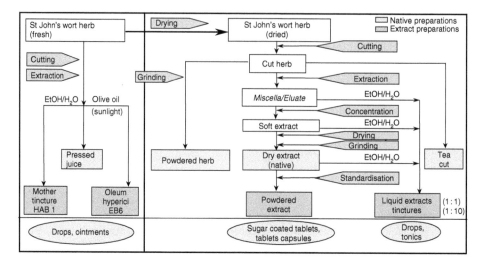

Figure 4.1 Manufacturing flow chart of St John's wort herbal drug preparations.

It can be deduced that according to the polarity of the extraction solvent, quite different constituents of St John's wort herb are transferred to tea infusions, extracts or oils. Consequently, not only different chemical and physical features but above all also different pharmacological and clinical effects result. In addition, it is obvious that the active principle of St John's wort herb and its preparations has not yet been clarified. Hypericin and pseudohypericin are not the only active substances, as it was assumed for years. It is evident that the synergism of numerous constituents is decisive for the therapeutic effect. This also explains why in the so-called 'Bühler Paper' of 1995 (Bühler 1995), total hypericin was no longer accepted to be the only 'active substance' but a 'pharmaceutically relevant constituent'. In Germany St John's wort extracts are, therefore, no longer permitted to be 'standardised' to a certain content of total hypericin (in German 'normiert'). It is more important for the therapeutic effect to standardise to a constant amount of native extract than to a constant amount of total hypericin. As a consequence the content of hypericin is no longer allowed to be labelled, but it is still an irreplaceable indicator for a good pharmaceutical quality of St John's wort herb and extracts. High contents of hypericin and pseudohypericin demonstrate the use of dried flowering tops (Table 4.5).

Besides hypericin the total composition of constituents must be included in the quality assessment. A reproducible composition from batch to batch (within realistic limits) is an essential precondition to guarantee a reproducible therapeutic success.

For St John's wort herb this is a difficult challenge due to the 'natural' variations and the variety of potentially active substances. In the concept of standardisation, not only should the quality of the herbal drug be included, but also the manufacturing process and the analytical characterisation.

Quality of the herbal drug

It is well known that the quality of St John's wort herb varies considerably due to numerous external influences such as, for example, the climate, the duration of solar radiation and the harvesting and drying conditions – even if the herb is cultivated.

Table 4.2 General classification of the constituents of herbal drugs and/or their preparations

No.	Classification	Definition	Examples	
			Substances present	*Extracts*
1	Active principles/ constituents or groups of constituents determining efficacy	Constituents showing equal or similar therapeutic effect being isolated or present in the whole extract	• Silymarin • Aescin • Sennosides	• Extr. Cardui mariae • Extr. Hippocastani • Extr. Sennae
2	Pharmaceutically relevant constituents or groups of constituents	Constituents present which do *not* show the same or similar therapeutic effect in an isolated state as the whole extract. They are, however, co-responsible and/or relevant for the total effect of the extract according to our present level of knowledge (see Bühler 1995).	• Hypercin/ Pseudohypericin/ Hyperforins • Procyanidins/ Flavonoids • Ginsenosides	• Extr. Hyperici • Extr. Crataegi • Extr. Ginseng
3	Markers	Constituents present which make *no* contribution to the stated effect. According to our present level of knowledge		
	3.1 Characteristic	Constituents which are characteristic for the individual species or plant family and are suitable for *identification* and *quantification* (batch specific control)	• Valerenic acids • Echinacoside	• Extr. Valeriannae off. • Extr. Echinaceae pall./ang.
	3.2 Ubiquitous	Constituents which are suitable only for *quantification* (batch specific control)	• Rutoside • Chlorogenic acid	• Extr. Hyperici/ Extr. Solidaginis • Extr. Hyperici/ Extr. Cynarae

The aim must be to compensate these natural fluctuations of St John's wort herb constituents as far as possible by 'quality standardisation' to achieve the required standards. These standards (specifications) are deduced from the quality profile of the herbal drug preparation which has been accepted by the registration authorities. Its reproduction is an indispensable precondition for the claimed therapeutic effect.

'Quality standardisation' can be achieved by

1 Switching from wild collection to cultivation (practised already for several years by large-scale manufacture).
2 Blending of herbal drug batches (as far as possible):

- from different crops
- from different sites
- from wild collection and cultivation.

Table 4.3 Constituents of St John's wort herb and their classification

No.	Substance group	Examples	Location	Lipophilic/ Hydrophilic	Classification			Assumed effect	Extracts
					Markers	Pharmaceutically relevant substance	Active substance		
1	Phloroglucin derivates	Hyperforin (Berghöfer 1987, Maisenbacher et al. 1992)	Blossom (fresh), buds, capsules (Berghöfer 1987, Berghöfer et al. 1988)	Very lipophilic	–	+++	–	• Antibacterial (Hagenström 1955) • Antidepressant • Antibiotic (Gurevich et al. 1971)	• Oil • CO_2 extract • Extract*
2	Essential oil	2-Methyl-3-buten-2-ol (Wohlfart 1983)	Blossom, herb, fruit	Lipophilic	+	–	–	• Sedative (Wohlfart 1983)	• Oil • CO_2 extract • Tincture • Extract*
3	Xanthones	Norathyriol (Berghöfer 1987, Sparrenberg 1993)	Root, blossom (Berghöfer et al. 1987)	Lipophilic	–	++	–	• Antidepressant (Berghöfer 1987, Sparrenberg 1993)	
4	Biflavanoids	Biapigenin (Berghöfer et al. 1988) (Amentoflavon)	Only blossom	Lipophilic	–	++	–	• Sedative (Nielsen et al. 1988) • Antiphlogistic (Schimmer 1986)	• Oil • Tincture • Extract*

No.	Group	Constituents	Plant part	Solubility				Effects	Forms
5	Naphthodianthrones	Hypericin Pseudohypericin (Berghöfer 1987)	Blossom, buds	Lipophilic	—	+++	—	• Antiviral (Meruelo et al. 1988) • Antidepressant (Butterweck et al. 1996) • Photosensitising/phototoxic (Knox et al. 1985)	• Tincture • Extract • Tea infusion
6	Flavonoids	• Quercetin • Quercetin-glycoside (hyperoside, rutoside, quercetrin)	Blossom, leaves	Lipophilic/hyrophilic	—	+++	—	• Antidepressant (Sparrenberg 1993) • Antiphlogistic	
7	Tannins/Polyphenols	Catechinic acid, Epicatechol	Stem, leaves, fruit	Hydrophilic	++	—	—	• Antiphlogistic • Antioxidative	
8	Vegetable plant acids	Chlorogenic acid. Caffeic acid	Blossom, herb	Hydrophilic	+++	—	—		• Hydrophilic extract • Tea infusion
9	Amino acids	GABA (Lapke et al. 1996)	Blossom, leaves	Hydrophilic	+	—	—	• Antidepressant (Lapke et al. 1996)	

Note
* Aqueous-alcoholic, strength > 50% (V/V).

Table 4.4 Disadvantages of the conventional assay method for dianthrones according to the DAC Monograph 86, 3rd Supp. 91 for St John's wort herb

1	The reproduciblility of the method is not sufficient
2	The calculated dianthrone content is too high due to the determination of matrix constituents and protocompounds
3	The minimum level of 0.04% dianthrones is too low

Table 4.5 Standard specification for St John's wort herb

Test parameters	Requirements	Remarks
1 Plant part used	Dried 'flowering tops'	Pharmaceutically relevant constituents/ potential active substances are found predominantly in the blossom/leaves, *not* in the stems
2 Identity		
• TLC fingerprint on hypericins/flavonoids	Method acc. to 'Apothekengerechte Prüfvorschriften'	Method acc. to DAC 86, 3rd Supp. 91 has little relevance
3 Purity		
• Portion of stems/leaves	• Max. 70%	See remark under 1
• Ash content	• Max. 7%	
4 Assay		
• Dianthrones acc. to DAC 86, 3rd Supp. 91	At least 0.08%	
• Σ flavonoids	Minimum values or ranges (HPLC)	Values must be derived from several harvests and manufacturing years
• Σ hypericin/pseudohypericin		
• Σ hyperforins		

Unfortunately, in Germany, there is still no Pharmacopoeia Monograph available that has been adapted to the present level of scientific knowledge. The current DAC Monograph must be considered as obsolete. The most important reasons for this are listed in Table 4.4 taking into account only the assay.

Even the draft monograph which was published recently in 'Pharmeuropa' (Pharmeuropa 1996) is not adequate to meet the high requirements of today's St John's wort herb preparations, because of its very few quality specifications and the omission of any assay. In most laboratories an HPLC method has therefore been developed for quality control, which selectively determines the hypericin and pseudohypericin content after exposure of the sample solution to light (Krämer *et al.* 1992).

Based on daily experience the following standard specification (Table 4.5) can be deduced for a monograph.

Extract manufacturing

Apart from a standardised herbal drug, the parameters listed in Figure 4.2 are relevant for a reproducible extraction process.

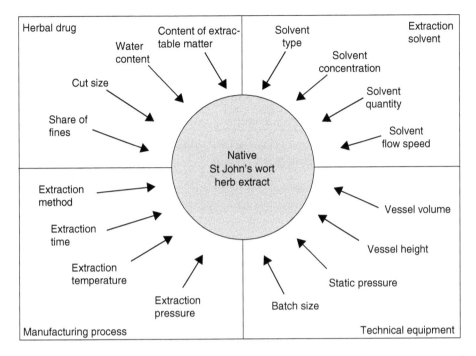

Figure 4.2 Dependence of yield and composition of the native St John's wort herb extract on the manufacturing and quality parameters.

The following figures demonstrate the dependence

1 of the extraction solvent on

- the content of extractable matter (Figure 4.3),
- the extraction rates of pseudohypericin and hypericin (Figure 4.4), and
- the extraction rates of flavonoids, biapigenin and hyperforin (Figure 4.5).

2 of the extraction temperature and time on

- the total hypericin content (Figure 4.6) and
- the flavonoid/biapigenin contents (Figure 4.7).

The validation data obtained from these experimental results lead to the optimised extraction conditions that are shown in Table 4.6. They are essential for an exhaustive extraction of the pharmaceutically relevant constituents of St John's wort herb.

The content of the naphthodianthrones is a good indicator for proof of success of extraction due to the low solubility of naphthodianthrones in ethanol 60% and methanol 80%. If the naphthodianthrones are extracted nearly quantitatively an exhaustive extraction of the herbal drug can be assumed.

To avoid the decomposition of the thermolabile naphthodianthrones and hyperforins the solvent (methanol/ethanol) has to be removed under mild conditions.

The soft extracts obtained contain resins and chlorophyll. They are inhomogeneous and even separate into two layers. At this stage it is almost impossible to take representative samples for

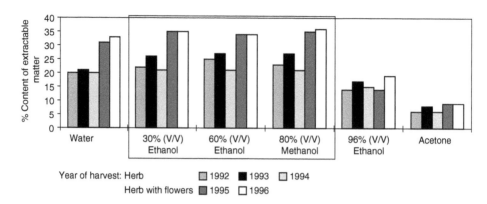

Figure 4.3 Dependence of the extractable matter content (DER_{native}) on the extraction solvent and year of harvest (laboratory preparations): optimal extraction solvents: 30–80% (V/V) alcohol/water mixtures.

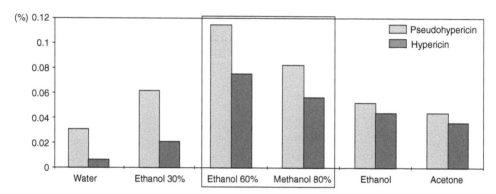

Figure 4.4 Dependence of the extraction rates of hypericin/pseudohypericin on the extraction solvents (laboratory preparations); optimal extraction solvents: 60% (V/V) ethanol/80% (V/V) methanol.

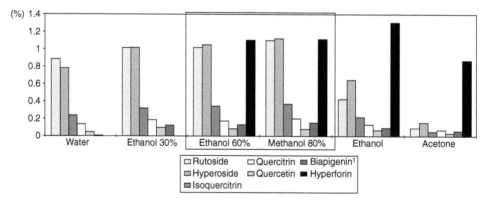

Figure 4.5 Dependence of the extraction rates of flavonoids/biapigenin and hyperforin (HPLC) on the extraction solvent (laboratory preparations); optimal extraction solvents: 60–80% (V/V) alcohol/water mixtures.

Note
1 Biapigenin calc as amentoflavon.

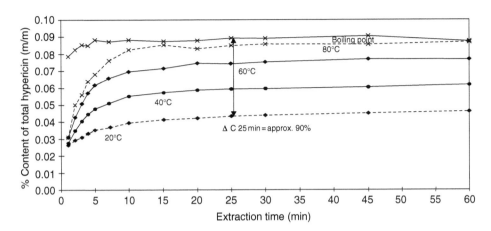

Figure 4.6 Dependence of the extraction rates of total hypericin (HPLC) on the extraction temperature and time; extraction solvent: 80% (V/V) methanol; optimal temperatures: 60–80°C (laboratory preparations).

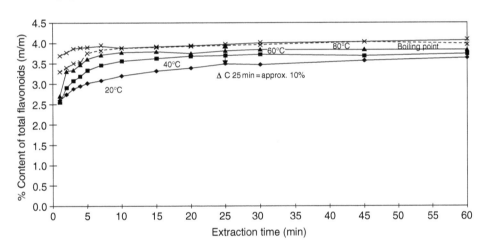

Figure 4.7 Dependence of the extraction rates of the flavonoids and the biapigenin on the extraction temperature and time; optimal temperatures: 60–80°C (laboratory preparations).

Table. 4.6 Optimised manufacturing conditions for St John's wort herb extracts

Manufacturing conditions	Requirements	Remarks
1 Herbal drug cut size	e.g. 3–6 mm	Dependent on the type of extraction/extractor
2 Extraction solvent	• Ethanol 60% (w/w) • Methanol 80% (V/V)	Extraction solvents of the clinically tested extracts
3 Extraction temperature	$> 50°C; < 80°C$	Higher extraction temperatures are generally not feasible in production
4 Extraction method	Exhaustive percolation/special maceration	Dianthrones are less soluble
5 DER$_{native}$	e.g. 3–6 : 1 (60 % (w/w) ethanol) 3–7 : 1 (80% (V/V) methanol)	Precision should be possible after production of several harvests and manufacturing years

Table 4.7 General possibilities for the standardisation of St John's wort extracts

	Standardisation 1	*Standardisation 2* (in German 'Normierung')
1 *General*		
Therapeutically relevant composition of the extract	Total native extract	Total hypericin
• Constant	• Amount of native extract • Amount of excipients	• Amount of total hypericin (pseudohypericin/hypericin)
• Variable	• Amount of total hypericin (pseudohypericin/hypericin)	• Amount of native extract • Amount of excipients
2 *Specification (example)*		
• Constant	• 80% native extract • 17% maltodextrin • 3% silica	• 0.30–0.33% total hypericin
• Variable	• Amount of total hypericin (e.g. at least 0.24% in the extract) • Batch-specific determination	• 80–97% native extract • 17–0% maltodextrin • 3% silica
3 *Labelling of the dianthrone content*	No	Yes

analysis. Thus, analytical testing of the soft extract can only serve for first orientation. The upper, solid layer that contains the chlorophyll and other lipophilic constituents (hyperforins) makes the following drying process rather difficult. In some special manufacturing processes the lipophilic layer is therefore eliminated either mechanically or with the aid of lipophilic solvents. In this case, however, a substantial loss of hyperforin must be accepted.

In most cases the soft extracts are homogenised while adding technical auxiliaries (e.g. maltodextrin, silica) and dried under mild conditions (Meier *et al.* 1996), for example, by means of spray, vacuum-belt or spray-belt driers. These preparations are less hygroscopic and show improved flowability compared with native extracts.

The standardisation with inert auxiliaries can be performed either at the stage of the soft extract (spissum) or at the stage of the dry extract (siccum). In principle, two different methods have to be distinguished, depending on whether the total native extract (standardisation 1) or the total hypericin (standardisation 2) are assumed to be responsible for the therapeutic effect. Depending on the respective interpretations the differences in the specification and the labelling are shown in Table 4.7. In Germany only standardisation 1 is allowed since the publication of the so-called 'Bühler-paper' of 8 September 1995 (Bühler 1995).

For the manufacturing of St John's wort extracts with a reproducible spectrum of constituents a detailed quality assurance system is required, as illustrated in Figure 4.8.

Analytical testing of St John's wort herb extracts

Apart from the standardisation of the herbal drug and the validation of the manufacturing process, the analytical methods used are also of significant importance. Only with the aid of selective and reproducible procedures an objective assessment of the quality of the herbal drug and extracts is possible. As mentioned before (Table 4.4), the DAC method does not meet these requirements. The inaccuracies are even more significant for the extract than for the herb due to the lack of chlorophyll elimination in the sample preparation of the extract assay. As a consequence, the values are too high (Gaedcke 1997b, Schütt *et al.* 1993). As an alternative, the HPLC method for the determination of naphthodianthrones, published by Krämer and Wiartalla (1992), is used. It determines the hypericin and pseudohypericin content after

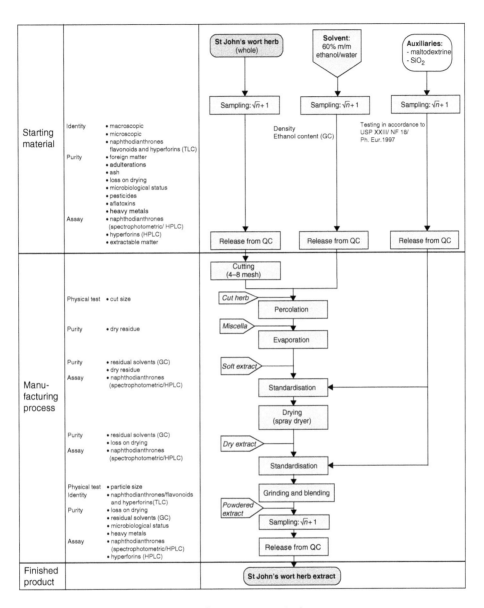

Figure 4.8 Quality assurance programme for St John's wort herb extracts.

exposure to light (Figure 4.9) and has been proved to be selective, robust and reproducible (Gaedcke 1997b).

To be able to compare the HPLC values of pseudohypericin and hypericin with the corresponding DAC values Table 4.8 shows the differences in the values according to both methods. By dividing the HPLC contents by the given factors (1.1 or 0.8) the values of dianthrones according to DAC can be calculated in the herbal drug and extracts. The advantage is that it is possible to relate the recommended daily dosages of 0.2–1.0 mg of total hypericin (according to DAC 1979) given in the Commission E monograph (Bundesanzeiger 1989) to the HPLC contents as demonstrated in Table 4.9.

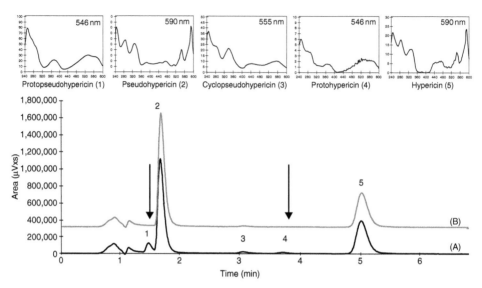

HPLC conditions: Eluent: Ethyl acetate/methanol/sodium dihydrogen phosphate phosphoric acid buffer pH 2.1
(isocratic)
Column: RP18–125 × 4 mm
Detection: 590 nm

Figure 4.9 HPLC chromatogram of the naphthodianthrones in St John's wort herb extract before (A) and after (B) exposure to light.

Table 4.8 Comparison between the determination of hypericin/pseudohypericin (HPLC) after Exposure of the sample solution to light (HPLC) and the dianthrone determination according to DAC 86 3rd supp. 91 in the herb and extract of St John's wort

Batch	Dianthrones DAC 86 – 3rd supp. 91 (%)	Pseudohypericin/Hypericin (HPLC) after light exposure (%)	DAC/HPLC factor
Herbal drug			
1	0.094	0.100	0.94
2	0.096	0.098	0.98
3	0.224	0.254	0.88
4	0.224	0.237	0.95
5	0.055	0.069	0.80
Average			approx. 0.9
Extract (standardisation 1)			
1	0.266*	0.239	1.11
2	0.264*	0.224	1.18
3	0.223*	0.170	1.31
4	0.220*	0.187	1.18
5	0.500*	0.320	1.56
Average			approx. 1.3

Note
* (mod.) methanolic solution of extract.

Table 4.9 Calculation of the correlating dianthrone and hypericin/pseudohypericin contents in St John's wort herb and extract related to the daily dosage of 0.2 to 1.0 mg total hypericin according to the Monograph of the Commission E

Daily dosage	Analytical method	Preparation
0.2 to 1.0 mg total hypericin	DAC 79	Herbal drug
0.17 to 0.83 mg dianthrones	DAC 86, 3rd Supp. 91	Herbal drug
about 0.19 to 0.92 mg pseudohypericin/ hypericin	HPLC (after exposure to light)	Herbal drug
about 0.13 to 0.64 mg pseudohypericin/ hypericin	HPLC (after exposure to light)	Extract

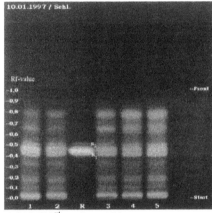 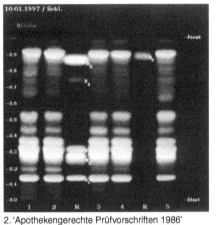

1. DAC 86-3rd supplement 91
Reference substances:
R = R$_1$ = caffeic acid R and R$_2$ = hypericin HF-HO2

2. 'Apothekengerechte Prüfvorschriften 1986'
Reference substances:
R = R$_1$ = rutin R, R$_2$ = chlorogenic acid RN
R$_3$ = hyperoside RN, R$_4$ = hypericin HF-HO2,
R$_5$ = caffeic acid R, and R$_6$ = amentoflavon

Figure 4.10 TLC fingerprint chromatograms on pharmaceutically relevant constituents of St John's wort herb extracts (extraction solvent: 60% (V/V) Ethanol – 5 production batches). (See Colour Plate XII.)

For the assessment of the qualitative composition of constituents and simultaneously for the control of batch to batch consistency TLC and HPLC methods are generally used. Concerning this, for St John's wort the TLC fingerprint chromatogram described in 'Apothekengerechte Prüfvorschriften' (Rohdewald *et al.* 1994) is much better and basically gives more information about quality than the one of the DAC 86 (Figure 4.10).

For the quantitative assessment the HPLC method described by Hölzl *et al.* (1987) shows a good separation of the pharmaceutically relevant constituents and has been proved to be a valuable method for monitoring the batch to batch consistency (Figure 4.11).

The evaluation of various batches of St John's wort extract, standardised on 0.3% total hypericin (according to DAC) leads, for example, to the following specification for the pharmaceutically relevant constituents:

- rutin: 2–4%
- hyperoside: 1.5–3%
- isoquercitrin: about 1%

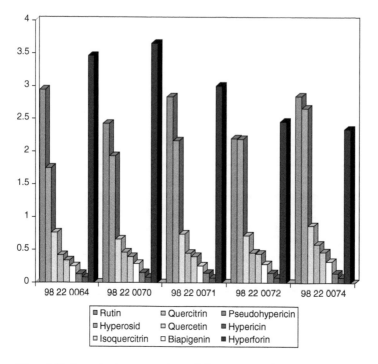

Figure 4.11 Quantitative comparison of pharmaceutically relevant constituents of five batches of St John's wort herb extract (extraction solvent: 60% (V/V) Ethanol – 5 production batches).

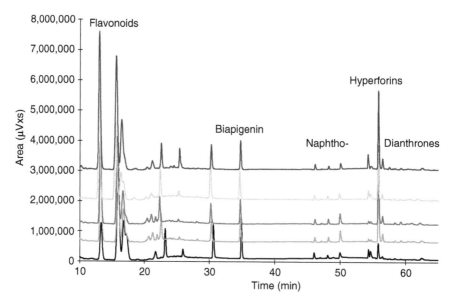

Figure 4.12 HPLC fingerprint chromatogram on pharmaceutically relevant constituents of St John's wort herb extracts (extraction solvent: 60% (V/V) Ethanol – 5 production batches). (See Colour Plate XIII.)

- quercitrin: about 0.5%
- biapigenin: about 0.3%
- hyperforin: 2–4%
- quercetin: not more than 0.5%

Summary

The herbal drug, the type and concentration of the extraction solvent and the manufacturing process have a great influence on the spectrum of constituents present in St John's wort herb extracts. In the past few years enormous efforts have been made to optimise and validate the manufacturing process of St John's wort and to achieve a reproducible composition of the pharmaceutically relevant constituents – such as hypericins, flavonoids, biapigenins and hyperforins – with the aid of a detailed quality assurance programme. Taking into account the peculiarities of phytotherapy Figures 4.11 and 4.12 illustrate that it is possible nowadays to assure the quality of St John's wort herb extracts in such a way that batch to batch consistency and, as a result, reproducible therapeutical effects are guaranteed (Meier *et al.* 1996, Gaedcke 1997a).

References

Bundesanzeiger (Federal Gazette) No. 228 dated 05.12.1984, in the version dated 02.03.1989.

Berghöfer, R. (1984) Analytik und Isolierung phenolischer Inhaltsstoffe von *Hypericum perforatum* L. aus Anbau und Wildvorkommen und Vergleich mit anderen heimischen Hypericum-Arten (Analysis and isolation of phenolic substances in cultivated or wild *Hypericum perforatum* L., and comparison with other native Hypericum species). *Dissertationes Botanicae*, Verlag J. Cramer, Berlin, Stuttgart, vol. 106.

Berghöfer, R. and Hölzl, J. (1986) Johanniskraut (*Hypericum perforatum* L.). Prüfung auf Verfälschung. (St John's wort herb, testing on adulterations). *Dtsch. Apoth. Ztg.* 126, 2569–73.

Berghöfer, R. and Hölzl, J. (1988) Biflavonoids in *Hypericum perforatum*. Part I: Isolation of I3, II8-biapigenin. *Planta Med.* 53, 219–21.

Bühler, W. Letter from the BfArM dated 07.09.1995.

Butterweck, V., Winterhoff, H. and Nahrstedt, A. (1996) An extract of *Hypericum perforatum* decreases immobility time of rats in the force swimming test. *Second International Congress on Phytomedicine*, Munich, 1996.

Gaedcke, F. (1997a) Herstellung, Qualität, Analytik und Anwendung von Johanniskrautextrakten. (Manufacturing, Quality, Analytical testing and use of St John's Wort Extracts.) *Z. Arznei. Gewürzpfl.* 2, 63–72.

Gaedcke, F. (1997b) Johanniskraut und dessen Zubereitungen. Qualitätsbeurteilung über ein selektives reproduzierbares HPLC-Verfahren. (St John's Wort and its preparations; assessment of quality by a selective and reproducible HPLC method.) *Dtsch. Apoth. Ztg.* 137, 3753–7.

Gurevich, A.I. *et al.* (1971) *Antibiotiki* 6, 510. C.A. 1971, 75: 5625.

Hagenström, U. (1955) Das Vorkommen einer antibakteriellen Fraktion in den Kapseln einiger Hypericum-Arten (Presence of an antibacterial fraction in the capsules of certain Hypericum species). *Arzneim. Forsch. (Drug Research)* 5, 155.

Hölzl, J. and Ostrowski, E. (1987) Johanniskraut (*Hypericum perforatum* L.), HPLC-Analyse der wichtigen Inhaltsstoffe und deren Variabilität in einer Population (St John's wort (*Hypericum peforatum* L.), HPLC analysis of the relevant substances present and their variability in a population). *Dtsch. Apoth. Ztg.* 127, 1227–330.

Knox, J.P. and Dodge A.D. (1985) Isolation and activity of the photodynamic pigment hypericin. *Plant Cell Environ.* 8, 19–25.

Krämer, W. and Wiartalla, R. (1992) Bestimmung von Naphtodianthronen (Gesamthypericin) in Johanniskraut (*Hypericum perforatum* L.) (Determination of naphthodianthrones (total hypericin) in St John's wort (*Hypericum perforatum* L.)). *Pharm. Ztg. Wiss.* 137, 202–7.

Lapke, C., Schilcher, H. and Riedel, E. (1996) GABA in *Hypericum perforatum* L. and in other herbal drugs. *Planta Med Abstracts Supp. 44th Annual GA Congress*, Prague, 3–7 Sept. 1996.

Maisenbacher, P. and Kovar, K.-A. (1992) Adhyperforin: a homologue of hyperforin from *Hypericum perforatum*. *Planta Med.* **58**, 291–3.

Meier, B. and Linnenbrink, N. (1996) Status und Vergleichbarkeit pflanzlicher Arzneimittel. *Dtsch. Apoth. Ztg.* **136**, 4205–20.

Meruelo, D., Lavie, G. and Lavie, D. (1988) Therapeutic agents with dramatic antiretroviral activity and little toxicity at effective doses; aromatic polycyclic diones hypericin and pseudohypericin. *Proc. Natl. Acad. Sci. USA* **85**, 5230–4.

Nielsen, M., Froekjaer, S. and Braestrup, C. (1988) High affinity of the naturally-occurring biflavonoid amentoflavon to brain benzodiazepine receptors *in vitro*. *Biochem. Pharmac.* **37**, 3285–7.

Niesel, S. and Schilcher, H. (1990) Johanniskraut (*Hypericum perforatum* L.); Vergleich der Freisetzung von Hypericin und Pseudohypericin in Abhängigkeit verschiedener Extraktionsbedingungen (St John's wort (*Hypericum perforatum* L.); Comparison of the hypericin and pseudohypericin release under varying extraction conditions). *Arch. Pharm.* **323**, 755.

Pharmeuropa vol 8 No 1, March 1996, PA/PH/ EXP. 13/T (95) 84, ANP.

Rohdewald, P., Rücker, G. and Glombitza, K.-W. (1994) Apothekengerechte Prüfvorschriften: Prüfung von Arzneistoffen, Chemikalien, Drogen und Zubereitungen im Apothekenlabor Deutscher Apotheken (Pharmacy-adapted Testing Methods: Investigation of active substances, chemicals, herbal drugs and preparations in German pharmacy laboratories). Deutscher Apotheker Verlag, Stuttgart, 6th supp. edn, pp. 803–6.

Rote Liste 1997: Arzneimittelverzeichnis des BPI, VFA, BAH und VAP (Red List 1997: Drug list of the BPI, VFA, BAH and VAP); Rote Liste Service GmbH, Frankfurt.

Schimmer, O. (1986) Flavonoide – Ihre Rolle als biologisch aktive Naturstoffe (Flavonoids: their function as biologically active natural substances). *Dtsch. Apoth. Ztg.* **126**, 1811–6.

Schütt, H. and Hölzl, J. (1994) Vergleichende Qualitätsuntersuchungen von Johanniskraut-Fertigarzneimitteln unter Verwendung verschiedener quantitativer Bestimmungsmethoden. (Analytical quality comparison of St John's wort dosage forms by means of different quantitative methods of determination). *Pharmazie* **49**, 206–8.

Schwabe, U. and Paffrath, D. (1996) *Arzneiverordnungsreport 1996*. Gustav Fischer Verlag, Stuttgart, Jena, Lübeck, Ulm.

Sparrenberg, B. (1993) MAO – inhibierende Eigenschaften von Hypericum Inhaltsstoffen und Untersuchungen zur Analytik und Isolierung von Xanthonen aus *Hypericum perforatum* L. (MAO inhibiting properties of St John's wort herb constituents and analytical tests and isolation of xanthones from *Hypericum perforatum* L.) Dissertation, Marburg.

Wohlfart, R. (1983) Hopfen-Mite-Sedativum oder Placebo (Hop medium sedative or placebo). *Dtsch. Apoth. Ztg.* **123**, 1637–8.

5 Culture and biotechnology of *Hypericum*

Eva Čellárová

Introduction

The important objective in breeding of the species within the *Hypericum* genus, especially *Hypericum perforatum*, is a selection of genotypes with a high content of different potential active principles with significant pharmaceutical activity, namely naphthodianthrones, hypericin and pseudohypericin, acylphloroglucinols, hyperforin and adhyperforin, and flavonoids. Natural populations and breeding lines of *H. perforatum* show very high variability of the active principles. With regard to the flavonglycosides two distinct forms have been reported, one with a high content of rutine and hyperoside and the other with a very low content of rutine but high hyperoside and isoquercitrine content. The latter is mostly linked with a very high content of total hypericins (Franke *et al*. 1999). A high degree of variability was observed in the routes of seed formation and ploidy level. Eleven divergent reproductive pathways were reconstructed from the results of the flow cytometric seed screen analysis (Matzk *et al*. 2001).

Tissue cultures, which provide an important tool for plant breeders have begun to be used in *Hypericum* research from the beginning of the 1990s in several laboratories in Europe, Brazil and Japan. The establishment of an experimental system and the optimisation of culture conditions for *in vitro* regeneration and multiple shoot formation in order to broaden genetic variability has been reported for several *Hypericum* species (Mederos Molina 1991, 1997, Čellárová *et al*. 1992, 1994, Oliviera 1996, Čellárová and Kimáková 1999, Murch *et al*. 2000, Pretto and Santarém 2000). Much effort has recently been concentrated on the study of biosynthetic capacity of cell cultures and isolation of productive callus and cell lines in *H. erectum* (Yazaki and Okuda 1990, 1994) and in several other *Hypericum* species (Zdunek and Alfermann 1992, Kartnig *et al*. 1996, Ishiguro *et al*. 1996, 1999, Dias *et al*. 1998, 1999, 2000, Kirakosyan *et al*. 2000a,b).

In vitro culture of *Hypericum perforatum* L.

Source of explants and culture conditions

Most of the tissue cultures of *Hypericum* species were isolated from the whole young seedlings (Čellárová *et al*. 1992, 1994, 1999, Kartnig *et al*. 1996), seedling segments (Alfermann 1992, Yazaki and Okuda 1994, Zdunek and Murch *et al*. 2000), shoot apices and axillary buds of mature plants (Mederos Molina 1991, 1997) or nodal segments with attached leaves (Cardoso and Oliviera 1996) or leaf halves (Pretto and Santarém 2000). The explants isolated from different *Hypericum* species are successfully cultured on basal full or half strength Murashige-Skoog's (1962), Linsmaier-Skoog's (1965), Gamborg's B5 (Gamborg *et al*. 1968) or WPM (Lloyd and McCown 1981) culture medium supplemented with different plant growth regulators (Table 5.1).

Table 5.1 Establishment of tissue and cell cultures of some *Hypericum* species

Species	Culture medium	Growth regulators	Morphogenetic response	References
H. brasiliense	MS	None	Multiple shoots	(Cardoso and de Oliviera 1996)
H. brasiliense	MS/B5	$4.5–9.0 \times 10^{-6}$ M 2,4-D	Callus	(Cardoso and de Oliviera 1996)
H. canariense	MS	BAP and NAA	Shoots	(Mederos Molina 1991)
H. canariense	1/2 MS	IBA or NAA	Roots	(Mederos Molina 1991)
H. canariense	MS, B5, WPM, QL.4	10^{-7} M BAP and 10^{-8} M NAA	Shoots	(Mederos et al. 1997)
H. erectum	LS	10^{-5} M BAP and 10^{-5} M IAA	Callus	(Yazaki and Okuda 1990, 1994)
H. erectum	LS	10^{-5} M BAP and 10^{-5} M IAA	Shoot primordia	(Yazaki and Okuda 1990, 1994)
H. erectum	LS	10^{-5} M BAP and 10^{-5} M IAA	Etiolated multiple shoots	(Yazaki and Okuda 1990, 1994)
H. erectum	LS	10^{-5} M BAP and 10^{-5} M IAA	Green multiple shoots	(Yazaki and Okuda 1990, 1994)
H. maculatum	1/2 MS	10^{-6} M BAP and 10^{-7} M NAA	Callus cell cultures	(Kartnig et al. 1996)
H. tomentosum	1/2 MS	10^{-6} M BAP and 10^{-7} M NAA	Callus cell cultures	(Kartnig et al. 1996)
H. bithynicum	1/2 MS	10^{-6} M BAP and 10^{-7} M NAA	Callus cell cultures	(Kartnig et al. 1996)
H. glandulosum	1/2 MS	10^{-6} M BAP and 10^{-7} M NAA	Callus cell cultures	(Kartnig et al. 1996)
H. balearicum	1/2 MS	10^{-6} M BAP and 10^{-7} M NAA	Callus cell cultures	(Kartnig et al. 1996)
H. olympicum	1/2 MS	10^{-6} M BAP and 10^{-7} M NAA	Callus cell cultures	(Kartnig et al. 1996)
H. perforatum	1/2 MS	10^{-6} M BAP and 10^{-7} M NAA	Callus cell cultures	(Kartnig et al. 1996)
H. perforatum	1/2MS	$10^{-6}–10^{-5}$ M BAP and $10^{-7}–10^{-6}$ M IAA	Shoots	(Zdunek and Alfermann 1992)
H. perforatum	LS	$4.4 \times 10^{-7}–4.4 \times 10^{-6}$ M BAP	Green multiple shoots	(Čellárová et al. 1992, 1994)
H. perforatum	LS	$4.6 \times 10^{-7}–2.3 \times 10^{-6}$ M KIN	Shoots	(Čellárová et al. 1992)
H. perforatum	LS	$4.9 \times 10^{-7}–2.5 \times 10^{-6}$ M 2iP	Shoots	(Čellárová et al. 1992)
H. perforatum	LS	4.6×10^{-6} M KIN and 5.4×10^{-7} M NAA	Green multiple shoots	(Čellárová et al. 1992)
H. perforatum	LS	$0.57 \times 10^{-6}–10^{-4}$ M IAA	Roots	(Čellárová and Kimáková 1999)
		$0.49 \times 10^{-6}–10^{-4}$ M IBA	Roots	
		$0.45 \times 10^{-6}–10^{-4}$ M 2,4-D	Callus	
		$0.54 \times 10^{-6}–10^{-4}$ M NAA	Callus	
		$0.44 \times 10^{-6}–10^{-4}$ M BAP	Multiple shoots	
		$0.49 \times 10^{-6}–10^{-4}$ M 2iP	Shoots	
		$0.46 \times 10^{-6}–10^{-4}$ M KIN	Multiple shoots	
		$0.74 \times 10^{-6}–10^{-4}$ M ADE	Shoots	
H. perforatum		5×10^{-6} M thidiazuron	Shoots	(Murch et al. 2000)

(Continued)

Table 5.1 (Continued)

Species	Culture medium	Growth regulators	Morphogenetic response	References
H. perforatum	MS	0.44×10^{-6}–10^{-5} M BAP	Shoots	(Pretto and Santarém 2000)
		0.45×10^{-6}–10^{-5} M 2,4-D	Callus	
		0.46×10^{-6}–10^{-5} M KIN	Callus	
	1/2 MS	0.49×10^{-5} M IBA	Roots	

Notes
MS – Murashige–Skoog's (1962) culture medium; LS – Linsmaier–Skoog's (1965) culture medium; B5 – Gamborg's culture medium (Gamborg *et al.* 1968); WPM – Lloyd–McCown's (1981) culture medium; QL.4 – Mederos 1981; BAP – 6-benzylaminopurine; KIN – kinetin; 2iP – 6-(γ, γ-dimethylallylamino)-purine; IAA – indole-3-acetic acid; NAA – naphthaleneacetic acid; IBA – indole-3-butyric acid; 2,4-D – 2,4-dichlorophenoxyacetic acid; ADE – adenin.

Morphogenetic response

Morphogenetic response of *Hypericum* tissue cultures depends on the type and concentration of plant growth regulators and light conditions. The use of cytokinins alone or in combination with low auxin concentration under a 12 or 16 h photoperiod or continuous light results in development of multiple green shoots in *H. perforatum* (Figure 5.1) (Čellárová *et al.* 1992, 1994, Zdunek and Alfermann 1992, Pretto and Santarém 2000) and in *H. erectum* (Yazaki and Okuda 1994). Thidiazuron supplementation to the culture medium for a 9 day induction period resulted in prolific growth of viable plantlets (Murch *et al.* 2000). The bioassay study showed that not all cytokinins led to multiple shoot formation. Positive morphogenetic response was observed under the effect of BAP or kinetin but no stimulating effect on shoot formation was observed under the influence of adenine or 2iP. The stimulation of rooting ability was observed under the effects of IAA and IBA. The induced morphogenetic response was concentration-dependent (Čellárová and Kimáková 1999). Cardoso and Oliviera (1996) achieved differentiation of shoots of *H. brasiliense* on the basal medium without plant growth regulators. Kartnig *et al.* (1996) reported on the isolation of callus and cell suspension cultures of seven *Hypericum* species under the same culture conditions (Table 5.1). Brutovská *et al.* (1994) found that the supplementation of culture medium containing BAP with 0.001% (w/v) of the non-ionic surfactant, Pluronic F-68, enhanced biomass production from cultured *H. perforatum* seedlings, as reflected by increases in both the overall fresh weight of the regenerated shoots and the number of adventive shoots produced per seedling. The growth of leaf-derived callus was unaffected by culture in the presence of Pluronic F-68; however, there was a tendency for callus produced from leaf explants grown in the presence of Pluronic to be highly pigmented with anthocyanins. In *H. brasiliense* the optimal callus induction occurred when the basal medium was supplemented with 2,4-D (Cardoso and Oliviera 1996). Combination of auxins and cytokinins was beneficial for callus induction in *H. perforatum* (Dias *et al.* 1998, Ishiguro *et al.* 1999, Kirakosyan *et al.* 2000a,b), in *H. patulum* (Ishiguro *et al.* 1996) and *H. androsaemum* (Dias *et al.* 2000). Browning exudates were successfully eliminated in tissue culture of *H. canariense* by rosmanol, a natural diterpenic antioxidant (Mederos *et al.* 1997).

Adaptation to ex vitro *conditions*

Regenerants with developed root system were adapted to *ex vitro* conditions by transferring them to perlite and keeping at 90% relative humidity. After 2 weeks cultivation in perlite, regenerants can be transferred to a hotbed or to field conditions (Čellárová *et al.* 1992).

Figure 5.1 Multiple shoot formation on seedling explant of *Hypericum perforatum* L. on the LS medium supplemented with BAP (Photo: M. Urbanová).

Production of secondary metabolites by callus and cell cultures

Data on the production of secondary metabolites by callus and cell suspension cultures of *Hypericum* species are still limited. Yazaki and Okuda (1990, 1994) isolated six procyanidins from callus and multiple shoot cultures of *H. erectum* and Kartnig *et al.* (1996) determined hypericin, pseudohypericin and flavonoids in cell cultures of seven *Hypericum* species and their chemotypes. Dias *et al.* (1999) found that calli and suspended cells of *H. perforatum* and *H. androsaemum* produce mainly xanthones while in *in vivo* plants these compounds were not detected. Similarly, the production of xanthones in *H. patulum* was reported by Ishiguro *et al.* (1996, 1999) and in *H. androsaemum* by Dias *et al.* (2000). A new naturally occurring compound, 6-prenyl luteolin was isolated from callus cultures of *H. perforatum* by Dias *et al.* (1998). As shown by Kirakosyan *et al.* (2000a), cell cultures of *H. perforatum* contain only trace amounts of hypericin and pseudohypericin as biosynthesis of these naphthodianthrones is correlated with the degree of cell differentiation and depends on the formation of specific morphological structures. These authors also reported on elicitation of hypericin production in shoot cultures of *H. perforatum* (Kirakosyan *et al.* 2000b) (Table 5.2).

Somaclonal variation in *Hypericum perforatum* L. – a source of genetic variability

The potential of somaclonal variation to contribute genetic variation in the improvement of plants has been widely studied (see Bajaj 1990). The tissue culture derived regenerants exhibit a range of altered characteristics which can be inherited. The maintenance of desirable properties

Table 5.2 Production of secondary metabolites by different cultures of some Hypericum species

Species	Type of culture	1	2	3	4	5	6	7	8	9	10	11	12	13	14	15	References
H. erectum	Callus	+	+	+	+	+	t										(Yazaki and Okuda 1990, 1994)
	Shoot primordia	+	+	+	+	+	t										(Yazaki and Okuda 1990, 1994)
	Etiolated multiple shoots	+	+	+	+	+	t										(Yazaki and Okuda 1990, 1994)
	Green multiple shoots	+	+	+	+	+	+										(Yazaki and Okuda 1990, 1994)
H. perforatum	Cell cultures															+	(Dias *et al.* 1999, 2000)
H. perforatum	Cell cultures													t	t		(Kirakosyan *et al.* 2000a)
H. perforatum	Shoot cultures													+	+		(Kirakosyan *et al.* 2000a)
H. perforatum	Shoot cultures													+	+		(Zdunek and Alfermann 1992)
H. perforatum	Cell cultures						+	+	+	+	+	+	+	+	+		(Kartnig *et al.* 1996)
H. maculatum	Cell cultures						+	+	+	+	+	+	+	+	+		(Kartnig *et al.* 1996)
H. tomentosum	Cell cultures						+	-	-	-	-	-	+	+	+		(Kartnig *et al.* 1996)
H. bithynicum	Cell cultures						+	+	+	+	+	+	+	+	+		(Kartnig *et al.* 1996)
H. glandulosum	Cell cultures						+	+	+	+	+	+	+	+	+		(Kartnig *et al.* 1996)
H. balearicum	Cell cultures						+	+	+	+	+	-	+	+	+		(Kartnig *et al.* 1996)
H. olympicum	Cell cultures						-	-	-	-	-	-	+	-	-		(Kartnig *et al.* 1996)
H. patulum	Cell cultures															+	(Ishiguro *et al.* 1996, 1999)

Notes

1 (−)-epicatechin; 2 procyanidin B2; 3 procyanidin C1; 4 cinnamtannin A2; 5 hyperin; 6 quercitrin; 7 rutin; 8 hyperoside; 9 isoquercitrin; 10 quercetin; 11 I3, II8-biapigenin; 12 amentoflavone; 13 hypericin; 14 pseudohypericin; 15 xanthone. t trace; + secondary metabolite present; − secondary metabolite not present.

in variants created from tissue cultures in *H. perforatum* may be expected if apomictic psedogamy is predicted as the prevalent mode of reproduction. Since Noack's observations (1939) it has been considered that *H. perforatum* is a facultative apomict of the *Hieracium* type in which about 97% of plants develop aposporically without meiotic reduction and only 3% reproduce sexually. Recently, on the basis of artificial crossing between diploid and tetraploid tissue culture derived plants followed by chromosome counting and flow cytometry analysis of progenies it has been found that 81% of the progeny were pentaploids (B_{III} hybrids), 12% were tetraploids (apomictic plants) and 7% were triploids (B_{II} hybrids). Both B_{III} hybrids and apomictic plants were derived from aposporous initials as revealed by histological observations (Brutovská *et al.* 1998). However, the results from cytogenetic analysis of the progeny obtained from tetraploid plants after self-pollination confirmed Noack's assumption (Brutovská *et al.* 1998). These authors assume that the decisive factor in the reproduction process of this species is the presence of fertile pollen at the time of egg cell maturity. The analyses of seed samples of diploid R_0 somaclones and their three subsequent generations revealed that the majority of plants reproduced sexually forming B_{II} hybrids. In addition to sexually developed seeds, some seeds have probably arisen by the endopolyploidisation process (Koperdáková *et al.* unpublished).

In vitro regeneration from young seedlings under the effect of BAP provides hundreds of somaclones which show a high degree of variability (Čellárová *et al.* 1992) which has been studied at different levels.

Biomass production and morphological alterations

The R_0 somaclones differ in several characters such as fresh and dry weight of herbage, height, number of branches per plant and number of dark glands containing dianthrones per leaf area. A comparison with the control plants which were derived *in vitro* from seeds without plant growth regulators showed a significant decrease in fresh and dry weight of regenerants. However, no significant differences between the control and *in vitro* regenerated plants were found in the number of branches and number of hypericin-containing multicellular glands per leaf area. The concentration of BAP within the range of 0.44–4.40 μM did not influence the observed characteristics (Čellárová *et al.* 1992).

Great variability occurred also in the leaf shape of regenerated plants. On the basis of the index determined by the ratio of the width to length of a leaf, broad-leaved, intermediate and narrow-leaved plants were found among the somaclones which originated from one genotype although the density of hypericin glands was comparable (Čellárová *et al.* 1994).

In the first year of cultivation, the R_0 somaclones as well as control seed derived plants had an unusual trailor-plant habit and actually reached the stage of flowering. However, in the second year of cultivation the habit of regenerants and control plants resembled that of plants from natural populations.

Cytogenetic changes and ploidy-dependent traits

The ploidy level analysis of the R_0 somaclones which originated from tetraploid maternal plants by both chromosome counting and flow cytometry methods revealed the presence of diploid ($2n = 2x = 16$), triploid ($2n = 3x = 24$), tetraploid ($2n = 4x = 32$) as well as mixoploid plants. The analyses of seed samples of diploid somaclones revealed 93% of diploids that were formed by sexual process. In addition to sexually developed seeds, some plants exhibited different chromosome numbers, which might be explained by the endopolyploidisation process.

The tetraploid plants were prevalently facultative apomicts producing B_{III} hybrids, parthenogenetic seeds or twin embryos as well and some produced aneuploid seeds.

On the other hand, seed samples of triploid, pentaploid and hexaploid plants showed an extensive variation in the chromosome number (Koperdáková *et al.* unpublished).

Karyotype analysis of haploid, diploid, triploid and tetraploid cells from mixoploid somaclones revealed that the chromosomes are morphologically very similar, median or submedian. In the basic chromosome set, the most distinguishable is chromosome number 1, which was subjected to detailed analysis. It was found that there are two types of this chromosome which contribute differentially in diploid, triploid and tetraploid plants (Brutovská *et al.* 2000a).

The chromosomal position of 5S/25S rRNA genes of three *Hypericum* species (*H. perforatum, H. maculatum* and *H. attenuatum*) were comparatively determined by FISH. The rDNA loci between *H. perforatum* and *H. maculatum* seem to be identical indicating that *H. perforatum* probably arose by autopolyploidisation from an ancestor related to *H. maculatum* (Brutovská *et al.* 2000b).

Study of somaclones of different ploidy showed that some morphological, physiological and biochemical traits are ploidy-dependent. Positive correlation was found between ploidy and biomass production or leaf shape (Čellárová and Bruňáková 1996). However, the number of hypericin glands per leaf blade was found not to be correlated with ploidy. Total hypericin content was negatively correlated with ploidy (Čellárová *et al.* 1995, 1997). Ploidy also affects the mode of reproduction in *H. perforatum* (Koperdáková *et al.* unpublished).

Biochemical level

Determination of hypericin content in R_0 somaclones and R_1–R_3 progenies revealed great variability within and between different genotypes, between the first and second year of cultivation and differed in plants in relation to ploidy level. As shown in Figure 5.2, great differences were detected between individual generations of progenies. With regard to these differences, at least

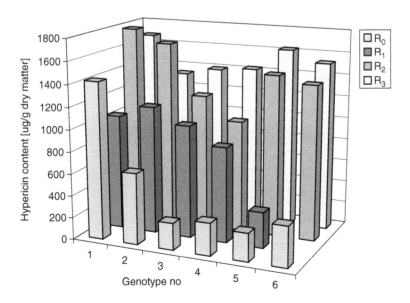

Figure 5.2 Comparison of hypericin content in the R_0 somaclones and their progenies.

two important reasons have to be taken into account. First, several possible routes of seed formation, and second, environmental factors affecting quantitative traits. Analyses of hypericin content revealed negative correlation between ploidy and hypericin content (Čellárová et al. 1995, 1997).

DNA polymorphism

DNA variation among the control seed derived plants and the R_0 H. perforatum somaclones as well as their R_1 progenies was studied by RFLP analysis with several rDNA probes from Lycopersicon esculentum (Halušková and Čellárová 1997). Two RFLP patterns were identified in both somaclones and seed-derived plants when digested with EcoRV and hybridised to 25SrDNA probe. The RFLP patterns of either seed-derived plants or somaclones were transmitted to most of their progenies. Identical RFLP pattern in somaclones and their progenies indicate the prevalence of an apomictic mode of reproduction in tetraploids while differences indicate that some individuals may reproduce sexually.

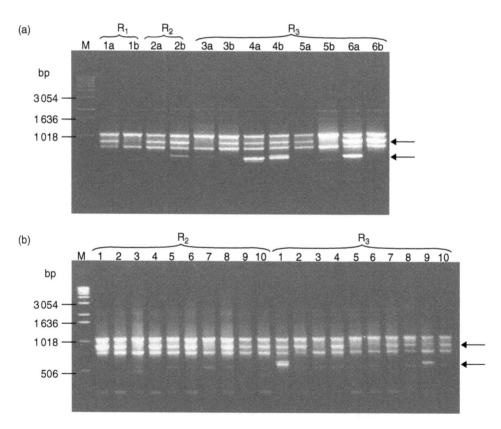

Figure 5.3 (a) Comparison of the amplification profiles of the R_1–R_3 progenies of Hypericum perforatum somaclones with the minisatellite primer. (5'-AGGGCTGGAGGAGGGC-3') (Jeffreys et al. 1985) Arrow indicates polymorphic band. a, b – plants originated from the same genotype. (b) Polymorphism generated with the minisatellite primer (5'-AGGGCTGGAGGAGGGC-3') (Jeffreys et al. 1985) among the R_2 and R_3 somaclones originated from one genotype. Arrow indicates polymorphic band (Photo: J. Košuth).

Seed progeny of $R_0–R_3$ generations selected on the basis of diploid chromosome number and flow-cytometry analysis, which confirmed sexual mode of reproduction, was screened by RAPD with mini- and microsatellite primers. Variability was determined not only between different genotypes but also within a particular genotype (Figure 5.3a,b) (Košuth *et al.* unpublished).

Cryopreservation of *Hypericum perforatum* L. meristems

Shoot tips of *H. perforatum* were cryoconserved in liquid nitrogen by the slow freezing method. The ability of cryopreserved meristems to produce multiple shoots as a proof of their recovery was determined. Plants regenerated *in vitro* from cryopreserved meristems were subjected to chromosomal analysis in order to ascertain genetic stability. The study of pre-culture and cryo-protection on the morphogenetic capacity of cryopreserved meristems showed that three day pre-culture of meristems on the medium supplemented with the pre-culture additives lead to direct shoot regeneration in about 10% of cryopreserved meristems. The highest survival of meristems after cryopreservation was detected on day 14 when the meristems were pre-cultured for 10 or 14 days on basal medium supplemented with 0.076 μM ABA. The prolonged pre-culture period resulted in a slight callus formation followed by shoot differentiation. Shoot tips of *H. perforatum* were treated with a cryoprotective mixture consisting of Glycerol–Sucrose–DMSO (10 : 20 : 10, v/w/v). The samples were cooled at a rate of 0.3 and 0.5°C/min up to −10°C and then at a rate 1°C/min up to −40°C. When the cryopreserved shoot tips were placed on the medium supplemented with 0.5 mg/l BA, they started to differentiate after a 2 week lag phase. During the next subculture, the growth characteristics of cryopreserved cultures were comparable with the control unfrozen meristems. Morphogenetic reaction was genotype-dependent and varied between 10% and 48%.

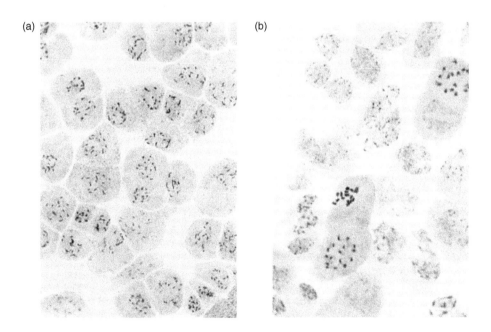

(a) (b)

Figure 5.4 (a) Mitotic activity in cryopreserved *H. perforatum* meristematic cells. (b) Mitotic activity in meristematic cells of the control plants (Photo: M. Urbanová).

Cytogenetic analyses of almost 1480 metaphases of apical bud meristematic cells of both directly and indirectly regenerated plants from cryopreserved as well as control non-cryopreserved meristems did not show any alterations of the chromosome number. Meristems after cryopreservation contained mostly dividing cells. The mitotic index varied between 96% and 99% and was comparable with the control (Figure 5.4a,b) (Urbanová *et al.* 2002).

The biotechnological potential of *Hypericum* sp. provides several possibilities for further research and use in different ways:

(i) widening the genetic variability followed by the selection of genotypes with desirable properties;
(ii) high degree of polymorphism as to several morphological, physiological and biochemical characters;
(iii) availability of genotypes with different ploidy that can be used in the study of ploidy-dependent gene expression related to some important traits such as genetic basis of apomixis;
(iv) potential of callus and cell suspension cultures for the production of some secondary metabolites.

References

Bajaj, Y. P. S. (1990) *Biotechnology in Agriculture and Forestry*, vol. 11. *Somaclonal Variation in Crop Improvement I*, Springer, Berlin.

Brutovská, R., Čellárová, E., Davey, M. R., Power, J. B. and Lowe, K. C. (1994) Stimulation of multiple shoot regeneration from seedlings of *Hypericum perforatum* L. by Pluronic F-68. *Acta Biotechnol.* 14, 347–53.

Brutovská, R., Čellárová, E. and Doležel, J. (1998) Cytogenetic variation in tissue culture-derived *Hypericum perforatum* L. plants and their seed progenies. *Plant Sci.* 133, 221–9.

Brutovská, R., Kušniriková, P., Bogyiová, E. and Čellárová, E. (2000a) Karyotype analysis of *Hypericum perforatum* L. *Biol. Plant.* 43, 133–6.

Brutovská, R., Čellárová, E. and Schubert, I. (2000b) Cytogenetic characterization of three *Hypericum* species by *in situ* hybridization. *Theor. Appl. Genet.* 101, 46–50.

Cardoso, N. A. and de Oliviera, D. E. (1996) Tissue culture of *Hypericum brasiliense* Choisy: Shoot multiplication and callus induction. *Plant Cell Tiss. Org. Cult.* 44, 91–4.

Čellárová, E., Kimáková, K. and Brutovská, R. (1992) Multiple shoot formation and phenotypic changes of R_0 regenerants in *Hypericum perforatum* L. *Acta Biotechnol.* 12, 445–52.

Čellárová, E., Daxnerová, Z., Kimáková, K. and Halušková, J. (1994) The variability of hypericin content in the regenerants of *Hypericum perforatum*. *Acta Biotechnol.* 14, 267–74.

Čellárová, E., Kimáková, K., Daxnerová, Z. and Mártonfi, P. (1995) *Hypericum perforatum* (St. John's Wort): *In vitro* culture and the production of hypericin and other secondary metabolites. In *Biotechnology in Agriculture and Forestry 33, Medicinal and Aromatic Plants* VIII. Y. P. S. Bajaj (Ed.), Springer, Heidelberg, pp. 261–75.

Čellárová, E. and Bruňáková, K. (1996) Somaclonal variation in *Hypericum perforatum* (St. John's Wort). In *Biotechnology in Agriculture and Forestry 36, Somaclonal Variation in Crop Improvement* II. Y. P. S. Bajaj (Ed.), Springer, Heidelberg, pp. 267–79.

Čellárová, E., Brutovská, R., Daxnerová, Z., Bruňáková, K. and Weigel, R. C. (1997) Correlation between hypericin content and the ploidy of somaclones *of Hypericum perforatum* L. *Acta Biotechnol.* 17, 83–9.

Čellárová, E. and Kimáková, K. (1999) Morphoregulatory effect of plant growth regulators on *Hypericum perforatum* L. seedlings. *Acta Biotechnol.* 19, 163–9.

Dias, A. C. P., Tomás-Barberán, F. A., Fernandes-Ferreira, M. and Ferreres F. (1998) Unusual flavonoids produced by callus of *Hypericum perforatum*. *Phytochemistry* 48, 1165–8.

Dias, A. C. P., Seabra, R. M., Andrade, P. B. and Fernandes-Ferreira, M. (1999) The development and evaluation of an HPLC-DAD method for the analysis of the phenolic fractions from *in vivo* and *in vitro* biomass of *Hypericum* species. *J. Liq. Chrom. & Rel. Technol.* 22, 215–27.

Dias, A. C. P., Seabra, R. M., Andrade, P. B., Ferreres, F. and Fernandes-Ferreira, M. (2000) Xanthone biosynthesis and accumulation in calli and suspended cells of *Hypericum androsaemum*. *Plant Sci.* 150, 93–101.

Franke, R., Schenk, R., Bauermann, U., Giberti, G., Craker, L., Lorenz, M., Mathe, A. and Giulietti, A. (1999) Variability in *Hypericum perforatum* L. breeding lines. Proc. 2nd World Congress on Medicinal and Aromatic Plants for Human Welfare, WOCMAP-2, Mendoza, Nov. 1997.

Gamborg, O. L., Miller, R. A. and Ojima, K. (1968) Nutrient requirements of suspension cultures of soybean root cells. *Exp. Cell Res.* 50, 151–8.

Halušková, J. and Čellárová, E. (1997) RFLP analysis of *Hypericum perforatum* L. somaclones and their progenies. *Euphytica* 95, 229–35.

Ishiguro, K., Fukumoto, H., Suitani, A., Nakajima, M. and Isoi, K. (1996) Prenylated xanthones from cell suspension cultures of *Hypericum patulum*. *Phytochemistry* 42, 435–7.

Ishiguro, K., Oku, H. and Isoi, K. (1999) *Hypericum patulum*: *In vitro* culture and production of xanthones and other secondary metabolites. In *Biotechnology in Agriculture and Forestry 43, Medicinal and Aromatic Plants* XI. Y. P. S. Bajaj (Ed.), Springer, Heidelberg, pp. 199–212.

Jeffreys, A. J., Wilson, V. and Thein, S. L. (1985) Hypervariable "minisatellite" regions in human DNA. *Nature* 314, 67–73.

Kartnig, T., Göbel, I. and Heydel, B. (1996) Production of hypericin, pseudohypericin and flavonoids in cell cultures of various *Hypericum* species and their chemotypes. *Planta Med.* 62, 51–3.

Kirakosyan, A. B., Vardapetyan, R. R. and Charchoglyan, A. G. (2000a) The content of hypericin and pseudohypericin in cell cultures of *Hypericum perforatum*. *Russ. J. Plant Physiol.* 47, 270–3.

Kirakosyan, A., Hayashi, H., Inoue, K., Charchoglyan, A. and Vardapetyan, H. G. (2000b) Stimulation of the production of hypericins by mannan in *Hypericum perforatum* shoot cultures. *Phytochemistry* 53, 345–8.

Linsmaier, E. M. and Skoog, F. (1965) Organic growth factor requirement of tobacco tissue cultures. *Physiol. Plant.* 18, 100–27.

Lloyd, G. and McCown, B. (1981) Commercially feasible micropropagation of mountain laurel, *Kalmia latifolia* by use of shoot-tip culture. *Combined Proc. Int. Propag. Soc.* 30, 421–6.

Matzk, F., Meister, A., Brutovská, R. and Schubert, I. (2001) Reconstruction of reproductive diversity in *Hypericum perforatum* L. opens novel strategies to manage apomixis. *Plant J.* 26, 275–82.

Mederos Molina, S. (1991) *In vitro* growth and multiplication of *Hypericum canariense* L. *Acta Hort.* 289, 133–4.

Mederos, S., San Andrés, L. and Luis, J. G. (1997) Rosmanol controls explants browning of *Hypericum canariense* L. during the *in vitro* establishment of shoots. *Acta Soc. Bot. Poloniae* 66, 347–9.

Murashige, T. and Skoog, F. (1962) A revised medium for rapid growth and bioassays with tobacco tissue cultures. *Physiol. Plant.* 15, 473–9.

Murch, S. J., Choffe, K. L., Victor, J. M. R., Slimmon, T. Y., KrishnaRaj, S. and Saxena, P. K. (2000) Thidiazuron-induced plant regeneration from hypocotyl cultures of St John's wort (*Hypericum perforatum*, cv. 'Anthos'). *Plant Cell Rep.* 19, 576–81.

Noack, K. L. (1939) Über *Hypericum*-Kreuzungen VI. Fortpflanzungsverhältnisse und Bastarde von *Hypericum perforatum* L. *Z. Indukt. Abstamm. Verebungslehre* 76, 569–601.

Pretto, F. R. and Santarém, E. R. (2000) Callus formation and plant regeneration from *Hypericum perforatum* leaves. *Plant Cell Tiss. Org. Cult.* 62, 107–13.

Urbanová, M., Čellárová, E. and Kimáková, K. (2002) Chromosome number stability and mitotic activity of cryopreserved *Hypericum perforatum* L. meristems. *Plant Cell Rep.* 20, 1082–6.

Yazaki, K. and Okuda, T. (1990) Procyanidins in callus and multiple shoot cultures of *Hypericum erectum*. *Planta Med.* 56, 490–1.

Yazaki, K. and Okuda, T. (1994) *Hypericum erectum* Thunb. (St John's Wort): *In vitro* culture and production of procyanidins. In *Biotechnology in Agriculture and Forestry 26, Medicinal and Aromatic Plants* VII. Y. P. S. Bajaj (Ed.), Springer, Heidelberg, pp. 167–78.

Zdunek, K. and Alfermann, A. W. (1992) Initiation of shoot organ cultures of *Hypericum perforatum* and formation of hypericin derivatives. *Planta Med.* **58** supplement, 621–2.

6 Chemical constituents of *Hypericum* ssp.

Josef Hölzl and Maike Petersen

Introduction

The genus *Hypericum* belonging to the family Clusiaceae includes more than 400 species growing worldwide (Robson 1977, 1981). For a better classification the genus is subdivided into 30 sections; for this purpose also the appearance of specific natural products is used. *Hypericum* contains a wide range of different natural product classes. Best known are the naphthodianthrones with hypericin and pseudohypericin as well as their precursors protohypericin and protopseudohypericin as the main compounds. Also important are the prenylated phloroglucinols with hyperforin and adhyperforin and their oxygenated derivatives. Typical natural products are moreover xanthones (e.g. 1,3,6,7,-tetrahydroxyxanthone), flavonoids (e.g. hyperosid, rutin, quercitrin, isoquercitrin) and biflavonoids (I3,II8-biapigenin, I3',II8-biapigenin), tannins and proanthocyanidins as well as phenolic acids. The essential oil of *Hypericum* contains hydrocarbons as well as terpenoids. From a medicinal point of view only one *Hypericum* species, *Hypericum perforatum*, is important.

Hypericin and related compounds

Hypericins

Hypericins are natural plant products belonging to the group of naphthodianthrones. This group includes the compounds hypericin, pseudohypericin, protohypericin and protopseudohypericin (Figure 6.1) (Brockmann *et al.* 1939, 1942, Brockmann 1957, Brockmann and Sanne 1953, 1957, Brockmann and Pampus 1954, Brockmann *et al.* 1974) and derivatives occurring in lower concentrations such as isohypericin, demethylpseudohypericin, hypericodehydrodianthrone, pseudohypericodehydrodianthrone (Brockmann 1957) and cyclopseudohypericin (Brockmann and Pampus 1954, Häberlein *et al.* 1992). Not all of the minor compounds have been proved to occur genuinely in the plant – they might be artefacts of the isolation. Protohypericin is a precursor for hypericin and protopseudohypericin for pseudohypericin; both conversions take place in light.

Occurrence of hypericins in the plant kingdom

Hypericins mainly occur in plants of the genus *Hypericum* belonging to the family Clusiaceae. Formerly *Hypericum* was placed into the families Guttiferae and then Hypericaceae. Although anthrone derivatives are reported from related subfamilies, *Hypericum* is currently the only plant taxon containing condensed anthrones such as the hypericins (Kitanov 2001). *Harungana madagascariensis* belonging to the same family as *Hypericum* was reported to contain hypericin and

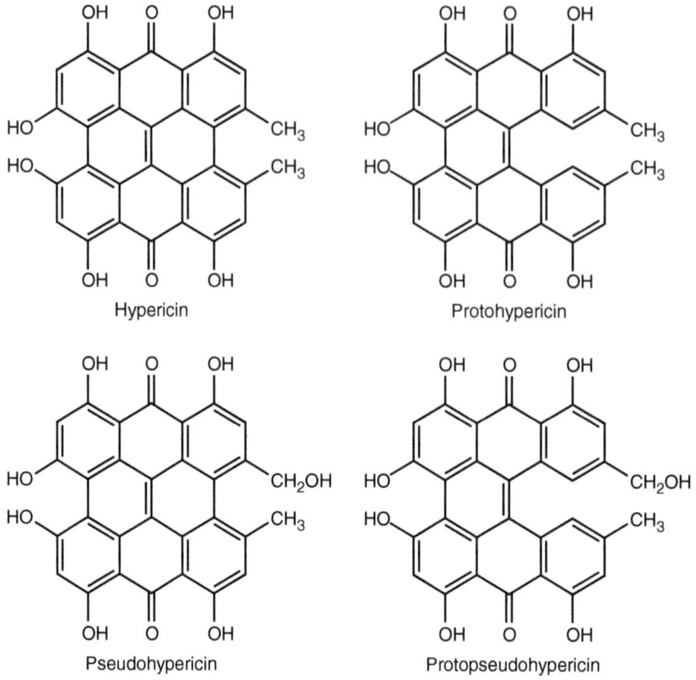

Figure 6.1 Structural formulae of hypericin, pseudohypericin, protohypericin and protopseudohypericin.

pseudohypericin in leaves (Fisel *et al.* 1966) and *Porospermum guineense* contains a red photosensitizing pigment in the root bark which is similar or identical to hypericin (Hegnauer 1966).

More than 400 *Hypericum* species are reported from all over the world, 60% of which are known to contain hypericins as constituents (Robson 1977, 1981). According to the Flora Europaea, 59 species are native to Europe. The genus *Hypericum* is subdivided into several sections and it was reported that only the more advanced sections contain hypericin and pseudohypericin (Kitanov 2001). The only species with medicinal importance is *H. perforatum*. Several subspecies exist of this species, two of which are pharmaceutically important: *H. perforatum* ssp. *angustifolium* (DC) GAUDIN is mainly native to Southern Europe, whereas the subspecies *perforatum* predominantly occurs in Northern Europe. These two subspecies are morphologically as well as phytochemically distinct. The natural products differ qualitatively and quantitatively. The narrow-leaved subspecies *angustifolia* has a much higher hypericin content than the broad-leaved subspecies *perforatum*. On the other side, the subspecies *perforatum* additionally contains the flavonol glycoside rutin (Southwell and Campbell 1991). Also seasonal variation of the hypericin content in these subspecies was reported (Southwell and Bourke 2001).

Distribution of hypericin in plant organs

Hypericin is localized in small glandular structures in plants. These glands are present in all above-ground parts (flowers, capsules, leaves, stems), but not in roots. The content of hypericins in the dry herb is 0.1–0.15% and in flowers and flower buds 0.2–0.3% (Kaul 2000). Seeds and the cotyledons of seedlings do not contain hypericins, but they can already be found in the first true leaves after germination. The distribution of hypericin in plant organs of *H. perforatum*

Table 6.1 Distribution of hypericin in plant organs (Berghöfer 1987)

Plant part	Hypericin content (% of dry material)
Small flower buds	0.360
Large flower buds	0.436
Flowers	0.187
Decayed flowers	0.070
Capsules	0.116
Leaves	0.080
Shoot, upper part	Traces
Shoot, lower part	Not detectable
Roots	Not detectable

was investigated by Berghöfer (1987) and is shown in Table 6.1. Pseudohypericin usually is found concomitantly with hypericin. However, pseudohypericin does not occur in *H. hirsutum* and *H. empetrifolium*. Up to now, the only species containing only pseudohypericin is *H. formosissimum* (Kitanov 2001).

In vitro culture experiments have shown that light is not essential for the biosynthesis and accumulation of hypericin. Shoot cultures of *H. perforatum* grown in the light and in the dark both showed the typical dark red oil glands. The contents of hypericin and pseudohypericin were a little lower in the dark-grown shoot culture, but both were drastically lower than in the natural plant (Zdunek and Alfermann 1992, Zdunek 1993). The same hypericins as well as flavonoids were found in cell cultures of several *Hypericum* species (Kartnig *et al.* 1996) and the content of hypericins could be stimulated (Kirakosyan *et al.* 2000a,b).

Hypericin in fungi and insects

Protohypericin and hypericin were isolated from the fruit body of the Australian fungus *Dermocybe austroveneta* (Gill *et al.* 1988) as well as other *Dermocybe* species. Another pigment of this toadstool was identified as austrovenetin (Figure 6.2), a 5,5'-coupled dimeric anthracenone pigment (Gill and Gimenez 1991). This was the only report of hypericin occurring in an organism other than the family Clusiaceae. Skyrin (Figure 6.2), an orange pigment, could be isolated from the same fresh fungal material. Skyrin was shown to be present in *H. perforatum* as well (Berghöfer 1987). It most probably is a precursor of protohypericin. Skyrin, hypericin and pseudohypericin were found in members of the Coccoidea (Banks *et al.* 1976, Cameron and Raverty 1976). Insects do not often use *Hypericum* as a food plant, except specialists feeding on this species exclusively. The hypericins in these organisms, therefore, most probably genuinely stem from the food plant.

Compounds related to hypericin

Fagopyrin (Figure 6.3) in a concentration of 0.01–0.03% is present in flowers and leaves of buckwheat (*Fagopyrum esculentum*; Brockmann *et al.* 1952). Fagopyrin is a naphthodianthrone coupled with two molecules of pyrine.

The dimeric anthraquinone skyrin (see Figure 6.2) was found to occur concomitantly with hypericins in *Hypericum* and the fungus *Dermocybe* (Gill *et al.* 1988). Moreover, skyrin and related derivatives were reported from *Penicillium* species and some ascomycetes as well as some lichens

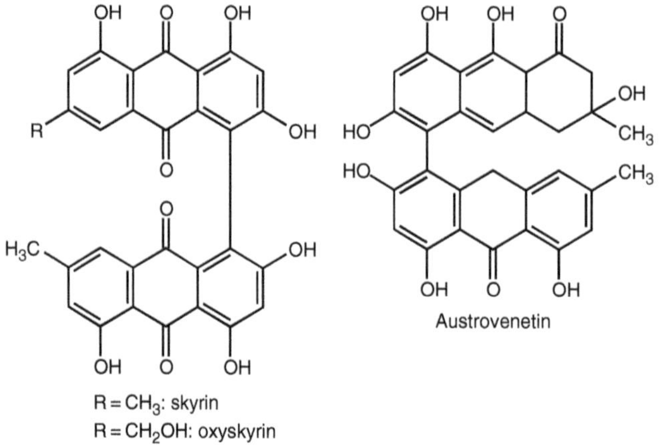

Figure 6.2 Structural formulae of skyrin, oxyskyrin and austrovenetin.

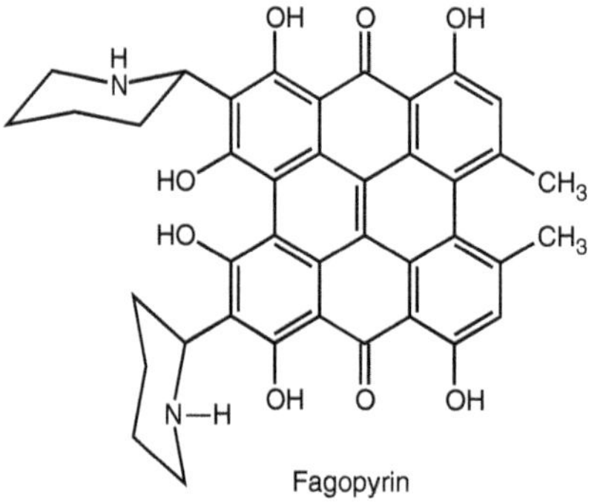

Figure 6.3 Structural formula of fagopyrin.

(e.g. *Cladonia, Trypetheliopsis, Physcia, Pyxine*) (Steglich *et al.* 1997). *Penicillium* moreover contains the yellow 5,5'-di-emodine anthrone penicilliopsin (Brockmann and Eggers 1955a).

Compounds structurally related to hypericin as well as to other naphthodianthrones are known to be pigments of photoreceptors in some lower animals (Kuhlmann 1998), for example, stentorin in the ciliate *Stentor coeruleus* (Song and Walker 1981, Häder and Häder 1991, Tao *et al.* 1994), zoopurine (blepharismin) from the ciliate *Blepharisma japonicum* (Giese 1981).

Chemical characteristics of hypericin

Hypericin (10,11-dimethyl-1,3,4,6,8,12-hexahydroxynaphthodianthrone, $C_{30}H_{16}O_8$, molecular mass 504.45) was first isolated as a pure compound from *H. hirsutum* by Brockmann *et al.* (1939). They also elucidated the structure of hypericin and reported the first total chemical synthesis. Further reports of chemical syntheses of hypericin followed starting from the natural anthraquinones emodin and its reduction product emodin anthrone (Rodewald *et al.* 1977). Hypericin could be crystallised from pyridine with methanolic HCl as blue–black needles. The melting point cannot be determined since the substance decomposes at 320°C.

Pseudohypericin ($C_{30}H_{16}O_9$; $\lambda_{max} = 589$ nm) behaves similarly (Falk and Schmitzberger 1992). Absorbance data for hypericin and pseudohypericin were determined and collected recently by Wirz *et al.* (2001). Hypericin is soluble in pyridine or other organic bases, but nearly insoluble in most organic solvents. The solution is red and exhibits brick-red fluorescence.

An X-ray diffraction spectrum of the pyridinium salt of hypericin crystallised from pyridine was published in 1993 (Etzlstorfer *et al.* 1993). The spectrum shows a distorted molecule with a helical twist. pH-dependent absorption spectra indicate that hypericin is an acidic compound. The first pK_a-value is at pH 1.7 and 1.2 in ethanol (80%) or DMSO, respectively, the second pK_a is at pH 12 in both solvents. Hypericin forms monobasic salts with inorganic bases in pH ranges of 4–11. Salts of hypericin are generally soluble in polar solvents in contrast to free hypericin which is poorly soluble in these solvents. Sodium salts of hypericin are used in biological assays and in clinical trials, in the plant hypericin is reported to occur as the potassium salt (Etzlstorfer *et al.* 1993).

Organic solutions of hypericin salts are red; they exhibit a high light absorption (λ_{max} in ethanol at 545 and 590 nm, $\varepsilon = 52,000$ for the latter peak) and exhibit red fluorescence. Monobasic salts of hypericin disperse in water giving slightly transparent non-fluorescent and high-molecular aggregates. Aqueous hypericin colloids precipitate in phosphate-buffered saline, whereas in organic solvents they dissociate to monomers. Hypericin interacts with components of biological fluids. It binds to plasma proteins, for example, human serum albumin or lipoproteins. Complexes of hypericin and human serum protein in the ratio 1:1 are formed. Hypericin interacts with the lipid moiety of lipoproteins, especially with phospholipids. The binding constant of the hypericin–lipoprotein complex is higher than in the complex with human serum protein. Hypericin is not bound to human immunoglobulin G. In lower concentrations hypericin is incorporated as monomers into detergent micelles (e.g. Tween 80, Triton X-100); in higher concentrations, hypericin forms aggregates. Hypericin and its salts can act as electron acceptors as well as electron donors as shown by electrochemical and EPR experiments; they are oxidizing as well as reducing agents (Lavie *et al.* 1995).

Solubility of hypericin in organic and aqueous solvents

Experiments for the determination of the solubility of hypericin were performed for 5 days at 30°C under stirring. Besides methanol and water, biological buffers such as Krebs–Ringer and phosphate buffer pH 7.4 (PBS, phosphate buffered saline) were also used as solvents. The solubility of hypericin in methanol is at 500 μg/ml. Hypericin has to be considered as insoluble in water (app. 40 μg/ml). In the indicated buffers the solubility is even lower with 10 μg/ml. The solubility in aqueous systems could be enhanced by the addition of 2-hydroxypropyl-β-cyclodextrin yielding 30% of the concentration in methanol (Sattler 1997).

Isolation of hypericin and pseudohypericin from **H. perforatum**

There are several reports about the isolation of pure hypericin from plant material. Mostly, these methods only yield milligram quantities of hypericin or the isolated compounds are not sufficiently pure.

The following protocol can be performed in every laboratory with standard equipment (Sattler 1997):

Extraction Dry flowers and flower buds weighing 160 g of *H. perforatum* were extracted in a soxhlett apparatus with dichloromethane (4 l) in order to remove very lipophilic constituents (e.g. chlorophyll). The drug is further extracted with acetone (3 l) in the same soxhlett apparatus.

The acetone extract is evaporated to near dryness yielding a viscous residue (appr. 11 g) and 3 g of the residue are redissolved in methanol (250 ml). The crude methanolic extract is fractionated using a polyamide column (diameter 5 cm, height 3 cm). Flavonoids and phenolic acids are desorbed by washing with methanol (3 l). Hypericins still adhere to the column by H-bonds between the phenolic hydroxyl groups of the hypericin and the amide groups of the stationary phase. They can be desorbed by methanol/25% ammonia 99:1 (1 l). The eluate is evaporated to dryness under vacuum at 40°C. This yields a residue (app. 170 mg) which is redissolved in 150 ml methanol. The solution is illuminated (strong day light) in order to transform precursors to hypericin.

Preparative HPLC   HPLC fractionation is performed at room temperature on a Hibar-RT LiChrosorb® RP-18 (length 25 cm, diameter 25 mm, particle size 7 μm) column with methanol : acetonitrile : *o*-phosphoric acid (85%) 89 : 10 : 0.5 as eluent. A flow gradient (0–2 min: 10 ml/min; 2–15 min: 12 ml/min, elution of pseudohypericin; 15–30 min: flow 15 ml/min, elution of hypericin) is applied for the fractionation.

In order to remove the phosphoric acid from the hypericin-containing fraction, this fraction is again applied to a polyamide column (as above) and the acid is eluted with approximately 2 l water as eluent. Hypericins are eluted with methanol/25% ammonia 99:1 (2 l). The hypericin has a purity of approximately 95% (Sattler 1997).

Chemical synthesis of protohypericin and hypericin

Chemical synthesis of protohypericin and hypericin from emodin anthrone was achieved by, for example, Brockmann and Eggers (1955b) and Spitzner (1977). For the latter method of synthesis of protohypericin, emodin is heated at 100°C together with 1.8 mol equivalents 0.6 N KOH and 1 mol equivalent hydroquinone for 20 days under nitrogen and exclusion of light. The violet mixture is acidified with 0.1 N HCl, the precipitate is collected by suction filtration, washed and dried and further exhaustively extracted with ethylacetate. The crude pigment is chromatographed on a cellulose column with DMF/H_2O/butylacetate 1 : 1 : 2 as eluent. A first orange coloured band contains anthraquinone derivatives and is followed by a zone containing protohypericin. A second column chromatography is necessary to remove DMF from the fraction. After addition of 0.1 N HCl the precipitate is again filtered, washed, dried and extracted with ethylacetate and then dissolved in acetone. The acetone solution of the crude pigment is applied to a Sephadex LH 20 column and eluted with acetone.

Hypericin is synthesised accordingly but without exclusion of light.

Hyperforin and related compounds

Discovery of hyperforin and related compounds in H. perforatum

The first indications to hyperforin emerged from the finding that constituents with good antibacterial activity are present in acetone extracts of St John's wort. The highest activities were shown in extracts from fruits (capsules) of the plant. Russian scientists developed enriched extracts which were used for treatment of festering infections. Gurevich *et al.* (1971) succeeded in the isolation and characterisation of hyperforin (Figure 6.4) as a compound with strong antibiotic activity. The final elucidation of the structure was published by Russian and Norwegian scientists (Bystrov *et al.* 1975, Brondz *et al.* 1982). According to Berghöfer and Hölzl (1986) hyperforin can mainly be found in buds, flowers and capsules of *H. perforatum*. They also suspected other compounds with hyperforin-like properties to be present. In 1992,

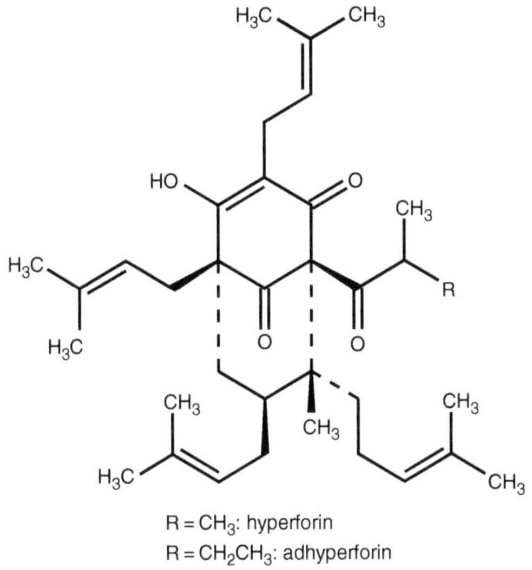

R = CH₃: hyperforin
R = CH₂CH₃: adhyperforin

Figure 6.4 Structural formulae of hyperforin and adhyperforin.

Maisenbacher and Kovar succeeded in the isolation and structural elucidation of adhyper-forin (Figure 6.4) and they qualitatively demonstrated the presence of additional polar acylphloroglucinols. These oxidised hyperforin derivatives might be genuine plant compounds or might be formed during the extraction (Verotta *et al.* 2000). Orthoforin was described as the main degradation product of hyperforin (Orth *et al.* 1999). Recently, the dihydrofuran furohyperforin was reported by Trifunovic *et al.* (1998) and Verotta *et al.* (1999) after super-critical CO_2 extraction. Further three oxygenated hyperforin derivatives (33-deoxy-33-hydroperoxyfurohyperforin, oxepahyperforin, 8-hydroxyhyperforin 8,1-hemiacetal) were published by Verotta *et al.* (2000) and the hyperforin analogues pyrano[7,28-*b*]hyperforin, (2R,3R,4S,6R)-6-methoxy-carbonyl-3-methyl-4,6-di(3-methyl-2-butenyl)-2-(2-methyl-1-oxopropyl)-3-(4-methyl-3-pentenyl)cyclohexanone   and   (2R,3R,4S,6S)-3-methyl-4,6-di(3-methyl-2-butenyl)-2-(2-methyl-1-oxopropyl)-3-(4-methyl-3-pentenyl)cyclohexanone by Shan *et al.* (2001). Hydroperoxycadiforin, a combination of cadinan, a sesquiterpene and hyper-forin, was isolated from the shoots and leaves of *H. perforatum* in very low amounts (0.0006%; Rücker *et al.* 1995). *H. brasiliense* contains the phloroglucinols hyperbrasilols A, B and C as well as isohyperbrasilol (Rocha *et al.* 1995, 1996). Phloroglucinols and rare O-prenylated phloroglucinol derivatives were found in the aerial parts of *H. japonicum* (Ishiguro *et al.* 1994, Hu *et al.* 2000).

Primarily, hyperforins were supposed to occur only in buds, flowers and capsules of *H. perforatum*. Schelosky (1997), however, could demonstrate the presence of this class of compounds in leaves and stems as well, although in low or very low concentrations, respectively.

The quantitative determination of hyperforin in different plant organs has only been possible after the establishment of a HPLC method (Hölzl and Ostrowski 1987). Maisenbacher and Kovar (1992) showed an increase in hyperforin and adhyperforin during the ontogenetic devel-opment from flowers to ripe fruits, the latter containing up to 1.8% adhyperforin and 4.4% hyperforin in the dry mass.

Properties of hyperforin

Hyperforin is an extremely lipophilic compound, soluble in organic solvents, but insoluble in water. In solution it is very unstable, which was already known from hyperforin-containing extracts (Berghöfer 1987). This instability is supposed to be due to the numerous double bonds which can react with singlet oxygen to form oxidation products. After illumination of a solution of hyperforin, more than 13 polar decomposition products were found with hyperforin-like electron spectra (Maisenbacher 1991). In ethanolic extracts from dry plant material hyperforin decomposed in a few days; in ethanolic extracts from fresh plants, however, hyperforin was still detectable after several months (Berghöfer 1987). It was supposed that this might be due to stabilising agents extracted from the fresh plants together with hyperforin. Pure solvent-free hyperforin as well as methanolic solutions of the pure compound are stable in the dark (Schelosky 1997). Patents have been filed for the production of stable hyperforin-containing plant extracts. The rapid decomposition of hyperforin in extracts is at least partially due to the photochemical properties of hypericin which induces the formation of singlet oxygen and which is always present in plants and plant extracts.

The stability of pure hyperforin (5 mg/ml) in different solvents was investigated by Granzow (2000) who showed that it was stable for more than 8 h in acetonitrile, methanol and chloroform, whereas in cyclohexane 50% of the hyperforin was decomposed in only 3 h.

2-Methyl-3-buten-2-ol is found in the essential oil of *H. perforatum* and it is supposed to be formed by oxidation of isoprenoid side-chains of hyperforin by oxygen radicals (Chialva 1982).

Besides the extreme lipophilicity, hyperforin also shows a strong acidity. The low pK_a value of 4.8 is due to the vinologous carboxylic acid structure in the molecule (Gurevich *et al.* 1971).

Isolation of hyperforin from H. perforatum

Extraction and purification of hyperforin by preparative TLC
(according to Berghöfer 1987)

Fresh flowers or buds weighing 30 g are homogenised for 2 min in 300 ml acetone with an Ultra-Turrax under constant cooling with ice. Alternatively, 2.5 g lyophilised flowers and fruits are homogenised for 2 min in 200 ml heptane. The filtrate is concentrated to dryness and the residue dissolved in acetone and used for preparative thin layer chromatography (TLC). Preparative TLC is performed on silica gel 60 F_{254} plates as stationary phase and hexane/diethylether 80 : 20 as solvent system. Hyperforin exhibits blue fluorescence under UV (360 nm) at an R_f value of 0.4–0.5. The same fluorescence colour is shown by adhyperforin at $R_f = 0.6$ just above hyperforin. Both compounds intensively quench short wave UV (254 nm). The silica gel containing the respective substances is scraped off the plates and extracted with acetone. The silica gel is sedimented by centrifugation and the clear extract evaporated to dryness. This yields a white to slightly yellow oil with a slight hop-odour. Crystallisation is achieved from the acetone solution.

Purification of hyperforin by preparative HPLC *(according to Granzow 2000)*

Plant extracts containing methanol-soluble material is chromatographed on a LiChrospher® RP-18 100 column (particle size 5 μm, length 25 cm, diameter 10 mm) with methanol/acetonitrile/H_2O/H_3PO_4 62.5 : 20 : 16.5 : 1 as eluent at a flow rate of 14 ml/min and detection at 274 nm. Hyperforin-containing fractions are diluted 1 : 1 with water and the organic solvents are completely evaporated under vacuum at 40°C. Hyperforin precipitates as a yellow waxy substance and is redissolved in methanol and washed with water several times to remove phosphoric acid. Hyperforin with a purity of 98% can be isolated using this method.

Influence of hypericin on the decomposition of hyperforin

In order to elucidate the influence of hypericin on the decomposition of hyperforin, hyperforin solutions alone or together with hypericin were illuminated by an electric bulb (100 W). In the solution containing hyperforin together with hypericin the hyperforin was decomposed nearly completely after 2 h. In the dark as well as under illumination, hyperforin is stable for at least 2 h when hypericin is not present (Schelosky 1997).

Biosynthesis of phloroglucinol derivatives

Acylphloroglucinols are polyketides that are formed from a coenzyme A-activated aliphatic starter acid (acetyl-, propionyl-, butyryl-, isobutyryl-, 2-methylbutyryl-CoA) and malonyl-CoA as extender acid from which an acetyl-moiety is transferred under release of CO_2. For the formation of acylphloroglucinols from the primary polyketo acids, ring-formation is established by C-acylation. The molecules are further transformed by reductions, C-methylations, formation of intramolecular C–C bridges, prenyl transfer or other modifications. Other members of the acylphloroglucinols are, for example, the hop bitter acids from *Humulus lupulus* or the cannabi-noids from *Cannabis sativus*. Hyperforin is highly substituted with isoprenoid moieties and therefore is very lipophilic. In its biosynthesis isobutyryl-CoA is the starter acid which is elongated with three acetyl-moieties from malonyl-CoA by phlorisobutyrolactone synthase, a chalcone synthase-like enzyme (Beerhues, personal communication). In *Humulus lupulus*, two phloroglucinol derivatives, 2-(3-methylbutanoyl)-1,3,5-benzenetriol (phloroisovalerophenone) and 2-(2-methylpropanoyl)-1,3,5-benzenetriol (phloroisobutyrophenone) as hop bitter acids, are synthesised by a chalcone synthase-like enzyme from malonyl-CoA and isovaleryl-CoA or isobutyryl-CoA, respectively (Zuurbier *et al.* 1995). Furthermore, the enzymatic oxidation of deoxyhumulone to the hop α-acid humulone was demonstrated in cell-free extracts from hop (Fung *et al.* 1997). Both, hyperforin and the hop bitter acids are very unstable.

Flavonoids and biflavonoids

Hypericum perforatum contains flavonoid aglyca, flavonoid glycosides as well as biflavonoids in a concentration of 2–4% of the dry mass. The most prominent flavonoid aglyca are the flavonols quercetin, kaempferol, myricetin, the dihydroflavonol dihydroquercetin and the flavone lute-olin, whereas nearly only quercetin was shown as the aglycon of flavonol glycosides (Figure 6.5). Dorossiev (1985) determined rutin, hyperoside and isoquercitrin as the main compounds. Jürgenliemk and Nahrstedt (2002) thoroughly re-investigated the phenolic compounds of *H. perforatum* and were able to detect quercetin and a number of quercetin glycosides (rutin, hyperoside, isoquercitrin, miquelianin, astilbin, guaijaverin, quercitrin, quercetin-3-O-(2″-acetyl)-β-D-galactoside) as well as isoorientin (6β-glucosylluteolin) and cyanidin-3-O-α-L-rhamnoside as anthocyanin.

The most prominent biflavonoids of *H. perforatum* are the biapigenins ($C_{30}H_{18}O_{10}$, M_r 538,47) I3,II8-biapigenin and amentoflavone = I3′,II8-biapigenin (Figure 6.6) (Berghöfer and Hölzl 1987, 1989). The concentrations range between 0.1 and 0.5% for I3,II8-biapigenin and 0.01–0.05% for amentoflavone (Berghöfer 1987). I3,II8-biapigenin was only found in flowers and buds and in traces in fruits of *H. perforatum*, but was not restricted to this species since the

Flavonol agylca:

Quercetin: $R_1 = H$, $R_2 = R_3 = OH$

Kaempferol: $R_1 = R_3 = H$, $R_2 = OH$

Myricetin: $R_1 = R_2 = R_3 = OH$

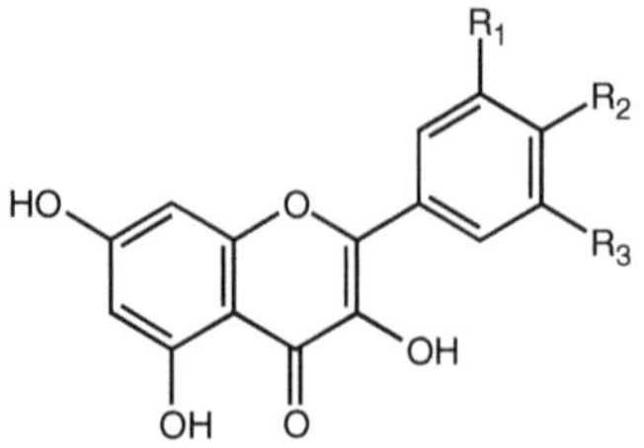

Quercetin glycosides:

R = α-1-rhamnosyl   Quercitrin

R = β-D-glucosyl   Isoquercitrin

R = β-D-galactosyl Hyperoside

R = β-rutinosyl   Rutin

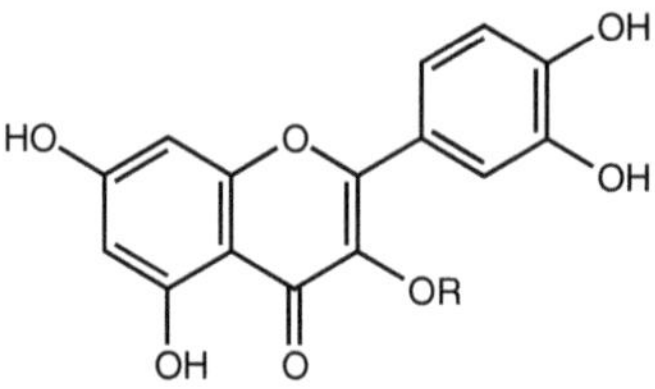

Figure 6.5 Structural formulae of flavonols and flavonol glycosides from *H. perforatum.*

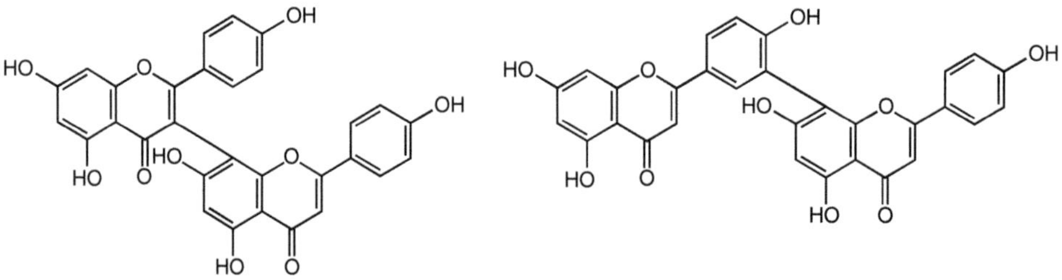

I3,II8-biapigenin

I3′,II8-biapigenin (amentoflavone)

Figure 6.6 Structural formulae of biflavonoids from *H. perforatum.*

compound was also present in flowers of *H. hirsutum, H. barbatum* and *H. montanum* (Berghöfer and Hölzl 1987).

Xanthones

Xanthones in Hypericum *species*

It is important to note with respect to chemotaxonomic classification that xanthones are generally only found in the families Gentianaceae and Clusiaceae. They mainly occur in the roots (Kaul 2000). Simple oxygenated xanthones are found in both families, whereas prenylated xanthones are commonly found in Clusiaceae (Schmidt *et al.* 2000a), but only rarely in Gentianaceae. Xanthones described for *H. perforatum* are 1,3,6,7-tetrahydroxyxanthone (norathyriol) in a concentration of 0.4 mg/100 g dry herb and mangiferin, isomangiferin and the xanthonolignoid kielcorin (Figure 6.7) (Kitanov and Blinova 1978, Nielsen and Arends 1978, Berghöfer and Hölzl 1986, Berghöfer 1987, Bennett and Lee 1989, Sparenberg 1993, Sparenberg *et al.* 1993, Kitanov and Nedialkov 1998).

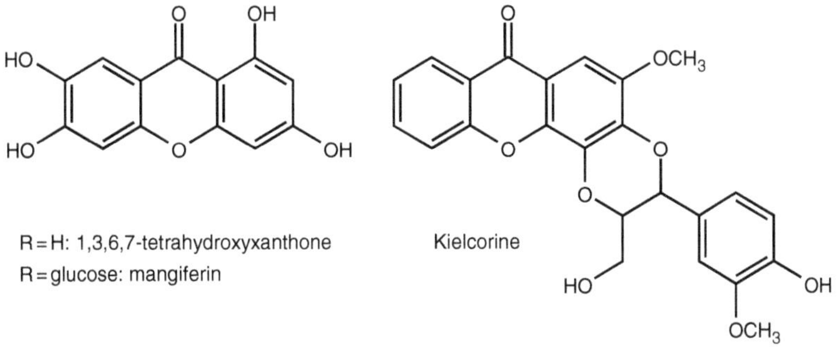

R = H: 1,3,6,7-tetrahydroxyxanthone
R = glucose: mangiferin

Kielcorine

Figure 6.7 Structural formulae of xanthones from *H. perforatum*.

Biosynthesis of xanthones

The biosynthesis of xanthones is currently examined in cell cultures of *Centaurium erythraea* (Gentianaceae) and *H. androsaemum* (Clusiaceae) accumulating 3,5,6,7,8-pentamethoxy-1-O-primverosylxanthone and prenylated and/or C-glucosylated derivatives of 1,3,6,7-tetrahydroxyxanthone, respectively (Peters *et al.* 1998, Schmidt *et al.* 2000b). Benzophenone synthase (Beerhues 1996), a chalcone synthase-like enzyme, couples 3-hydroxybenzoyl-CoA, formed from 3-hydroxybenzoic acid and coenzyme A by 3-hydroxybenzoate : CoA ligase (Barillas and Beerhues 1997), with acetyl-moieties from 3 malonyl-CoA leading to 2,3',4,6-tetrahydroxybenzophenone. This precursor molecule is differently coupled by specific xanthone synthases either to 1,3,5-trihydroxyxanthone in *C. erythraea* or to 1,3,7-trihydroxyxanthone in *H. androsaemum*. The xanthone synthases are cytochrome P450-oxidases catalysing different regiospecific oxidative phenol coupling reactions (Peters *et al.* 1998). Xanthone 6-hydroxylases, cytochrome P450 monooxygenases, further hydroxylate xanthones in position 6. The enzyme from *C. erythraea* (Gentianaceae) is specific for 1,3,5-trihydroxyxanthone, whereas the enzyme from *H. androsaemum* (Clusiaceae) prefers 1,3,7-trihydroxyxanthone (Schmidt *et al.* 2000b).

Isolation of norathyriol from shoots of H. perforatum

The determination and isolation of xanthones from the herb of *H. perforatum* (Sparenberg 1993) is quite tedious, since these compounds only occur in minute amounts in the plant.

Lipophilic components are removed from the lyophilised herb (150 g) in a soxhlett apparatus with dichloromethane (3.5 l, 15 h). Xanthones are then extracted from the pre-extracted herb with methanol (3.5 l, 9 h). Methanol is evaporated and the extract is redissolved in 500 ml water and partitioned between ethylacetate (5 × 500 ml) and water. The ethylacetate phase is collected and evaporated to dryness. The dry extract is then redissolved in methanol and fractionated on a Sephadex LH-20 column (Ø 4.2 cm, 11.5 cm) with methanol as eluent. Xanthone containing fractions are again chromatographed on a Sephadex LH-20 column (Ø 4.4 cm, 75.5 cm) with methanol as eluent. A third chromatography on Sephadex LH-20 (Ø 2.4 cm, 36.5 cm) was performed with methanol : water 80 : 20. Preparative HPLC was run on a LiChrosorb® 100 RP-18 column (particle size 7 μm, Ø 25 mm, length 250 mm) with 90% methanol (9 ml/min) as eluent. Preparative TLC on silica gel is performed as the last purification step with the solvent system ethylacetate/formic acid/water 30 : 2 : 3. Fluorescent zones are scraped off the plates and the silica gel eluted with methanol.

According to Sparenberg (1993) the concentrations of norathyriol (1,3,6,7-tetrahydroxyxan-thone) in flowers of *H. perforatum* are at 1.28 mg/100 g dry plant material followed by 0.95 mg/100 g in dry roots and 0.22 mg/100 g in dry leaves.

Chemical synthesis of norathyriol

Synthesis of norathyriol is extensively described by Lin *et al.* (1992) and Schelosky (1997).

Tannins and proanthocyanidins

Condensed tannins were shown to be present in *H. perforatum* as early as 1925 (Porodko 1925) and were described in more detail by Neuwald and Hagenström (1953) who found 3.8% condensed tannins in shoots, 12.4% in leaves and 16.2% in flowers. Usually tannin contents with catechin and epicatechin as main building moieties range from 6 to 15% in *H. perforatum*. Racz and Fuzi (1959) first detected proanthocyanidins in *H. perforatum*. Several condensed tannins (Akhtardzhiev and Kitanov 1975, Akhtardzhiev *et al.* 1984) and procyanidins (Melzer *et al.* 1989, 1991, Melzer 1990, Hölzl *et al.* 1994, Hölzl and Münker 1995) were described from *H. perforatum* and their pharmacological activities have been evaluated.

Using the method of Porter *et al.* (1986) a content of 7.36% procyanidins was determined by acid hydrolysis in *H. perforatum*. By using the methods according to PharmEur (2001) a total tannin content of 8.35% was determined. Since procyanidins as well as condensed tannins show binding to powdered skin, condensed tannins will account for 1% of the tannins in *H. perforatum*, since they cannot be transformed to anthocyanidins by acid hydrolysis.

Dimeric as well as trimeric procyanidins were found in *H. perforatum* with procyanidin B2 (epicatechin-(4β→8)-epicatechin; Figure 6.8) as the main compound (Hölzl and Münker 1985,

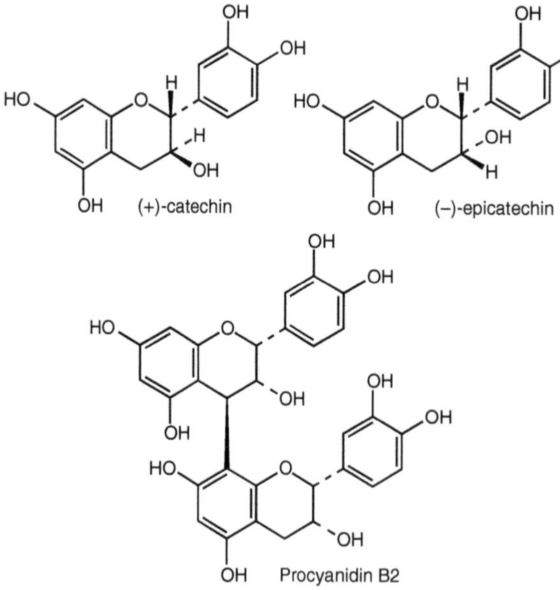

Figure 6.8 Structural formulae of catechin, epicatechin and procyanidin B2 (epicatechin-(4β→8)-epicatechin).

Melzer 1990). Recent investigations by Ploss *et al.* (2001) showed the occurrence of procyanidins A2, B1, B2, B3, B5, B7 and C1 in aqueous extracts of *H. perforatum* besides catechin and epicatechin (Figure 6.8).

Phenolic acids and other phenolic compounds

Widely occurring phenolic acids such as chlorogenic acid and caffeic acid have been found to be present in *Hypericum* species (Hölzl and Ostrowski 1987). A more detailed study further lists neochlorogenic acid (2-O-[*E*]-caffeoylquinic acid), 3-O-[*E*]-4-coumaroylquinic acid, 3-O-[*Z*]-4-coumaroylquinic acid, cryptochlorogenic acid (4-O-[*E*]-caffeoylquinic acid) and protocatechuic acid (Jürgenliemk and Nahrstedt 2002).

Coumarins like scopoletin and umbelliferone were described for *H. perforatum, H. elongatum, H. helianthemoides* and *H. scabrum* (Kitanov and Blinova 1987).

Anthraquinone derivatives

After TLC separation of extracts from *H. perforatum* two yellow to orange bands are visible which turn red after treatment with alkaline reagents; thus, they were suggested to be 1.8-dihydroanthraquinone derivatives. According to the polarity (R_f value) of the compounds the bands were supposed to correspond to dimeric structures. The compounds were tentatively identified as skyrin and/or oxyskyrin (Figure 6.2; Berghöfer 1987). It is supposed that these dimeric anthraquinones are precursors for the hypericins.

Terpenes and *n*-alkanes/*n*-alkanols

Hypericum perforatum has macroscopically visible translucent oil glands in all above-ground plant organs containing essential oil. The content of essential oil is highly variable ranging from 0.1 to 1% (Mathis and Ourisson 1964). Mathis and Ourisson (1963, 1964) determined essential oil components of 35 *Hypericum* species and found the following compounds: 2-methyloctan (isononan), *n*-nonan, *n*-undecan, *n*-octanal, *n*-decanal, myrcene, geraniol, α-pinene, β-pinene, limonene, α-terpineol and caryophyllene. Only flower buds and flowers, but not green shoot material contain essential oil. Cakir *et al.* (1997) reported the terpenoids α-pinene, (+)-3-carene, β-myrcene and β-caryophyllene as components of the essential oil. Other compounds are saturated hydrocarbons such as 2-methyl-octane (16.4%) as well as dodecanole, nonane, 3-methyl-nonane, isoundecane and undecane in appreciable concentrations. 2-methyl-3-buten-2-ol is found in the essential oil of *H. perforatum* and it is supposed that this is formed by oxidation of an isoprenoid side-chain of hyperforin by oxygen radicals (Chialva *et al.* 1982, 1983). 2-Methyl-3-buten-2-ol is also present in the essential oil of *Humulus lupulus*. Other compounds are myrcene, caryophyllene and humulene. A total of 29 compounds have been identified by gas chromatography as essential oil components of *H. perforatum* (Chialva *et al.* 1982, 1983). C_9–C_{13} saturated aliphatic hydrocarbons are components of the essential oil; furthermore, C_{16}–C_{29} alkanes and alkanols are present in *H. perforatum* (Chialva *et al.* 1981, Brondz and Gleibrokk 1983, Brondz *et al.* 1983).

References

Akhtardzhiev, K. and Kitanov, G. (1975) Catechin composition of *Hypericum perforatum. Farmatsiya (Sophia)* 24, 17–20.

Akhtardzhiev, K., Koleva, M. and Kitanov, S. (1984) Pharmacognostic study of representatives of *Arum, Althea* and *Hypericum* species. *Farmatsiya (Sophia)* 34, 1–6.

Banks, H.J., Cameron, D.W. and Raverty, W.D. (1976) Chemistry of the Coccoidea II – Condensed polycyclic pigments from two Australian pseudococcids (Hemiptera). *Aust. J. Chem.* 29, 1509–21.

Barillas, W. and Beerhues, L. (1997) 3-Hydroxybenzoate:coenzyme A ligase and 4-coumarate: coenzyme A ligase from cultured cells of *Centaurium erythraea*. *Planta* 202, 112–16.

Beerhues, L. (1996) Benzophenone synthase from cultured cells of *Centaurium erythraea*. *FEBS Lett.* 383, 264–6.

Bennett, G.J. and Lee, H.X. (1989) Xanthones from *Guttiferae*. *Phytochemistry* 28, 967–98.

Berghöfer, R. (1987) Analytik und Isolierung phenolischer Inhaltsstoffe von *Hypericum perforatum* L. aus Anbau und Vergleich mit anderen *Hypericum*-Arten. *Dissertationes Botanicae* vol. 106, Verlag J. Cramer, Berlin.

Berghöfer, R. and Hölzl, J. (1986) Johanniskraut (*Hypericum perforatum* L.) – Prüfung auf Verfälschung. *Dtsch. Apoth. Ztg.* 126, 2569–73.

Berghöfer, R. and Hölzl, J. (1987) Biflavonoids in *Hypericum perforatum*; Part 1. Isolation of I3,II8-biapigenin. *Planta Med.* 53, 216–17.

Berghöfer, R. and Hölzl, J. (1989) Isolation of I3′,II8-biapigenin (amentoflavone) from *Hypericum perforatum*. *Planta Med.* 55, 91.

Brockmann, H. (1957) Photodynamisch wirksame Pflanzenfarbstoffe. *Progr. Chem. Org. Nat. Prod.* 14, 141–77.

Brockmann, H. and Eggers, H. (1955a) Umwandlung von Penicilliopsin in Protohypericin und Hypericin. *Angew. Chemie* 67, 706.

Brockmann, H. and Eggers, H. (1955b) Partialsynthese des Proto-hypericins und Hypericins aus Emodin-anthron-(9). *Angew. Chemie* 67, 706–7.

Brockmann, H., Franssen, U., Spitzner, D. and Augustiniak, H. (1974) Zur Isolierung und Konstitution des Pseudohypericins. *Tetrahedron Lett.* 23, 1991–4.

Brockmann, H., Haschad, M.N., Maier, K. and Pohl, F. (1939) Über das Hypericin, den photodynamisch wirksamen Farbstoff aus *Hypericum perforatum*. *Naturwissenschaften* 27, 550.

Brockmann, H. and Pampus, G. (1954) Die Isolierung des Pseudohypericins. *Naturwissenschaften* 41, 86–7.

Brockmann, H., Pohl, F., Maier, K. and Haschad, M.N. (1942) Über das Hypericin, den photodynamischen Farbstoff des Johanniskrautes (*Hypericum perforatum*). *Ann. Chemie* 553, 1–53.

Brockmann, H. and Sanne W. (1953) Pseudohypericin, ein neuer *Hypericum*-Farbstoff. *Naturwissenschaften* 40, 461.

Brockmann, H. and Sanne, W. (1957) Zur Kenntnis des Hypericins und Pseudohypericins. *Chem. Ber.* 90, 2480–91.

Brockmann, H., Weber, E. and Pampus, G. (1952) Protofagopyrine and fagopyrine, photodynamic pigments of buckwheat (*Fagopyrum esculentum*). *Ann. Chem. Pharm.* 575, 53–83.

Brondz, I. and Greibrokk, T. (1983) *n*-1-Alkanols of *Hypericum perforatum*. *J. Nat. Prod.* 46, 940–1.

Brondz, I., Greibrokk, T., Groth, P.A. and Aasen, A.J. (1982) The relative stereochemistry of hyperforin – an antibiotic substance from *Hypericum perforatum*. *Tetrahedron Lett.* 23, 1299–300.

Brondz, I., Greibrokk, T. and Aasen, A.J. (1983) n-Alkanes of *Hypericum perforatum*: a revision. *Phytochemistry* 22, 295–6.

Bystrov, N.S., Chernov, B.K., Dobrynin, V.N. and Kolosov, M.N. (1975) The structure of hyperforin. *Tetrahedron Lett.* 32, 2791–4.

Cakir, A., Duru, M.E., Harmadabrar, M., Ciriminna, R., Passananti, S. and Piozzi, F. (1997) Comparison of the volatile oils of *Hypericum scabrum* L. and *Hypericum perforatum* L. from Turkey. *Flavour and Fragrance Journal* 4, 285–7.

Cameron, D.W. and Raverty, W.D. (1976) Pseudohypericin and other Phenanthroperylene Quinones. *Aust. J. Chem.* 29, 1523–33.

Chialva, F., Doglia, G., Gabri, G. and Ulian, F. (1983) Direct headspace gas chromatographic analysis with glass capillary columns in quality control of aromatic herbs. *J. Chromatogr.* 279, 333–40.

Chialva, F., Gabri, G., Liddle, P.A.P. and Ulian, F. (1981) Study on the composition of the essential oil from *Hypericum perforatum* L. and *Teucrium chamaedrys* L. *Riv. Ital. EPPOS* 63, 286–8.

Chialva, F., Gabri, G., Liddle, P.A.P. and Ulian, F. (1982) Qualitative evaluation of aromatic herbs by direct headspace GC analysis. Applications of the method and comparison with the traditional analysis of essential oils. *J. High Resolut. Chromatogr. Chromatogr. Commun.* 5, 182–8.

Dorossiev, K. (1985) Determination of flavonoids in *Hypericum perforatum. Pharmazie* 40, 585–6.

Etzlstorfer, C., Falk, H., Müller, N., Schmitzberger, W. and Wagner, U.G. (1993) Tautomerism and stereo-chemistry of hypericin: force field, NMR, and x-ray crystallographic investigations. *Monatsh. Chem.* 124, 751–61.

Falk, H. and Schmitzberger, W. (1992) The Nature of 'Soluble' Hypericin in *Hypericum* Species. *Monatsh. Chem.* 123, 731–9.

Fisel, J., Gäbler, H., Schwöbel, H. and Trunzler, G. (1966) *Haronga madagascariensis.* Botanik, Pharmakognosie, Chemie und therapeutische Anwendung. *Dtsch. Apoth. Z.* 106, 1053–60.

Fung, S.Y., Zuurbier, K.W.M., Paniego, N.B., Scheffer, J.J.C. and Verpoorte, R. (1997) Conversion of deoxyhumulone into the hop α-acid humulone. *Phytochemistry* 44, 1047–53.

Giese, A.C. (1981) The photobiology of *Blepharisma. Photochem. Photobiol. Rev.* 6, 139–80.

Gill, M. and Gimenez, A. (1991) Austrovenetin, the principal pigment of *Dermocybe austroveneta. Phytochemistry* 30, 951–5.

Gill, M., Gimenez, A. and McKenzie, R.W. (1988) Pigments of fungi. Part 8. Bianthraquinones from *Dermocybe austroveneta. J. Nat. Prod.* 51, 1251–6.

Granzow, D. (2000) Untersuchungen zu *Hypericum perforatum* L.: Anbau und Selektion, analytische und präparative Arbeiten. Thesis, University of Marburg, Germany.

Gurevich, A.I., Dobrynin, V.N., Kolossov, M.N., Popravko, S.A., Ryabova, I.D., Chernov, B.K., Derbentseva, N.A., Aizenman, B.E. and Garagulya, A.D. (1971) Hyperforin, an antibiotic from *Hypericum perforatum* L. *Antibiotici (Moskow)* 16, 510–13.

Häberlein, H., Tschiersch, K.P., Stock, S. and Hölzl, J. (1992) Johanniskraut (*Hypericum perforatum* L.): Nachweis eines weiteren Naphthodianthrons. *Pharm. Ztg. Wiss.* 5/137, 169–74.

Häder, D.P. and Häder, M.A. (1991) Effects of solar radiation on motility in *Stentor coeruleus. Photochem. Photobiol.* 54, 423–8.

Hegnauer, R. (1966) *Chemotaxonomie der Pflanzen*, Birkhäuser, Basel.

Hölzl, J. and Münker, H. (1985) Vorkommen von Procyanidinen in *Hypericum perforatum. Acta Agron. Hung.* supplement 34, 52.

Hölzl, J. and Ostrowski, E. (1987) Johanniskraut-HPLC-Analyse der wichtigen Inhaltsstoffe und deren Variabilität in einer Population. *Dtsch. Apoth. Ztg.* 127, 1227–30.

Hölzl, J., Sattler, S. and Schütt, H. (1994) Johanniskraut – eine Alternative zu synthetischen Antidepressiva. *Pharm. Ztg.* 139, 3959–77.

Hu, L.H., Khoo, C.W., Vittal, J.J. and Sim, K.Y. (2000) Phloroglucinol derivatives from *Hypericum japonicum. Phytochemistry* 53, 705–9.

Ishiguro, K., Nagata, S., Fukumoto, H., Yamaki, M. and Isoi, K. (1994) Phloroglucinol derivatives from *Hypericum japonicum. Phytochemistry* 35, 469–71.

Jürgenliemk, G. and Nahrstedt, J. (2002) Phenolic compounds from *Hypericum perforatum. Planta Med.* 68, 88–91.

Kartnig, T., Gobel, I. and Heydel, B. (1996) Production of hypericin, pseudohypericin and flavonoids in cell cultures of various *Hypericum* species and their chemotypes. *Planta Med.* 62, 51–3.

Kaul, R. (2000) *Johanniskraut*, Wissenschaftliche Verlagsgesellschaft, Stuttgart.

Kirakosyan, A., Hayashi, H., Inoue, K., Charchoglyan, A. and Vardapetyan, H. (2000a) Stimulation of the production of hypericins by mannan in *Hypericum perforatum* shoot cultures. *Phytochemistry* 53, 345–8.

Kirakosyan, A.B., Vardapetyan, R.R. and Charchoglyan, A.G. (2000b) The content of hypericin and pseudohypericin in cell cultures of *Hypericum perforatum. Russ. J. Plant Physiol.* 47, 270–3.

Kitanov, G.M. (2001) Hypericin and pseudohypericin in some *Hypericum* species. *Biochem. Syst. Ecol.* 29, 171–8.

Kitanov, G.M. and Blinova, K.F. (1978) Mangiferin in some species of *Hypericum* genus. *Khim. Prirod. Soedin* 524.

Kitanov, G.M. and Blinova, K.F. (1987) Modern state of the chemical study of species of the genus *Hypericum. Khim. Prirod. Soedin.* 1987, 185–203.

Kitanov, G.M. and Nedialkov, P.T. (1998) Mangiferin and isomangiferin in some *Hypericum* species. *Biochem. Syst. Ecol.* 26, 647–53.

Kuhlmann, H.W. (1998) Photomovements in ciliated protozoa. *Naturwissenschaften* 85, 143–54.

Lavie, G., Mazur, Y., Lavie, D. and Meruelo, D. (1995) The chemical and biological properties of hypericin – a compound with a broad spectrum of biological activities. *Medicinal Research Reviews* 15, 111–19.

Lin, C.N., Liou, S.S., Ko, F.N. and Teng, C.M. (1992) Gamma-pyrone compounds. II: Synthesis and antiplatelet effects of tetraoxygenated xanthones. *J. Pharm. Sci.* 81, 1109–12.

Maisenbacher, P. (1991) Untersuchungen zur Analytik von Johanniskrautöl. Thesis, University of Tübingen, Germany.

Maisenbacher, P. and Kovar, K.A. (1992) Adhyperforin: a homologue of hyperforin from *Hypericum perforatum. Planta Med.* 58, 291–3.

Mathis, C. and Ourisson, G. (1963) Etude chimio-taxonomique du genre *Hypericum* I. *Phytochemistry* 2, 157–70.

Mathis, C. and Ourisson, G. (1964) Etude chimio-taxonomique du genre *Hypericum* II, III, IV, V. *Phytochemistry* 3, 115–31, 133–41, 377–9.

Melzer, R. (1990) Untersuchungen zur Analytik und kardiovaskulären Wirkung der oligomeren Procyanidine von *Hypericum perforatum* L. Thesis, University of Marburg, Germany.

Melzer, R., Fricke, U., Hölzl, J., Podehl, R. and Zylka, J. (1989) Procyanidins from *Hypericum perforatum*: effects on isolated guinea pig hearts. *Planta Med.* 55, 655–6.

Melzer, R., Fricke, U. and Hölzl, J. (1991) Vasoactive properties of procyanidins from *Hypericum perforatum* L. in isolated porcine coronary arteries. *Arzneim. Forsch.* 41, 481–3.

Nielsen, A. and Arends, P. (1978) Structure of the xanthonolignoid kielcorin. *Phytochemistry* 17, 2040–1.

Neuwald, F. and Hagenström, U. (1953) Über eine Bestimmung des Gerbstoffgehaltes von *Hypericum perforatum* L. *Sci. Pharm. (Wien)* 21, 279–82.

Orth, H.C., Hauer, H., Erdelmeier, C.A. and Schmidt, P.C. (1999) Orthoforin: The main degradation product of hyperforin from *Hypericum perforatum* L. *Pharmazie* 54, 193–200.

Peters, S., Schmidt, W. and Beerhues, L. (1998) Regioselective oxidative phenol couplings of 2,3',4,6-tetrahydroxybenzophenone in cell cultures of *Centaurium erythraea* RAFN and *Hypericum androsaemum* L. *Planta* 204, 64–9.

Ploss, O., Pastereit, F. and Nahrstedt, A. (2001) Procyanidins from the herb of *Hypericum perforatum. Pharmazie* 56, 509–11.

Porodko, Z. (1925) *Arch. Pharm.* 263, 170 cited according to Neuwald, F. and Hagenström, U. (1953) Über eine Bestimmung des Gerbstoffgehaltes von *Hypericum perforatum* L. *Sci. Pharm. (Wien)* 21, 279–82.

Porter, L.J., Hrstich, L.N. and Chan, B.G. (1986) The conversion of procyanidins and prodelphinidins to cyanidin and delphinidin. *Phytochemistry* 25, 223–30.

Racz, G. and Fuzi, J. (1959) The presence of leucoanthocyanidins in herbs. *Acta Pharm. Hung.* 29, 64–70.

Robson, N.K.B. (1977) Studies in the genus *Hypericum* L. (Guttiferae). I. Infrageneric classification. *Bull. Br. Mus. Nat. Hist. (Botany)* 5, 293–355.

Robson, N.K.B. (1981) Studies in the genus *Hypericum* L. (Guttiferae). 2. Characters of the genus. *Bull. Br. Mus. Nat. Hist. (Botany)* 8, 55–226.

Rocha, L., Marston, A., Potterat, O., Kaplan, M.A.C., Stoeckli-Evans, H. and Hostettmann, K. (1995) Antibacterial phloroglucinols and flavonoids from *Hypericum brasiliense. Phytochemistry* 40, 1447–52.

Rocha, L., Marston, A., Potterat, O., Kaplan, M.A.C. and Hostettmann, K. (1996) More phloroglucinols from *Hypericum brasiliense. Phytochemistry* 42, 185–8.

Rodewald, R., Arnold, R., Giesler, J. and Steglich, W. (1977) Synthese von Hypericin und verwandten meso-Naphthodianthronen durch alkalische Dimerisierung von Hydroxyanthrachinonen. *Angew. Chem.* **89**, 56–7.

Rücker, G., Manns, D., Hartmann, R. and Bonsels, U. (1995) Peroxides as constituents of plants. Part 19. A C_{50}-hydroperoxide from *Hypericum perforatum*. *Arch. Pharm. (Weinheim)* **328**, 725–30.

Sattler, S. (1997) Naphthodianthrone aus *Hypericum perforatum* L.: Isolierung, Pharmakokinetik. Löslichkeitsverbesserung und Absorptionsstudien am Caco-2 Zellkulturmodell. Thesis, University of Marburg, Germany.

Schelosky, N. (1997) Separation of hypericins in extracts of *Hypericum perforatum* (St. John's wort). Thesis, University of Innsbruck, Austria.

Schmidt, W., Abd El-Mawla, A.M.A., Wolfender, J.L., Hostettmann, K. and Beerhues, L. (2000a) Xanthones in cell cultures of *Hypericum androsaemum*. *Planta Med.* **66**, 380–1.

Schmidt, W., Peters, S. and Beerhues, L. (2000b) Xanthone 6-hydroxylase from cell cultures of *Centaurium erythraea* RAFN and *Hypericum androsaemum* L. *Phytochemistry* **53**, 427–31.

Shan, M.D., Hu, L.H. and Chen, Z.L. (2001) Three new hyperforin analogues from *Hypericum perforatum*. *J. Nat. Prod.* **64**, 127–30.

Song, P.S. and Walker, E.B. (1981) Molecular aspects of photoreceptors in protozoa and other microorganisms. In *Biochemistry and Physiology of Protozoa*. M. Levandowsky and S.H. Hutner (Eds), Academic Press, New York, vol. 4, pp. 199–233.

Southwell, I.A. and Bourke, C.A. (2001) Seasonal variation in hypericin content of *Hypericum perforatum* L. (St. John's wort). *Phytochemistry* **56**, 437–41.

Southwell, I.A. and Campbell, M.H. (1991) Hypericin content variation in *Hypericum perforatum* in Australia. *Phytochemistry* **30**, 475–8.

Sparenberg, B. (1993) MAO-inhibitierende Eigenschaften von Hypericuminhaltsstoffen und Untersuchungen zur Analytik und Isolierung von Xanthonen aus *Hypericum perforatum* L. Thesis, University of Marburg, Germany.

Sparenberg, B., Demisch, L. and Hölzl, J. (1993) Untersuchungen über die antidepressiven Wirkstoffe von Johanniskraut. *Pharm. Ztg. Wiss.* **138**, 50–4.

Spitzner, D. (1977) Synthese von Protohypericin aus Emodin. *Angew. Chem.* **89**, 55–6.

Steglich, W., Fugmann, B. and Lang-Fugmann, S. (Eds) (1997) *Römpp-Lexikon Naturstoffe*, Thieme, Stuttgart.

Tao, N., Deforce, L., Romanowski, M., Meza-Keuthen, S., Song, P.S. and Furuya, M. (1994) *Stentor* and *Blepharisma* photoreceptors: structure and function. Acta Protozool. 33, 199–211.

Trifunovic, S., Vajs, V., Macura, S., Juranic, N., Djarmati, Z., Jankov, R. and Milosavljevic, S. (1998) Oxidation products of hyperforin from *Hypericum perforatum*. *Phytochemistry* **49**, 1305–10.

Verotta, L., Appendino, G., Belloro, E., Jakupovic, J., Bombardelli, E. and Morazzoni, P. (1999) Furohyperforin, a prenylated phloroglucinol from St. John's wort (*Hypericum perforatum*). *J. Nat. Prod.* **62**, 770–2.

Verotta, L., Appendino, G., Jakupovic, J. and Bombardelli, E. (2000) Hyperforin analogues from St. John's wort (*Hypericum perforatum*). *J. Nat. Prod.* **63**, 412–15.

Wirz, A., Meier, B. and Sticher, O. (2001) Absorbance data of hypericin and pseudohypericin used as reference compounds for medicinal plant analysis. *Pharmazie* **56**, 52–7.

Zdunek, K. (1993) Untersuchungen zur Bildung von Hypericinen in *in vitro*-Kulturen von Johanniskraut. Diploma thesis, University of Düsseldorf, Germany.

Zdunek, K. and Alfermann, A.W. (1992) Initiation of shoot organ cultures of *Hypericum perforatum* and formation of hypericin derivatives. *Planta Med.* **58**, 621.

Zuurbier, K.W.M., Fung, S.Y., Scheffer, J.J.C. and Verpoorte, R. (1995) Formation of intermediates in the biosynthesis of bitter acids in *Humulus lupulus*. *Phytochemistry* **38**, 77–82.

7 Determination of hypericins and hyperforin in herbal medicinal products

Astrid Michelitsch, Werner Likussar,
Manfred Schubert-Zsilavecz and Mario Wurglics

Over the past several years, St John's wort products have enjoyed a tremendous growth and increased acceptance for the treatment of mild to moderate depression, both in Europe and in USA. During this period, clinical studies comparing these products with placebo or reference antidepressants have shown St John's wort extract to be an effective medicine (Laakmann *et al.* 1998, Harrer *et al.* 1999, Kaul 2000).

Despite intensive research efforts, it has not yet been possible to identify all the active components of St John's wort extract. On the basis of current knowledge, it is certain that hyperforin is one of the active components (Müller *et al.* 1997, 1998) and that hypericins and possibly flavonoids also contribute to the antidepressant properties of St John's wort (Butterweck *et al.* 2000).

In the light of the rapidly increasing importance of St John's wort extract in the treatment of depression, our study (Wurglics *et al.* 2000) was devoted to assessing the quality of the St John's wort products currently available on the German market. We compared hypericin and hyperforin content both on a batch-to-batch as well as on a between manufacturer basis to determine the reliability of the dose-to-dose reproducibility and to assess the switchability of the eight products tested.

Since *Hypericum perforatum* contains hypericin as well as pseudohypericin and their precursors protohypericin and protopseudohypericin, it is usual to determine these substances in the form of total hypericin by applying hypericin as reference (Schmidt 1996, Gaedcke 1997). For this reason, the precursors protohypericin and protopseudohypericin must be converted into hypericin and pseudohypericin, respectively, by treating them with artificial light or daylight.

In the past, the total hypericin content of the crude drug was determined photometrically (DAC, 1991). Because of the poor selectivity of this method, the measured concentration of hypericin was too high. Therefore, chromatographic methods were proposed for the determination of total hypericin in *Herba hyperici* (Freytag 1984, Hölzel and Ostrowski 1987, Kartnig and Göbel 1992) and in herbal medicinal products (HMPs) (Krämer and Wiartalla 1992, Klein-Bischoff and Klumpp 1993, Wagner and Bladt 1994).

The electrochemical properties of hypericin (Redepenning and Nengbing 1993) suggested the application of a differential pulse polarographic (DPP) method to determine the total hypericin content in HMPs.

Hypericin and pseudohypericin were reduced in a single step (sample direct current polarography (SDC)) or a single peak (DPP) at the dropping mercury electrode (DME) (pH 3.5–10.0). In the pH range 3.5–5.5, their DPP peaks were dependent on H^+-concentration; the shift of the peak potentials was about $-90\,\text{mV/pH}$. Between pH 6.0 and 10.0, no change in the peak potential was registered. The intensity of the peak current of hypericin and pseudohypericin was influenced both by the pH value and the type of the buffer system (i.e. acetate buffer, McIlvain

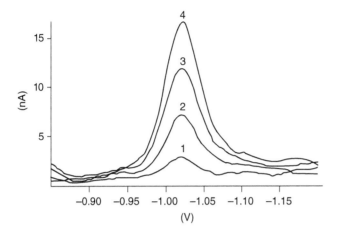

Figure 7.1 Differential pulse polarogram of hypericin in BR buffer: methanol solution (3 : 7 v/v; pH = 6.0). The concentrations of hypericin employed were: 1 : 1.0, 2 : 3.0, 3 : 5.0, 4 : 7.0 μg/ml.

buffer, Sörensen buffer and Britton–Robinson (BR) buffer). It was necessary to add methanol to the buffer solutions as a solubilizer.

It was found that the peak current was reproducible, with optimal sensitivity, in BR buffer: methanol solution (3 : 7 v/v) at pH 6.0 ± 0.1. At this pH value, the peak potential of hypericin was −1.02 V and that of pseudohypericin was −1.00 V vs. silver/silver chloride. Separation of hypericin and pseudohypericin was not possible because of the similarity in their electrochemical behaviour and their peak potentials. However, since the peak height vs. concentration correlations of hypericin and pseudohypericin were almost identical, the polarographic method may be considered suitable for the determination of total hypericin in HMPs (Figure 7.1) (Michelitsch *et al.* 2000).

The determination of hypericins and hyperforin in one run is not possible, due to different chemical properties of the compounds. Therefore the hyperforin content was analysed by an optimised HPLC method based on that of Maisenbacher and Kovar (1992).

At present there are more than 50 St John's wort products on the German market. Of these, eight were randomly selected for the study. Three of the eight products tested are capsules, four are film tablets and one is a sugar-coated tablet (Table 7.1). Products P2, P3 and P4 each contain 300 mg of St John's wort extract (on a dry basis), products P5, P6 and P7 each contain 425 mg, P1 contains 250 mg and product P8 612 mg. The dry extract used in most of the products was prepared using 60% ethanol, although in two products (P2 and P4) 80% methanol was used and in P8, 50% ethanol was used as the extraction medium. In the case of product P1, no information about the medium used to obtain the extract was given, even though this is obviously an important determinant of the final composition.

The total hypericin content also varied among manufacturers (Figure 7.2). Product P3 had the lowest hypericin content (0.16%), six products had hypericin contents between 0.2 and 0.3% and one, P7, contained more than 0.3%.

The product P1 contained hyperforin only in trace quantities (<0.2%) (Figure 7.3). A further product, P6, contained less than 2% hyperforin on average, five products had average hyperforin contents between 2% and 3% and product P3 had the highest (4.14%) hyperforin content.

Table 7.1 Characteristics of the eight St John's wort products studied

Product no.	Dosage form	DEV[a]	Extraction fluid	Trade name
Nominal amount of extract 250 mg				
P1	Film tablet	?	?	Remotiv
Nominal amount of extract 300 mg				
P2	Sugar-coated tablet	4–7 : 1	Methanol 80%	Jarsin 300
P3	Film tablet	2,5–5 : 1	Ethanol 60%	Neuroplant 300
P4	Film tablet	4–7 : 1	Methanol 80%	Texx 300
Nominal amount of extract 425 mg				
P5	Capsule	3,5–6 : 1	Ethanol 60%	Felis 425
P6	Capsule	3,5–6 : 1	Ethanol 60%	Futuran
P7	Capsule	3,5–6 : 1	Ethanol 60%	Helarium 425
Nominal amount of extract 612 mg				
P8	Film tablet	5–8 : 1	Ethanol 50%	Laif 600

Note
a Raw material: extract ratio (on a dry basis).

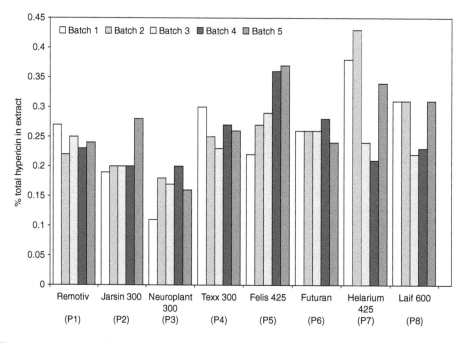

Figure 7.2 Average total hypericin content of individual batches for each of the eight products studied
($n = 10$ for each batch).

The batch-to-batch reproducibility also varied among manufacturers. In the case of hyper-forin, product P3 showed excellent batch-to-batch consistency, with a standard deviation (s.d.) from the average value of less than 4%. Product P2 also showed good reproducibility, with a s.d. of less than 15% from the overall average. Products P4, P5, P6 and P8 varied widely among

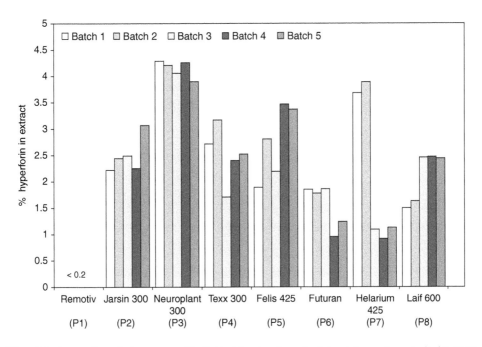

Figure 7.3 Average hyperforin content of individual batches for each of the eight products studied ($n = 10$ for each batch).

Table 7.2 Average total hypericin and hyperforin content of the products studies in mg per dosage unit and % of the extract

Product	Ø total hypericin content/dosage unit (mg)	% total hypericin in extract	s.d._{rel} (%)	Ø hyperforin content/dosage unit (mg)	% hyperforin in extract	s.d._{rel} (%)
P1	0.602	0.24	7.47	<0.5	<0.2	—
P2	0.643	0.21	16.01	7.48	2.49	13.76
P3	0.487	0.16	20.33	12.43	4.14	3.86
P4	0.788	0.26	10.28	7.51	2.50	21.22
P5	1.282	0.30	20.20	11.68	2.75	25.47
P6	1.100	0.26	4.73	6.53	1.54	24.26
P7	1.359	0.32	29.14	9.09	2.14	70.36
P8	1.704	0.28	16.97	12.81	2.10	24.00

batches (s.d. 20–25%), while product P7 had an unacceptably high variability in the hyperforin content (s.d. 70%). With respect to hypericin content, seven of the eight products had acceptably low batch-to-batch variability (s.d. <20% from the overall average). Here again, product P7 exhibited the highest variability (s.d. 30%) (Table 7.2).

According to a proposed monograph of St John's wort extract ('Powdered St John's wort Extract' USP), the hyperforin content of the extract should be at least 3% (*Pharmacopeial Forum* 1999). Of all the products we tested, only one (Neuroplant 300) would meet this criterion. On the other hand, seven of the eight products tested had a hypericin content exceeding the proposed monograph limit of 0.2% of the extract.

Comparison of eight St John's wort products currently available on the German market demonstrated that the content of hypericin and hyperforin varied significantly between products, indicating that they are not interchangeable for the treatment of mild to moderate depression. Instead, manufacturers should be required to state the content of these pharmacologically important components on the label, so that the doctors, pharmacists and patients are adequately informed. Furthermore, although several products exhibited excellent batch-to-batch reproducibility, not all products can be considered consistent.

Experimental

Sample preparation

To determine the content uniformity, individual dosage forms (tablets or capsules) were ground in a mortar and pestle and transferred quantitatively into 50 ml standard flasks using ethanol. After bringing the sample to required volume with 80% ethanol, the flask was stoppered and placed in an ultrasound bath for 10 min to ensure complete extraction of both the hyperforin and hypericin. Samples were then removed with a syringe, filtered through an 0.45-μm filter and filled directly into brown coloured HPLC vials. No further workup prior to analysis was necessary. For each batch from each manufacturer ten dosage forms were analysed for both hypericin (DPP) and hyperforin (HPLC).

Determination of hypericin by DPP

Polarographic measurements using DPP and SDC, including the polarographic analysis of herbal medicinal products, were carried out by using a 693 VA processor (Metrohm, Herisau, Switzerland) in combination with a VA stand 694 (Metrohm, Herisau, Switzerland). This stand consists of a multimode electrode (static mercury drop (SMDE), hanging mercury drop (HMDE) and DME) as the working electrode, a silver/silver chloride/potassium chloride (3M) reference electrode and a platinum wire is used as auxiliary electrode.

For DPP and SDC, the analyser operated and includes the following parameters: mode, SMDE; drop size, 9 (approx. $0.5\,mm^2$); t_{step}, 1.5 s; U_{step}, 4 mV; potential range, -0.85 to $-1.20\,V$; pulse amplitude, 20 mV (for DPP only). The Metrodata VA database was used to evaluate the data. In order to determine the concentration of total hypericin, standard addition method was used by analysing an aliquot of the solution described above after exposing it to a daylight lamp for 30 min. The data were evaluated by applying the tangent method.

Determination of hyperforin by HPLC

For the determination of hyperforin, an optimized HPLC-method based on that of Maisenbacher and Kovar (1992) was applied using two Dynamax Model SD-200 pumps, a Varian Pro Star Model 410 autosampler, a Dynamax Absorbance Detector Model UV-DII and a LiChrospher 100, RP8 (5 μm), 125 × 4 mm column.

Acknowledgements

We would like to acknowledge Fonds der Chemischen Industrie, AAI Deutschland GmbH & Co. KG, Wegenerstraße 13, 89231 Neu-Ulm.

References

Butterweck, V., Jurgenliemk, G., Nahrstedt, A. and Winterhoff, H. (2000) Flavonoids from *Hypericum perforatum* show antidepressant activity in the forced swimming test. *Planta Med.* 66 (1), 3–6.

Deutscher Arzneimittelcodex (1991) *DAC 86 3. Erg.*

Freytag, W.E. (1984) Bestimmung von Hypericin und Nachweis von Pseudohypericin in *Hypericum perforatum* L. durch HPLC. *Deutsch. Apoth. Ztg.* 124, 2383–6.

Gaedcke, F. (1997) Johanniskraut und dessen Zubereitungen. *Deutsch. Apoth. Ztg.* 137, 117–21.

Harrer, G., Schmidt, U., Kuhn, U. and Biller, A. (1999) Comparison of equivalence between the St. John's wort extract LoHyp-57 and fluoxetine. *Arzneim.-Forsch./Drug Res.* 49(1), 289–96.

Hölzel, J. and Ostrowski, E. (1987) Johanniskraut (*Hypericum perforatum* L.), HPLC-Analyse der wichtigen Inhaltsstoffe und deren Variabilität in einer Population. *Deutsch. Apoth. Ztg.* 127, 1127–30.

Kartnig, Th. and Göbel, I. (1992) Determination of hypericin and pseudohypericin by thin layer chromatography-densitometry. *J. Chromatogr.* 609, 423–6.

Kaul, R. (2000) *Johanniskraut, Handbuch für Ärzte, Apotheker und andere Natur-Wissenschaftler*, Wissenschaftliche Verlagsgesellschaft mbH Stuttgart.

Klein-Bischoff, U. and Klumpp, U. (1993) Hypericin und Fluoreszenz, eine quantitative fluorimetrische Bestimmungsmethode. *Pharm. Ztg. Wiss.* 138, 55–8.

Krämer, W. and Wiartalla, R. (1992) Bestimmung von Naphthodianthronen (Gesamthypericin) in Johanniskraut (*Hypericum perforatum* L.). *Pharm. Ztg. Wiss.* 137, 202–6.

Laakmann, G., Schüle, C., Baghai, T. and Kieser, M. (1998) St. John's wort in mild to moderate depression: the relevance of hyperforin for the clinical efficacy. *Pharmacopsychiatry* 31 (Suppl.), 54–9.

Maisenbacher, P. and Kovar, K.A. (1992) Analysis and stability of *Hyperici oleum*. *Planta Med.* 58(4), 351–4.

Michelitsch, A., Biza, B., Wurglics, M., Schubert-Zsilavecz, M., Baumeister, A. and Likussar, W. (2000) Determination of hypericin in herbal medicine products by differential pulse polarography. *Phytochem. Anal.* 11, 41–4.

Müller, W.E., Rolli, M., Schäfer, C. and Häfner, U. (1997) Effects of *Hypericum* extract (LI 160) in biochemical models of antidepressant activity. *Pharmacopsychiatry* 30 (Suppl.), 102–7.

Müller, W.E., Singer, A., Wonnemann, M., Häfner, U. and Rolli, M. (1998) Hyperforin represents the neurotransmitter reuptake inhibiting constituent of *Hypericum extract*. *Pharmacopsychiatry* 31 (Suppl.), 16–21.

Pharmacopeial Forum (1999) 25(2), 7758–60.

Redepenning, J. and Nengbing, T. (1993) Measurement of formal potentials for hypericin in dimethylsulphoxide. *Photochem. Photobiol.* 58, 532–5.

Schmidt, A.H. (1996) Online-Belichtung zur HPLC-Bestimmung von Gesamthypericin. *Lab. Praxis* 20, 58–60.

Wagner, H. and Bladt, S. (1994) Pharmaceutical quality of *Hypericum* extracts. *DJ. Geriatr. Pschiat. Neurol.* 7, 65–8.

Wurglics, M., Westerhoff, K., Kaunzinger, A., Wilke, A., Baumeister, A. and Schubert-Zsilavecz, M. (2000) Johanniskrautextrakt-Präparate: Vergleich aufgrund der Hyperforin- und Hypericingehalte. *Deutsch. Apoth. Ztg.* 140, 3904–10.

8 Secondary metabolites content of *Hypericum* sp. in different stages and plant parts

Katarzyna Seidler-Łożykowska

Introduction

The secondary metabolites content of *Hypericum* species is not only related to genetic and environmental factors, but also to its harvest period, drying process and storage (Bombardelli and Morazzoni 1995). Detailed analysis of the active compounds content of the species in its vegetative stages and in particular plant parts helps us to define the best harvest time and allows us to obtain high quality raw material.

Secondary metabolites content in the vegetative stages

The highest content of hypericin and pseudohypericin was recorded in the herb harvested in full blossom and it decreased as the plant overbloomed (Seidler *et al.* 1999). Similar results were reported by Bomme (1997), Brantner *et al.* (1994), Kartnig *et al.* (1997), Büter and Büter (2000). However, Mártonfi and Repčák (1994) in their previous investigation reported the highest content of hypercin and pseudohypericin while the herb was in bud and it decreased when flowers bloomed and overbloomed.

The largest amount of flavonoids contained in the herb was detected at the beginning of its bloom – the phase of yellow bud and then decreased as the plant bloomed then overbloomed (Seidler *et al.* 1999, Bomme 1997, Brantner *et al.* 1994). Bomme (1997) observed the maximum amount of flavonoids in plant leaves at the beginning of blossom stage. Kartnig's *et al.* (1997) study indicated that the largest amount of monoflavonoids was detected before flowering, while biflavonoids reached their highest level in blossom and then decreased when overbloomed (Figure 8.1). The content of I3,II8 biapigenin was higher in the phase of young opening bud than in blossom and later. Similar results were obtained by Tekel'ová *et al.* (2000) who found a progressive increase of biapigenin content in the developing buds and it reached its maximum just before blooming. The hyperosid content increased up to full bloom stage and the beginning of overblown flower stage and rapidly decreased when the plant was completely overblown. The quercitrin content was higher while flower was in young bud. The content of rutin reached its maximum in full bloom phase (Mártonfi and Repčák 1994). Tekel'ová *et al.* (2000) observed the highest content of quercetin when flowers overblown, while content of hyperosid and isoquercitrin reached their maximum in the bud phase. They also found a decrease in the content of quercitrin in older flowers (maximum – buds just before opening).

Broad analysis carried out by Mártonfi and Repčák (1994) showed the highest amount of hyperforin over the whole overblown flower stage and it was located mainly in capsules. The similar results were obtained by Büter and Büter (2000) who reported that hyperforin levels increased in advanced developmental stages. Also Tekel'ova's *et al.* (2000) in their study showed a gradual increase of hyperforin with the highest level in unripe fruits.

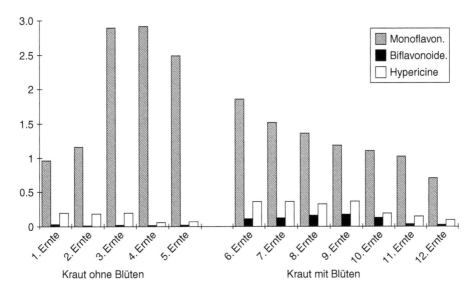

Figure 8.1 Total amount of monoflavonoids, biflavonoids and hypericin at different harvesting times in the herb of *H. perforatum* L. var. Topaz (based on the data of Kartnig *et al.* 1997).

Secondary metabolites content in different plant parts

Hypericin and pseudohypericin

Umek *et al.* (1999) who examined six *Hypericum* species collected around Slovenia stated that the level of hypericin and pseudohypericin in *Hypericum perforatum* flowers was the highest one compared with other *Hypericum* species. The largest content of hypericin and pseudohypericin was found in *H. perforatum* flowers (Seidler *et al.* 1999, Umek *et al.* 1999, Southwell and Campbell 1991, Tekel'ová *et al.* 2000). These results were proved by Bomme (1997) who observed that 90% of these compounds were located in 20–30 cm upper part of the plant, and by Hannig *et al.* (1995) who stated that 100% of hypericin and pseudohypericin was found in flowers and flower buds (Table 8.1). While cutting level was raised from 10 to 95 cm the increase in hypericins yield was observed, but the total herb yield decreased. The ratio of hypericin to pseudohypericin was 1 : 3, 1 : 4 (Hannig *et al.* 1995). Tekel'ová *et al.* (2000) reported that content of hypericin and pseudohypericin increased gradually during the development and growth of the blossom buds and reached their maximum when flowers were just opened. Repčák and Mártonfi (1997) who examined the distribution of the compounds in different flower parts reported that hypericin and pseudohypericin were located mainly in petals and stamens; there were no or only traces of them in pistils and sepals. Kartnig's *et al.* (1997) showed that the herb without flowers contained app.10 times less hypericin and pseudohypericin compared to flowers, while the herb with flowers – 4 times less. These results are similar to the observations made by Southwell and Campbell (1991), while Hannig *et al.* (1995) did not find any hypericin and pseudohypericin in stems and leaves.

Fornasiero *et al.* (1998) investigated the localization of active compounds in anatomical and cytochemical structure of leaves of *H. perforatum*. Black nodules appeared at the tip of young leaves and were located along leaf margins during their growth. Occasionally, nodules were found in the central lamina. Nodules differentiated and matured at the early stage of leaf

Table 8.1 Secondary compounds in *H. perforatum* plant parts (based on the data of Hannig *et al.* 1995)

Plant part	Herb mass (g)	Herb mass (%)	Extract content (%)	Hypericin content (%)	Σ Flavonoids (mg/g)	Σ Flavonoids (%)
Stem	98.00	57.4	13.0	0	1.74	4.0
Leaves	28.99	17.0	38.0	0	67.62	46.5
Flower buds	6.68	3.9	33.6	21	46.24	7.3
Flowers	36.95	21.7	33.2	79	48.26	42.2
Total	170.62	100.0	22.4	100	24.75	100.00

development, before the maturation of tissue. The pigments accumulated in enlarging secretion cells were hypericin and its derivatives. Secretion nodules of the leaf contained significantly more active substances than leaf portion without nodules. These results corresponded to those presented by Southwell and Campbell (1991) who found the positive correlation between the hypericin content and the size and density of glands. The top leaves contained more glands then down leaves. But the largest amount of hypericin was found in flowers and buds.

Flavonoids

Umek *et al.* (1999) found the highest amounts of rutin, I3,II8 biapigenin, amentoflavone in *H. perforatum*, while hyperoside, isoquercitrin and quercitrin were on the highest level in *H. maculatum* flowers. Another study (Brantner *et al.* 1994) demonstrated that *H. perforatum* had higher amounts of flavonoids and tiannins compared to *H.maculatum*. Similar effects were observed in cell cultures of various *Hypericum* species by Kartnig *et al.* (1996). The authors mentioned that from all the species investigated, the cell cultures of *H. perforatum* and *H. maculatum* showed the best production of flavonoids and hypericin.

Hannig *et al.* (1995) reported that about 50% of flavonoids were located in flowers, the rest of them were located in leaves, while Seidler *et al.* (1999) did not find any differences between the flavonoids content of flowers and 5-cm stem tops. The comparison of flowers and herbs of *H. perforatum* showed a larger amount of quercetin, I3,II8 biapigenin, amentoflavone in flowers, while the content of rutin was higher in the herb. Kartnig *et al.* (1997) also observed a larger amount of rutin, hyperosid, isoquercitrin in leaves, while quercitrin, quercetin, biapigein, amentoflavone were discovered mainly in flowers. Their study indicated that the rate of monoflavonoids to biflavonoids in the herb was 1 : 0,03 and in the flower − 1 : 0,15. Tekel'ová *et al.* (2000) reported that quercetin derivatives and apigenin–biflavonoids were the main constituent of the flowers. They noted the absence of rutin in the investigated samples.

In the flower, I3,II8 biapigenin was found mainly in stamens, while quercitrin, hyperosid, rutin and quercetin in sepals (Repčák and Mártonfi 1997).

Hyperforin

Umek *et al.* (1999) while examining various *Hypericum* species collected in the wild in Slovenia, found that *H. perforatum* flowers indicated the highest content of hyperforin. They also stated that hyperforin was detected only in *H. perforatum* and no other species. But Brantner *et al.* (1994) pointed out that the hyperforin content should not be used as a marker for *H. perforatum* identification, because of hyperforin variability in a dried herb.

Repčák and Mártonfi (1997) found the highest content of hyperforin in *H. perforatum* flower pistils.

Correlations

A strong positive correlation among all secondary metabolites was found in *H. maculatum* whereas this correlation was weaker in *H.perforatum* (Brantner *et al.* 1994). In Umek's *et al.* (1999) investigations there was a positive correlation between the content of other substances in flowers and in the herb except the amentoflavone content. The content of rutin was in a positive correlation with the altitude of a site, while that of quercitrin was in a negative one. These two substances are in a negative correlation themseves. A positive correlation of rutin content to the altitude was proved by Tekel'ová *et al.* (2000) and Brantner *et al.* (1994) who reported that the content of flavonoids had a positive correlation to the altitude. The results were proved by Wolf (1997) who showed that the high concentration of hypericin associated with the high content of other substances. Similar correlation was observed by Mártonfi and Repčák (1994) who found mostly positive correlation between the compounds. However, they did not discover any positive correlation between hypericin and I3,II8 biapigenin and rutin. Kartnig *et al.* (1996) in experiments on cell cultures of *H. perforatum* found a significant positive correlation between hypericin and pseudohypericin, and amentoflavone and biflavonoids. Positive correlation was also obtained between amentoflavone and rutin, and hyperosid and quercetin. Different results were given by Büter *et al.* (1998) who found a positive correlation only for amentoflavone and pseudohypericin, and hyperforin and rutin. They also reported that significant negative correlations between amentoflavone and quercetin, and biapigenin and hyperosid, and biapigenin and quercetin were found. No significant correlation was found between the content of hypericin and any other tested compounds. However, Pluhar *et al.* (2001) observed that *Hypericum* taxa with high content of hypericin rarely had high flavonoids accumulation.

Cellarova *et al.* (2001) found that hypericin content was negatively correlated with ploidy level.

Secondary metabolites content in the first and second year

Results obtained by Seidler *et al.* (1999) showed that hypericin, pseudohypericin and flavonoids content of the herb of var. Topaz was the same in the first and second year of cultivation with one exception. In the phase of flower buds the higher content of hypericin and pseudohypericin was recorded in the second year of harvest. In contrast to the results obtained by Seidler *et al.* (1999), Büter *et al.* (1998) found that the content of secondary metabolites in the first year were considerably lower compared to the second year (Figure 8.2). The smallest differences were found for hyperforin (80% of second year), the largest for hyperosid (only 13% of the second year). They also point out that in the first year the accession has strong effect on the tested compounds, whereas in the second year impact was weaker. Also Pluhár *et al.* (2000) observed the highest total content of hypericin derivatives in the second year of cultivation. Bomme (1997) observed relatively low content of hypericin in the second-cut of herb.

The environmental factors that may affect the quality of the herb are important for example accession can affect the contents of hypericin, pseudohypericin, rutin, quercetrin, amentoflavone (Büter *et al.* 1998). Although most authors (Bomme 1997, Büter *et al.* 1998, Umek *et al.* 1999) pointed out that the secondary metabolites content is determined by genetic factors. Büter *et al.* (1998) stated that investigation showed hypericin, pseudohypericin, rutin were strongly affected by genetic factors.

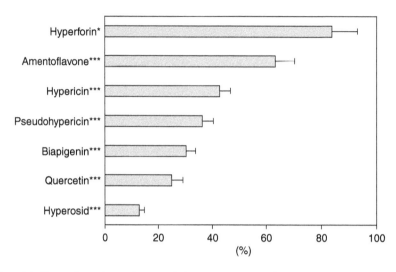

Figure 8.2 Comparison of secondary metabolite contents in the first and second year of cultivation. Second year (1996) results = 100%. Error bars represent standard errors; *, ***= significant difference between first (1995) and second (1996) year at $P = 0.05$ and $P = 0.001$ respectively (based on the data of Büter *et al.* 1998).

References

Berger Büter, K. and Büter, B. (2000) Ontogenetic variation regarding hypericin and hyperforin levels in four accessions of *Hypericum perforatum* L. *Abstracts of Second International Symposium*, Breeding research on medicinal and aromatic plants Chania, Greece PA4.

Bombardelli, E. and Morazzoni, P. (1995) *Hypericum perforatum*. Fitoterapia LXVI, 1: 43–68.

Bomme, U. (1997) Produktionstechnologie von Johanniskraut (*Hypericum perforatum* L.). *Z.Arzn. Gew.pfl.* 2, 127–34.

Brantner, A., Kartnig, Th. and Ouehenberger, F. (1994) Vergleichende phytochemische Untersuchungen an *Hypericum perforatum* L. und *Hypericum maculatum Crantz. Sci. Pharm.* 62, 261–76.

Büter, B., Orlacchio, C., Soldati, A. and Berger, K. (1998) Significance of genetic and environmental aspects in the field cultivation of *Hypericum perforatum*. *Planta Med.* 64, 431–7.

Čellárová, E., Koperdáková, J. and Brutovská, R. (2001) Genetic background of hypericin production in St. John's wort (*Hypericum perforatum* L.) somaclones and their progenies. *Abstracts of World Conference on Medicinal and Aromatic Plants*, Budapest, Hungary p. 101.

Fornasiero, R., Bianchi, A. and Pinetti, A. (1998) Anatomical and ultrastructural observations in *Hypericum perforatum* L. leaves. *J. Herbs, Spices Medicinal Plants* 5, 21–33.

Hannig, H.-J., Plescher, A. and Vollrath, G. (1995) Erfahrungen beim großflächigen Anbau von Johanniskraut – Anforderungen an die industrielle Verwertung. *Herba Germanica* 3, 96–103.

Kartnig, Th., Göbel, I. and Heydel, B. (1996) Production of hypericin, pseudohypericin and flavonoids in cell cultures of various *Hypericum* species and their chemotypes. *Planta Med.* 62, 51–3.

Kartnig, Th., Heydel, B. and Lässer, L. (1997) Johanniskraut aus Schweizer Arzneipflanzenkultur. *Agrarforschung* 4, 299–302.

Mártonfi, P. and Repčák, M. (1994) Secondary metabolites during flower ontogenesis of *Hypericum perforatum* L. *Zahradnictvi* 21, 37–44.

Pluhár, Z., Bernáth, J. and Németh, É. (2000) Investigations on the infraspecific variability of *Hypericum perforatum* L. *Abstracts of Second International Symposium*, Breeding research on medicinal and aromatic plants Chania, Greece, PA14.

Pluhár, Z., Bernáth, J. and Neumayer, E. (2001) Morphological, production-biological and chemical variability of St. John's wort (*Hypericum perforatum* L.). *Abstracts of World Conference on Medicinal and Aromatic Plants*, Budapest, Hungary, p. 23.

Repčák, M. and Mártonfi, P. (1997) The localization of secondary substances in *Hypericum perforatum* flower. *Biologia* 52, 91–4.

Seidler-Łożykowska, K., Dąbrowska, J. and Zygmunt, B. (1999) Content of active substances in herb of St. John's wort (*Hypericum perforatum* L.) cvar. Topaz in different vegetation phases. *Herba Pol.* **XLV**, 169–72.

Southwell, I. and Campbell, M. (1991) Hypericin content variation in *Hypericum perforatum* in Australia. *Phytochem.* **30**, 475–78.

Tekel'ová, D., Repčák, M., Zemková, E. and Tóth, J. (2000) Quantitative changes of dianthrones, hyperforin and flavonoids content in the flower ontogenesis of *Hypericum perforatum*. *Planta Med.* **66**, 778–80.

Umek, A., Kreft, S., Karting, Th. and Heydel, B. (1999) Quantitative phytochemical analyses of six *Hypericum* species growing in Slovenia. *Planta Med.* **65**, 388–90.

Wolf, E. (1997) 'Gewachsene' Qualität bei Phytopharmaka. *Pharm. Zgt.* **142**, 51.

9 Herbal medicinal products of St John's wort
Manufacturing and quality control

Beat Meier

Introduction

Medicinal plants are not in themselves healing agents. Some processing or preparation is required to produce a herbal drug preparation which can be used in herbal medicinal products (HMPs). For this reason the current terminology in the field of regulatory affairs distinguishes between the terms herbal drug and herbal drug preparation.

Herbal drug generally means the raw material. This is generally the dried, or more rarely the freshly harvested, whole or chopped drug, as we find it described in a pharmacopoeia. The European Pharmacopoeia included a monograph for St John's wort in the 2000 Addendum to Ph Eur 3. The monograph is fully integrated in the actual issue of Ph Eur 4. The Swiss Pharmacopoeia (eighth edition) has recently included the monograph for the freshly harvested plant (Herba hyperici recens) as the herbal drug for the production of the oily macerate. In the United States of America, a monograph for St John's wort and St Johns wort powder is part of the twentieth edition of the National Formulary included into the USP 25 published in 2002.

Herbal drugs may be processed into drug powders or extracts of various kinds: the resulting herbal drug preparations serve as active substances for a wide variety of dosage forms such as sugar-coated and film-coated tablets, capsules, soft gelatine capsules, drops, solutions and so on (Figure 9.1). In the case of St John's wort the standardised dried extracts are of greatest importance as active substances, but there are preparations with liquid or oily extracts as well.

The quality control of the finished product is based primarily on the specification of the herbal drug preparation. A fully comprehensive analysis of the product is generally not necessary, since only the dosage of the complex active substance according to the declaration has to be detected. With the principle of batch-specific controls, an adequate strategy was developed for HMPs. The Committee for Proprietary Medicinal Products (CPMP) of the EMEA has published a 'Note for Guidance on Quality of Herbal Medicinal Products' which came into operation in January 2002. The analytical demands for the control test on the finished product have been finalised as follows: 'The control tests on the finished product must be such as to allow the qualitative and quantitative determination of the composition of the active substances and a specification has to be given which may be done by using markers if constituents with known therapeutic activity are unknown. In the case of herbal drugs or herbal drug preparation with constituents of known therapeutic activity, these constituents must also be specified and quantitatively determined'. To date no substance in St John's wort can be classified as having a sole therapeutic activity, on which to base the proven pharmacological effects and the clinical efficacy of the herbal drug preparation (almost solely dried extracts were investigated) can be explained (Simmen *et al.* 2001). It is rather the case that a whole series of substances trigger an effect in

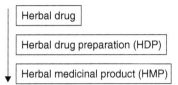

Figure 9.1 Standard terms for the production of HMPs according to the Note for Guidance on Quality of HMPs of the EMEA (The European Agency for the Evaluation of Medicinal Products), which came into operation in January 2002.

pharmacological test systems. For this reason the extract of St John's wort is considered to be the active substance. This leaves various strategies for control of the end product. In order to develop these it is necessary to look at the present analytical situation of *Hypericum perforatum* L.

Procedures for the analysis of *H. perforatum* and products thereof

Quantitative analyses of hypericin

Based on physicochemical properties, the hypericins are especially well-suited – probably more so than almost all other substance groups – for the quantitative level of standardisation: the hypericins are only slightly soluble in all currently used solvents and, furthermore, they are thermolabile. For the manufacturing of the extract this means that optimal conditions must be chosen should one wish to achieve a total hypericin concentration of 0.1–0.3% (which would lead to daily extract amounts of 333–1000 mg, with a daily dose of 1 mg total hypericin according to the proposals of the ESCOP-Monograph) (Meier and Linnenbrink 1996).

The hypericin complex primarily exists in herbal medicinal preparations as pseudohypericin and hypericin, which are specifically photometrically determined as the total hypericin or, using HPLC as the sum of hypericin and pseudohypericin. The protohypericins play a subordinate role. Even though they make up a considerable portion in the plant (Krämer and Wiartalla 1992), they, however, change under exposure to light (Figure 9.2) into the corresponding hypericins. Therefore, in extracts, they are considered among the 'minor compounds' (Gaedcke 1997).

For determining hypericins, there are two methods which are, in principle, equivalent: HPLC and photometry. For determining hypericin and pseudohypericin, RP-C18 systems with different mobile phases are used (Table 9.1). Peak symmetry, which requires a qualification of the column, is often poor. However, the separation of both compounds does not pose a problem. High α values are achieved. The availability and quality definition of the standards remain unresolved. They are imperative for the HPLC. Wirz *et al.* (2001) have analysed all published absorbance data of hypericin at 590 ± 2 nm and documented different ε values in ethanolic solution between 52,000 and 37,410 and in methanolic solution between 45,650 and 22,800. Self-isolated hypericin showed an ε value of 51,712 in methanol. Interestingly, the value for pseudohypericin was significantly lower with 43,486.

In practice, the situation is similar: so-called pure hypericin samples show a serious variation compared to the official absorption coefficients adopted by the European Pharmaocopeia of the original monograph in the German Drugs Code (Deutscher Arzneimittel Codex, DAC) with a value of 870 for $A_{1\,cm}^{1\%}$. Therefore, it is very difficult to get reference standards of equal quality. Table 9.2 shows some analysis of $A_{1\,cm}^{1\%}$ and of HPLC values of several commercially

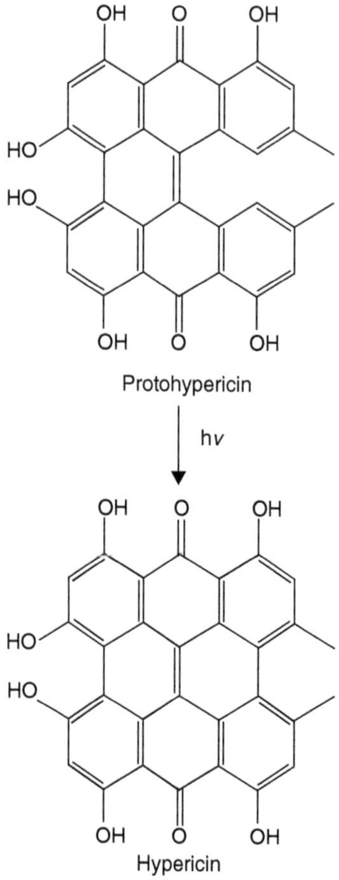

Figure 9.2 Transformation of protohypericin to hypericin under exposure to light.

available and isolated batches of so-called 'pure' hypericins compared to an 'origin' hypericin with the absorption coefficient 890, which is close to the official Ph Eur value.

The lack of a defined standard is a disadvantage in chromatography and the reason why the photometric determination has not yet been played out: even though the values which emerge with this process lie approximately 10–30% higher than the HPLC values when the same standard is used for HPLC and photometry (Wirz 2001), nevertheless, because the 870 coefficient provides a defined relative value, the comparability of lab to lab as well as the examination of declared values is significantly simpler with photometric analyses, still used in the mentioned monographs.

The hypericins are indeed physico-chemical difficult substances which continuously demonstrate unexpected behaviour. They have a great affinity for forming associates. The central problem has become the extraction, which also shows that Ph Eur uses the rather unusual solvent, tetrahydrofurane, for the extraction of the drug. Without knowing the ingredients and the manufacturing process, it is difficult to dissolve the hypericins completely from solid dosage forms. This presumably explains the largely diverging results achieved in serial investigations compared to those in the declaration (Schütt and Hölzl 1994). A methanolic accelerated solvent extraction with the ASE™-System at 80° C and more resulted in low levels of hypericins

Table 9.1 HPLC methods described in the literature for the quantification of hypericin (H) and pseudohypericin (P) in herbal drug preparation of Hyperici herba

Reference	Column[a]	Eluent[b] and flow	°C	Detection	H (min)	P (min)
Freytag (1984)	LiChrosorb RP-18 (250 × 4 mm I.D., 5 μm)	MeOH–EtOAc–0.1 M NaH₂PO₄ (pH 2.1) (5.6 : 1.6 : 1, m/m) Flow: 0.8 ml/min	RT	VIS absorption (590 nm)	8	—
Freytag (1984)	LiChrospher 100 CH-18/2 (250 × 4 mm I.D., 5 μm)	A: MeOH, B[d]: aqueous H₃PO₄ (pH 2.0) 0–15 min A–B (25 : 75) to A–B (5 : 95) Flow: 1 ml/min	RT	VIS absorption (590 nm)	9	18
Gaedcke (1997)	Kromasil RP-18 (125 × 4 mm I.D., 5 μm) (MZ-Analytical)	MeOH–EtOAc–NaH₂PO₄/H₃PO₄ (pH 2.1) (3.6 : 1 : 1.2, m/m) Flow: 1 ml/min	40	VIS absorption (590 nm)	5	1.8
Krämer and Wiartalla (1992)	LiChrosorb RP-18 (250 × 4 mm I.D., 5 μm) (Merck)	MeOH–EtOAc–0.1 M NaH₂PO₄/H₃PO₄ (pH 2.1) (3.6 : 1 : 1.2, m/m) Flow: 0.8 ml/min	25–28	VIS absorption (590 nm)/ Fluorimetric emission (excitation 254 nm, emission > 530 nm)	14.0	5.3
Maisenbacher (1991)	LiChrospher 100 RP-18 (125 × 4 mm I.D., 5 μm) (Merck)	A: MeOH–THF (315 : 185), B: 0.01 M NH₄H₂PO₄/H₃PO₄ (pH 2.2–2.7) Elution profile: 0–5 min A–B (50 : 50) to A–B (80 : 20), 5.01–15 min A–B (80 : 20) Flow: 0.6 ml/min	30	VIS absorption (588 nm)	14	11.9
Micali et al. (1996[e])	HS RP-18 (83 × 4 mm I.D., 3 μm) (Perkin Elmer)	A: MeOH–0.05 M Na₂HPO₄/KH₂PO₄ (pH 7) (7 : 3), B: H₂O–MeOH (3 : 7) Elution profile: 0–3 min 100% A, 3.01–5 min 100% A to 100% B, 5.01–18.0 100% B Flow: 1.5 ml/min	50	Fluorimetric emission (excitation 470 nm, emission 590 nm)	9	—
Niesel (1992)	MN Nucleosil 5 100 RP-18, (250 × 4 mm I.D.)	MeOH–0.01 M NaH₂PO₄/NaOH (75 : 25) Flow: 0.8 ml/min	35	VIS absorption (589 nm)	19.5	6.5
Ostrowski (1988)	Supersphere RP-18 (250 mm, 4 μm) (Merck)	MeOH–ACN–THF–H₂O–H₃PO₄ (19.8:48.3:11.4:19.5:1) Flow profile: 0–17 min 0.3 ml/min, 17.01–50 min 1.3 ml/min	c	UV absorption (254 nm)	23	14.4

(Continued)

Table 9.1 (Continued)

Reference	Column[a]	Eluent[b] and flow	°C	Detection	H (min)	P (min)
Piperopoulos et al. (1997)	LiChrosorb RP-18 (125 × 4 mm I.D., 5 μm) (Merck)	A: MeOH–ACN (5:4), B: 0.1 M triethylammonium acetate (pH 7) Elution profile: 0–8 min A–B (70:30) to A–B (90:10), 8.01–13 min A–B (90:10) to A–B (70:30) Flow: 0.6 ml/min	c	(1) VIS absorption (590 nm), (2) Electrospray mass spectrometry for qualitative analysis	11.9	8.2
Schütt and Hölzl (1994)	LiChrospher 100 RP-18 (125 mm, 5 μm) (Merck)	MeOH–EtOAc–0.1 M NaH$_2$PO$_4$/H$_3$PO$_4$ (pH 2.1) (3.6:1:1.2) Flow: 1 ml/min	c	VIS absorption (590 nm)	7.6	2.5
Schütt (1996)	LiChrospher 100 RP-18 (250 × 4 mm I.D., 5 μm)	MeOH–ACN–THF–H$_2$O–H$_3$PO$_4$ (18.5:49:4:28:0.5) Flow profile: 0–2 min 0.5–1 ml/min, 2.01–5 min 1 ml/min, 5.01–8 min 1–2 ml/min, 8.01–10 min 2–2.5 ml/min, 10.01–15 min 2.5 ml/min, 15.01–20 min 2.5 to 0.5 ml/min	RT	VIS absorption (590 nm)	12.3	4.8
Stock (1992[f])	LiChrosorb 100 RP-18 (250 × 4 mm I.D., 5 μm) (Merck)	MeOH–ACN–H$_3$PO$_4$ 85% (69.5:30:0.5) Flow: 1 ml/min	c	UV absorption (254 nm)/ Fluorimetric emission (excitation 580 nm, emission 600 nm)	13.9	3.4
Tateo et al. (1998[e])	Techsphere RP-18 (150 × 4.6 mm I.D., 3 μm)	A: ACN–MeOH–H$_3$PO$_4$ (59:40:1) B: ACN–H$_2$O–H$_3$PO$_4$ (19:80:1) Elution profile: 0–8 min 100% B, 8.01–28 min 100% B to 100% A, 28.01–55 min 100% A	c	VIS absorption (590 nm)	33.5	27
Zevakov et al. (1991)	Silasorb (250 × 4 mm I.D., 7 μm)	H$_2$O–EtOAc–MeOH–ACN–H$_3$PO$_4$ (41:25:22:9:3)	c	VIS absorption (590 nm)	16.0	13.7

Notes
a Precolumns are not listed in this table.
b Wash out and reequilibration are not considered.
c Not mentioned in the reference.
d Not clearly defined.
e Analysis of alcoholic beverages.
f The method was used for analysis of plasma samples as well (Table 9.6).

Table 9.2 Specific adsorption coefficient of several hypericin references declared as 'poor' by the supplier related to the peak area with HPLC. The 'original hypericin' supplied by Roth Company was used as 100% standard in HPLC. Hypericin was dissolved in a mixture of methanol and 1% pyridine.

Sample	$A_{1\,cm}^{1\%}$	Area % in HPLC compared to the origin hypericin
Standard, Supplier 1	890	100 (origin)
Supplier 1, Sample 1, TLC quality	792	64.9
Supplier 1, Sample 2	624	60.0
Supplier 2, Sample 1	844	82.6
Supplier 3, Sample 1	921	91.1
Supplier 4, Sample 1	789	81.2
Supplier 4, Sample 2	420	39.5
Supplier 4, Sample 3	612	50.5
Supplier 4, Sample 4	905	94.2
Supplier 4, Sample 5	292	26.0
Supplier 5, Sample 1	703	64.5
Supplier 5, Sample 2	769	65.8
Supplier 6, Sample 1	933	102.9
Supplier 7, Sample 1	896	83.5

compared to the theoretical amount calculated from the extract analysis (Morf *et al.* 1998). Constant and equivalent results could be obtained after a prepurification with methylene chloride. Interactions of solvent and ingredients, especially with magnesium stearate and with the hypercins has to be taken into consideration. The easiest and most promising sample preparation has been established with ultra-sonication.

The low solubility of pure hypericin is a further problem in the analyses of this compound with HPLC. Hypericin is only sparingly soluble in methanol and other common solvents. Addition of 1% pyridine, sonication and waiting time before dilution (about four days) are means to prevent incomplete dissolution leading to wrong calibration graphs. The results of Wirz (2001) showed that the solubility of hypericin in methanol–pyridine (99 : 1, v/v) is about nine times better than in pure methanol. The improved solubility of hypericin in the presence of pyridine may be explained by the formation of hypericin–pyridine complexes. As absorption spectra of hypericin in methanol and methanol–pyridine (99 : 1, v/v) were the same, the finding could not be explained by ionisation of hypericin or the accelerated breakdown of homoassociates by pyridine due to enhanced solvatisation (Kapinus *et al.* 1999). Stability tests proved that the addition of 1% pyridine and a waiting time of 4 days do not affect hypericin solutions, when excluded from light. The degradation of pseudohypericin, however, is slightly accelerated by pyridine and therefore not recommendable for standard solutions of pseudohypericin. Solubility is less a problem with pseudohypericin, as the additional hydroxyl group makes it better soluble in common solvents. Because the absorbance properties of hypericin are not influenced by the addition of 1% pyridine, it is possible to prepare standard solutions of hypericin with methanol–pyridine (99 : 1, v/v) and extract solutions with pure methanol.

For the sake of completeness, it must be mentioned that for determining the hypericin content, a fluorimetric method (Klein–Bischoff and Klumpp 1993) has also been proposed. Nevertheless, a wide-spread application for this process still remains unaccepted for the herbal drug and herbal drug preparation to date, even though the specificity and selectivity might be quite high.

Qualitative analyses by HPLC- and TLC-fingerprint

St John's wort herb has quite a constant and homogeneous spectrum of different kinds of flavonoids. With the HPLC-fingerprint method, several quercetin derivatives can be recognised: the diglyco-side rutin, the monoglycosides hyperoside, isoquercitrin (this is usually not separated of 3,3',4',5,7-pentahydroxyflavanone-7-O-rhamnopyranoside) quercitrin and the aglycon, quercetin, as well as additionally a biflavone, I3,II8-biapigenin (Brolis *et al.* 1998). These compounds have to be present in extracts and HMPs. (Figure 9.3, trace a) High quercetin peaks arise with extraction by enzymatic splitting from its glycosides. In well dried herbal St John's wort drugs, the quercetin peak is significantly smaller than that of hyperoside and rutin in HPLC-fingerprint. Several systems for the HPLC-analyses of St John's wort fingerprint are given in Table 9.3.

For analysing the fingerprints of St John's wort preparations, the RP-HPLC is particularly suited if quantitative or semi-quantitative results are desired (e.g. for comparing different batches and preparations as well as for extract validation and stability tests), while the TLC (see Figure 9.4) is especially appropriate for checking the identity of large quantities of raw materials and for in-process controls, since it is quickly performed. Several solvents show similar results. The best separation was achieved with the solvent 6 in Table 9.4 on HPTLC in a saturated twin trogue chamber after dipping the plate in a 0.5% solution of diphenylboric acid aminoethyl ester in ethyl acetate, drying and heating to 105°C (5 min). The detection was done immediately after derivatisation at 366 nm. Some problems with the detection of the hypericins have been observed. The detection with diphenylboric acid aminoethyl ester is very specific for flavonoids, but critical for the hypericin spots. Occasionally, they can be very weak. Therefore, it is proposed to detect the hypericins before spraying the plates with the reagents at UV 366 nm (Blatter 2001).

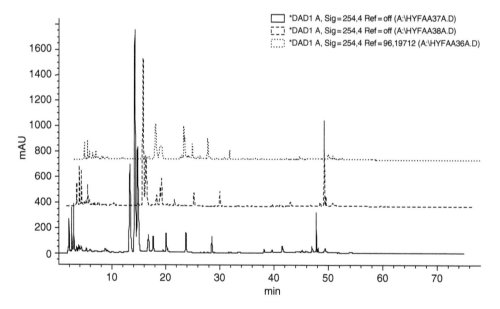

Figure 9.3 HPLC-fingerprint chromatogram of Hyperici herba. Most commercial and cultivated samples correspond to the reference chromatogram (a) (unbroken line); (b) (broken line) contains no rutin (first peak at a retention period of approximately 13 min), which sometimes occurs; (c) (dotted line) deviates completely from the basic sample. Such samples are, however, very rare. All samples of the species *Hypericum perforatum* L. were morphologically classified.

Table 9.3 HPLC methods described in the literature for the qualitative analysis of herbal drug and herbal drug preparation of Hyperici herba (Fingerprint analyses)

Reference	Column[a]	Eluent[b] and Flow	°C	Detection
Brantner *et al.* (1994)	Superspher RP-18 (250 × 4 mm I.D.) (Merck)	A: H_2O, B: H_3PO_4, C: ACN, D: MeOH Elution profile: 0–30 min A–B–C (82 : 1 : 17), 30.01–40 min A–B–C (67 : 1 : 32), 40.01–45 min A–B–C–D (30 : 1 : 49 : 20), 45.01–55 min A–B–C–D (30 : 1 : 49 : 20), 55.01–75 min A–B–C–D (10 : 1 : 49 : 40) Flow profile: 0–40 min 1 ml/min, 40.01–75 min 0.6 ml/min	c	UV absorption (254 nm)
Brolis *et al.* (1998)	201 TP 54 RP-18 (250 × 4.6 mm I.D., 5 μm) (Vydac Separation Group Hesperia)	A1 (Detection 1): H_2O–85% H_3PO_4 (99.7 : 0.3), A2 (Detection 2): H_2O–HCOOH (99.7 : 0.3), B: ACN, C: MeOH Elution profile: 0–10 min 100% A to A–B (85 : 15), 10.01–30 min A–B (85 : 15) to A–B–C (70 : 20 : 10), 30.01–40 min A–B–C (70 : 20 : 10) to A–B–C (10 : 75 : 15), 40.01–55 min A–B–C (10 : 75 : 15) to A–B–C (5 : 80 : 15) Flow: 1 ml/min	30	(1) UV absorption (270 nm), (2) Electrospray mass spectrometry for qualitative analysis
Butterweck (1997)	Prosep RP-18 (150 × 4 mm I.D., 5 μm) (Latek)	A: ACN, B: 0.5% TFA in H_2O, C: MeOH–ACN–0.5% TFA (60 : 39.5 : 0.5) Elution profile: 0–5 min A–B (1 : 9) to A–B (2 : 8), 5.01–7.0 min A–B (2 : 8) to B–C (7 : 3), 7.01–10 B–C (7 : 3) to B–C (5 : 5), 10.01–20 min B–C (5 : 5) to 100% C, 20.01–45 min 100% C Flow: 1 ml/min	25	UV absorption (284 nm)
Häberlein *et al.* (1992)	LiChrospher RP-18 (250 × 4 mm I.D., 5 μm) (Merck)	A: ACN–H_2O–H_3PO_4 (19 : 80 : 1) B: ACN–MeOH–H_3PO_4 (59 : 40 : 1) Elution profile: 0–8 min 100% A, 8.01–30 min 100% A to A–B (50 : 50), 30.01–40 min A–B (50 : 50) to 100% B, 40.01–60 min 100% B Flow profile: 0.6 to 0.8 ml/min (linear)	c	UV absorption (254 nm)

(*Continued*)

Table 9.3 (Continued)

Reference	Column[a]	Eluent[b] and Flow	°C	Detection
Hölzl and Ostrowski (1987)	Supersphere RP-18 (250 × 4 mm I.D.) (Merck)	A: ACN–H$_2$O–H$_3$PO$_4$ (19:80:1), B: ACN–MeOH–H$_3$PO$_4$ (59:40:1) Elution profile: 0–8 min 100% A, 8.01–30 min 100% A to A–B (50:50), 30.01–45 min A–B (50:50) to 100% B, 45.01–80 min 100% B Flow: 0.6 ml/min	RT	UV absorption (254 nm)
Mulinacci et al. (1999)	LiChrosorb RP-18 (250 × 4.6 mm I.D., 5 μm) (Merck)	A: H$_2$O/H$_3$PO$_4$ (pH 3.2), B: MeOH, C: ACN Elution profile: 0–10 min A–C (88:12), 10.01–15 min A–C (82:18), 15.01–30 min A–C (82:18) to A–C (55:45), 30.01–35 min A–C (55:45) to B–C (55:45), 35.01–42 min B–C (55:45) Flow: 1 ml/min	26	VIS absorption (590 nm)
Ostrowski (1988)	Supersphere RP-18 (250 mm, 4 μm) (Merck)	A: ACN–MeOH–H$_2$O–H$_3$PO$_4$ (36:20:44:1), B: ACN–MeOH–H$_3$PO$_4$ (59:40:1) Elution profile: 0–3 min 100% A, 3.01–6 min 100% A to 100% B, 6.01–45 min 100% B Flow: 0.6 ml/min	RT	UV absorption (254 nm)

Notes
a Precolumns are not listed in this table.
b Wash out and reequilibration are not considered.
c Not mentioned in the reference.

Table 9.4 Different solvents for TLC-fingerprint analyses. Solvent 6 showed the best results when the following criteria have been analysed: separation of flavonoids, separation of hypericin and pseudohypericin, resolution

Solvent	Method	Mobile phase composition
Solvent 1	Ph Eur	EtOAc, Formic acid, H_2O (90 : 6 : 9)
Solvent 2	Ph Helv 8	EtOAc, Formic acid, H_2O (30 : 2 : 3)
Solvent 3	Deutscher Arzneimittel Kodex 1999	EtOAc, Formic acid, H_2O (86 : 6 : 8)
Solvent 4	National Formulary 20	EtOAc, Formic acid, Acetic acid, H_2O (100 : 11 : 11 : 26)
Solvent 5	CAMAG Application Note A-69.2	EtOAc, Formic acid, H_2O (20 : 2 : 1)
Solvent 6	EM-Science (www.emscience.com)	EtOAc, Formic acid, Acetic acid, H_2O, DCM (100 : 10 : 10 : 11 : 25)

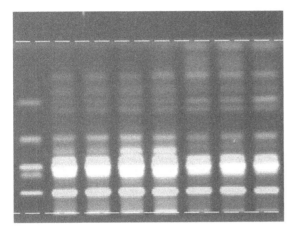

Figure 9.4 Typical TLC-Fingerprint of St John's wort herbal drug and herbal drug preparations. Yellow spots: flavonoids (rutin in the lowest position). White or blue spots: plant acids. Red spots: hypericin and pseudohypericin. (See Colour Plate XIV.)

Recently, deviations have been increasingly observed: in the dried herb of *H. perforatum* of oriental origin, especially China, with a botanically definitive identity, rutin is missing. Rutin has been included in the European monograph (Ph Eur) because in a whole series of possible adulterations (other *Hypericum* species) this rutin is missing. Figure 9.3 (traces (b) and (c)) shows two diverging flavonoid fingerprint traces with HPLC in comparison to the standard trace, which currently shows the majority of the sources.

The analyses of hyperforin

In the dried herb, hyperforin and its companion, adhyperforin, have been regularly measured in the fingerprint, in as far as the gradient was driven to the point that apolar substances could also be included, but, however, has not been measured in extracts (Hölzl and Ostrowski 1987). In the meantime, this has changed. Those substances which had long been considered as unstable, show, not unexpectedly, a large spread when batch-to-batch comparisons from the individual manufacturers are made. Different German solid dosage forms of St John's wort have been analysed with a hyperforin and adhyperforin specific method (Melzer *et al.* 1998, Wurglics *et al.* 2000).

Table 9.5 HPLC methods described in the literature for the qualitative and quantitative analyses of hyperforin (HF) and adhyperforin (A) in herbal drug and herbal drug preparation of Hyperici herba

Reference[a]	Column[b]	Eluent and flow	°C	Detection	HF (min)	A (min)
Biber et al. (1998)	Nucleosil RP-18 (125 × 3 mm, 5 μm) (Macherey + Nagel)	ACN–0.26 % H_3PO_4 in H_2O (75:25) Flow: 1 ml/min	c	UV absorption (273 nm)	c	c
Biber et al. (1998)	Intersil RP-8 (60 × 4 nm) (M&W)	MeOH–0.02 M NH_4COOCH_3 (6:4)	c	MS/MS	c	c
Butterweck (1997)	Prosep RP-18 (150 × 4 mm I.D., 5 μm) (Latek)	A: ACN, B: 0.5% TFA in H_2O, C: MeOH–ACN–0.5% TFA (60:39.5:0.5) Elution profile: 0–5 min A–B (1:9) to A–B (2:8), 5.01–7.0 min A–B (2:8) to B–C (7:3), 7.01–10 B–C (7:3) to B–C (5:5), 10.01–20 min B–C (5:5) to 100% C, 20.01–45 min 100% C Flow: 1 ml/min	25	UV absorption (284 nm)	24.9	c
Erdelmeier (1998)	Spherisorb ODS (250 × 4 mm I.D., 2.5 μm) (Knauer)	A: H_2O–H_3PO_4–triethylamine (995:3:2) B: ACN–H_3PO_4–triethylamine–H_2O (935:3:2:60) Elution profile: 0–5 min A–B (99:1), 5.01–55 min A–B (99:1) to A–B (60:40), 55.01–90 min A–B (60:40) to A–B (1:99) Flow: 1.2 ml/min	c	UV absorption (254 nm)	92	93

Reference	Column	Mobile phase / elution	Temp (°C)	Detection		
Hölzl and Ostrowski (1987)	Supersphere RP-18 (250 × 4 mm I.D.) (Merck)	A: ACN–H$_2$O–H$_3$PO$_4$ (19:80:1) B: ACN–MeOH–H$_3$PO$_4$ (59:40:1) Elution profile: 0–8 min 100% A, 8.01–30 min 100% A to A–B (50:50), 30.01–45 min A–B (50:50) to 100% B, 45.01–80 min 100% B Flow: 0.6 ml/min	RT	UV absorption (254 nm)	65	c
Maisenbacher; Maisenbacher/ Kovar (1991/1992)	LiChrospher 100 RP-8 (125 × 4 mm I.D., 5 µm) (Merck)	ACN–0.01 M NH$_4$H$_2$PO$_4$/H$_3$PO$_4$ (pH 2.2–2.7) (8:2) Flow: 0.8 ml/min	30	UV absorption (270 nm)	c.10	c.12
Melzer et al. (1998)	LiChrospher 100 RP-18 (250 × 4 mm I.D., 5 µm) (Merck)	ACN–0.01 M NH$_4$H$_2$PO$_4$ (pH 2.5) (85:15) Flow: 2 ml/min	30	UV absorption (270 nm)	10.1	12.2
Orth (1999)	Nucleosil 100 RP-18 (250 × 4 mm I.D., 5 µm) (Macherey + Nagel)	ACN–H$_2$O/H$_3$PO$_4$ (pH 4.5) (89.5:10.5) Flow: 1.2 ml/min	c	UV absorption (272 nm)	9.2	10.5
Ostrowski (1988)	Supersphere RP-18 (250 mm, 4 µm) (Merck)	MeOH–ACN–THF–H$_2$O–H$_3$PO$_4$ (19.8:48.3:11.4:19.5:1) Flow profile: 0–17 min 0.3 ml/min, 17.01–50 min 1.3 ml/min	c	UV absorption (254 nm)	c.47	c
Wirz (2001)	Spherisorb S5 ODS 2	ACN–THF–H$_2$O–H$_3$PO$_4$ (105:45:50:1.2) Flow 1.6 ml/min	25	UV absorption (274 nm)	14.5	17.3

Notes
a References in bold print are suited for the quantification of hyperforin as far as retention time is the decisive factor.
b Precolumns are not listed in this table.
c Not mentioned in the reference.

It is only recently that extracts have been manufactured with a high hyperforin content. Thus, hyperforin has found an entrance into the analysis of herbal preparations of St John's wort (Table 9.5) and will therefore become in the future, more homogeneous in some extracts than it has been to date. Interestingly enough, positive pharmacological results were presented on the one hand with extracts and fractions which contained hyperforin, on the other hand, where no hyperforin could be analysed. This also applies to clinical studies. In older, but also in newer studies (Schrader *et al.* 1998, Schrader 2000, Woelk 2000), extracts were used which contained – at best – small quantities of hyperforin. From this, one can conclude that hyperforin is a very interesting substance, but, however, it cannot currently be considered as being evident for the effect of HMPs of St John's wort.

Wirz (2001) has studied the factors influencing the extractability of hyperforin. Aqueous ethanol 70% to ethanol 100% turned out to be the most effective solvents among those investigated. Extraction at higher temperature decreased the yield (Figure 9.5). Among the methanol–water mixtures, the highest yields were achieved with methanol 90%. No hyperforin could be detected in the tea preparation. Light exclusion did not significantly influence the results, but an influence to the stability was observed. The content of hyperforin in the extracts prepared without light exclusion decreased up to 80% within 14 days, although they were stored at −20°C. Fastest degradation of hyperforin and adhyperforin could be observed in the solvent ethanol 100%, when extraction was done at room temperature. Methanol 100% provided the least stability of the methanolic extract solutions for hyperforin. Extraction at increased temperature led to extracts with improved stability. Best results regarding stability were achieved when light and oxygen were excluded during extraction. The contents of those extracts did not change significantly within 14 days in spite of being stored at 20°C. The addition of ascorbic acid turned out not to be necessary.

Furthermore, the drying process is of importance for the yield of hyperforin. Spray drying works not without the separation of chlorophylls as a step of the concentration process. The hyperforins are very lipophilic and therefore concentrated in the chlorophyll fractions.

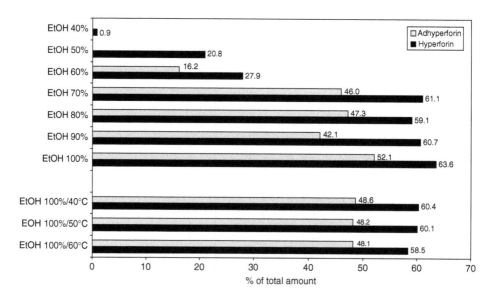

Figure 9.5 Influence of the extracting solvent and temperature on the yield of hyperforin and adhyperforin.

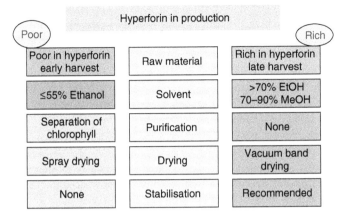

Figure 9.6 Flow diagram of two types of extraction technology for St John's wort herbal preparation: extracts poor in hyperforin; extracts rich in hyperforin.

In contrast, vacuum band drying is not sensitive to chlorophyll. A lot of manufacturers have changed the process of extraction within the last years (Figure 9.6).

Other compounds

Amentoflavon, kaempferol, luteolin, myricetin and xanthones (not approved) are present in the drug in such small quantities that no analytical methods have yet been developped for them. On the basis of current scientific knowledge, the essential oil, consisting primarily of *n*-alkanes, is not therapeutically relevant. There is no suitable method of analysis for procyanidines. To date, the procyanidine present in *Hypericum* has only been poorly investigated. In general, analytical research makes only slight progress in this area due to the complexity of the bonds (Rohr 1999) in spite of a multitude of publications available. Currently, only non-specific procyanidine determinations can be used based on standard procedures: extraction with ethylacetate, hydrolysis in butanol/HCl, measurement of the red colouring in comparison with cyanidol or with an absorption coefficient according to the references (Rehwald 1995, Hiermann *et al.* 1986, Porter *et al.* 1986). With this process, values greater than 5% are measured in extracts. St John's wort contains an extensive spectrum of amino acids (Lapke 2000), which can also be extracted with ethanol and methanol. The extraction rate is higher with more polar solvents compared to the solvents traditionally used. Lapke analysed approximately 80 µmol/g dry weight free amino acids into the herbal drug. Alanine, arginine, asparagine, γ-amino-butyric acid and glutamine are the major amino acids in St Johns wort (*c.*8–14 µmol/g dry weight). The same spectrum and similar amounts have been analysed in the extract Ze117 (Table 9.7). Interesting from a pharmacological point of view are γ-amino-butyric acid and tryptophan. Tryptophan does not belong to the major free amino acids in St John's wort. Generally, the additional daily ingestion of amino acids with St John's wort preparations is indeed not very significant in comparison to the amount ingested daily with food.

Concepts to analyse herbal drug products of *H. perforatum*

The testing of a herbal medicinal product includes parameters which characterise the dosage form (e.g. average weight, uniformity of weight, disintegration time), purity tests (especially microbial contamination) as well as phytochemical parameters. Only the latter will be discussed here. For the qualitative test (identity), thin layer chromatography has proved itself. The TLC-fingerprint (compare Figure 9.4) covers in the same run the flavonoids, the hypericins and the plant acids. The flavonoids appear as yellow spots, rutin is expressed as bands with the deepest Rf level. The plant acids exhibit light blue fluorescent bands between rutin and hyperoside. In the upper Rf field the hypericins appear with clear red fluorescent colours, and in the front region the aglycons, of which I3,II8 biapigenin and quercetin dominate. Using this procedure, the identity of St John's wort can be clearly demonstrated. Thin layer chromatography is consequently of the same value as HPLC as a technical procedure, but is considerably simpler. New documentation and evaluation systems using video cameras and subsequent image processing will increase the value of the procedure in future (see www.camag.com).

The quantitative testing must aim to prove the declared dosage of the active substance, in this case, an extract. The principle of batch-specific controls is suitable for this purpose. The content of a substance is determined in each extract and from this is calculated the amount which must be found in the herbal medicinal product, in order to ensure a dose accuracy of 90–110%. In principle any substance is suitable for this purpose. The simplest is the quantitative analysis of rutin, which is selectively separated from the other flavonoids in all the fingerprint methods documented in Table 9.3. Rutin is available without problems as a reference substance, but is, however, in no way specific as a component solely of St John's wort. For this reason the content of hypericin is determined much more often. Hypericin and pseudohypericin are the most characteristic compounds of St John's wort, which was mentioned by the Commission E in their monograph dosage guidelines, and later confirmed by ESCOP.

To obtain hypericin, especially in compressed forms such as tablets, an expensive extraction procedure is often necessary, in order to obtain a reproducible result corresponding to the extract used. In liquid products, especially in alcoholic tinctures, hypericin determination is the method of choice, since this does not normally require any sample preparation. The red colouring of the solutions by the hypericins can be determined immediately. The extent to which chlorophyll, which is extracted with it and not separated, affects the levels and thus limits the reproducibility of the data, is still a subject of discussion and investigation (Wirz 2001). If the drug is 'defatted' with methylene chloride to remove the chlorophyll, lower levels result when the hypericin is determined, without, however, achieving a complete congruence between the photometric and HPLC levels. With extracts the differences vary, depending on whether or not a chlorophyll separation has been undertaken in the extraction process.

The red colouring is also measured in oil macerates, but in these the structure of the staining hypericin compounds has not yet been explained. It is known that the 'oil hypericins' are neither hypericin nor pseudohypericin.

The standardisation of extracts using hyperforin has only recently been introduced. For such preparations it is advisable to test the dosage in the end product using this compound.

There is a tendency to make demands with regard to precision of the methods, which the latter cannot fulfil. In relation to the total weight of the herbal preparation the content of rutin or hyperforin is 1–4% at most. The hypericin is clearly lower at approximately 0.1%. The compounds are also present in a double complex matrix (extractive substances and excipients). This leads to expensive concepts of sample preparation, in which each step is liable to error, and to chromatographic procedures, in which a baseline separation, as is usual with synthetic compounds, cannot be achieved to the same extent, making reproducibility of the integration more difficult. A calculation of the sum of all the errors of the single steps of the analytical procedures leads immediately to errors, which may be expected to amount to almost ± 10%. The result of this is that reproducibility of the methods, either between laboratories or over time, is not guaranteed within the narrow range required, which for the analysis of synthetic drugs is understandably narrow. These facts are often not recognised until the stability testing, when actual fluctuations occur. Optimisation to prevent these problems is scarcely possible, because the reference substances needed for this are not available in sufficient amounts at reasonable cost, and because there is no 'blind' extract matrix. It is worth making error estimations for all analytical processes of medicinal herbal products which are intended for application, in order to dampen all too high expectations and avoid unpleasant surprises. The authorities will be asked to bear in mind these analytical limitations, if regulatory provisions which have been drawn primarily on the basis of synthetic drugs are to be used for HMPs. Generally, requirements of a dose accuracy of ±5%, as laid down by the German drugs law, are unrealistic for HMPs. This is not because such dose accuracy is not achieved technologically, but because the analysis procedures do not exhibit the necessary precision. If limits are imposed, this will lead to batches having to be destroyed although their quality does not differ from batches which do not fall below the specifications. A range of ± 10% is realistic. This is especially to be taken into consideration in the case of hypericin/pseudohypericin analyses.

Wirz (2001) asks, after a broad research in the field, several questions (Figure 9.7):

> There are open questions still waiting for clarification. Is it really only the difference in polarity of the extracting solvents tetrahydrofuran–water (8 : 2) and acetone, which led to significantly higher contents of naphthodianthrones applying tetrahydrofuran–water (8 : 2)? Why does prolonged extraction with acetone not give similar results as tetrahydrofuran–water (8 : 2)? The Soxhlet extraction of a drug sample with acetone is apparently completed after 3 hours. Why does resumed extraction the day after provide additional hypericin? Does tetrahydrofuran accelerate some transformation processes in contrast to acetone? Have reactions been completed that were assumed to happen during biosynthesis (Cameron *et al*. 1976, Gill and Giménez 1991), or is the transformation of austroventin to penicillopsin and the conversion of penicillopsin to protohypericin still going on when the drug is extracted? What do the transformation and degradation products of hypericin and pseudohypericin look like? Is pseudohypericin synthesized in a separate biosynthetic pathway or is it generated from hypericin by oxidation? Considering these questions it is probable that the content of hypericin in drug samples of *Hypericum perforatum* is not completely stable but part of a dynamic system.

Solid dosage forms

Tablets, capsules, effervescents

St John's wort is generally manufactured as a dried extract, and processed into solid pharmaceutical dosage forms. The handling of extracts is not entirely without problems. Most of the St John's wort extracts are generally highly hygroscopic. The permitted water content of extracts is normally limited to <5%. Storage of the extracts has to be carried out under controlled conditions, transport is in well-closed containers. Solid pharmaceutical forms must be protected from moisture. For blister packs it is worth using the best quality with barriers as strong as possible. This applies in particular to film-coated tablets, which are particularly susceptible to a dilatation of the core, because the coating can easily split.

Spray dried extracts generally exhibit less density and are accordingly very voluminous. The problem can be reduced by compacting the extracts. Analysis of 42 dried extract preparations offered as HMPs in Germany according to the Yellow List (Gelbe Liste, Pharmindex, Media–Media, Neu–Isenburg, 1998) shows that more than half (23) are offered as hard gelatine

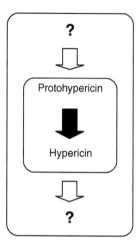

Figure 9.7 Potential 'Dynamic System' in *Hypericum perforatum* L.

capsules. There is generally a much lower proportion of excipients in capsules than in sugar- or film-coated tablets. This is important, because extract dosages of 250, 300 and 425 mg produce rather large dosage forms. Increased size reduces the ability to swallow these preparations. Film-coated tablets have an advantage compared to sugar-coated tablets (dragees), that the coating is of no importance with regard to weight; the thicker coating of dragees on the other hand, protects them from absorbing moisture. The use of sucrose, which is still normal when making dragees, is a disadvantage of this form for various reasons (diabetes, tooth decay). Uncoated cores look unattractive since the extract has an unpleasant reddish brown colour.

This slow onset of action means there is little point in manufacturing solid forms with sustained release of the active substance. These are offered on the American market: the extract is micro-encapsulated and coated in such a way that release occurs at various times. In Europe, corresponding pharmaceutical forms are not possible without proving the action and giving reasons, at least not in those countries where HMPs from St John's wort are registered as medicines. Transdermal systems with St John's wort extracts are really imaginative products, since there is no proof or experience of the efficacy of topical application. Occasionally also St John's wort powdered herbal drug is processed into tablets or filled into capsules. These uses are based on special traditions, such as those of Pastor Kneipp in Germany and Küenzli in Switzerland.

There are only a few systematic papers on the manufacture of solid pharmaceutical forms from extracts. For St John's wort, Rocksloh (1999) analysed five extracts from three not named manufacturers, for their technological properties and their behaviour when manufactured as tablets. The extracts were all prepared with methanol (80% v/v) as extraction agent and dried after thickening on a vacuum band-type drier. The substance was ground after drying and some small amounts of excipient (not more than 7%) were added. The physical tests showed, that all the extracts have poor fluidity and that they are hygroscopic and can quickly absorb water. On the whole, the analysis of the extracts showed no essential differences.

For the tabletting of dried extracts only direct pressing is possible, since there is no point in renewed moistening of the extract by moist granulation. In all other cases thickened extracts are suitable. But the handling of these is considerably more difficult than that of dried extracts.

The main problem in direct tabletting is the disintegration of the tablets, which according to the monograph of the European Pharmacopoeias must be <60 min for sugar coated tablets and <30 min for film-coated tablets. At a dosage of 300 mg extract and an extract proportion between 57.7% and 42.9% of the total composition, no tablets could be manufactured which disintegrate sufficiently swiftly. In spite of the use of modern dispersing agents the disintegration times were 45 min and higher.

Better results were achieved with compacted extracts. The magnesium stearate (3%) required for compacting also serves as a lubricant in the tabletting. It is not necessary to add another lubricant. With Avicel PH101 it was possible to alleviate demixing tendencies, the addition of flow regulators (1% high disperse silicon oxide, Cab-O-Sil M5) improved not only the fluidity but also the disintegration of the tablets. Cross-linked polyvinylpyrrolidone proved the best dispersing agent, but brings with it the risk of irreversible adsorption of the phenols contained in the extracts (flavones, hypericins). For this reason cross-linked carboxymethylcellulose (Ac-Di-Sol) is preferable. Tablets with a total mass of 700 mg showed a disintegration time of approximately 1 min at a fracture strength of 60–80 N. They contained 329 mg of St John's wort compactate (including 3% magnesium stearate), 336 mg Avicel PH101, 28 mg Ac-Di-Sol and 7 mg Cab-O-Sil M5.

The manufacture of effervescent tablets is associated with difficulty, since it is not possible to dissolve the clinically tested alkanolic extract completely in water. Solubilising agents have to be used to dissolve the extracts. These usually have a tendency to foam. Undissolved particles rise in this foam and leave unsightly traces on the glass. The solubilising agents can also inhibit the access of the water to the carbonates, so that there is no effervescent effect. To date no product has been developed in Germany and Switzerland which has reached the market place.

Dissolution

For solid oral dosage forms the dissolution is generally a firm part of the documentation for marketing authorisation in the parts IIA (Development Pharmaceutics) and IIF (Stability Tests). Especially for slightly soluble substances and substances with different crystal modifications influencing the solubility (polymorphism) the test is also required under IIE (Control Tests on the Finished Product). The question is, whether for herbal remedies, too, dissolution tests can contribute to the description of the quality or not.

In accordance with Bauer and Tittel (1996), herbal remedies can be subdivided into two groups:

1   HMPs, in which single substances or substance groups are known as active constituents;
2   HMPs, in which the active constituents are not known and the analytics are based on the so-called markers.

Group 1 includes products with extracts of rauwolfia roots (rauwolfia alkaloids, especially reserpine), kava-kava (kava pyrones), horse chestnut seeds (aescins), ammi visnaga (khellin), Silybum marianum (silymarins) and – considering the numerous particularities just concerning the dissolution – the anthraquinones. For most of these preparations a dissolution is required. The share of the active constituent is high – as isolated individual compounds or as purified fraction the substances are slightly soluble – and the pharmacokinetics of these substances is at least partly known. Therefore, the requirement of a dissolution is justified.

Group 2 includes most of the medicinal herbal products in use today. These preparations can be defined as follows: preparations with a pharmacologically broad effect and/or a therapeutically

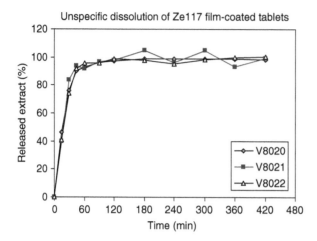

Figure 9.8 Extract dissolution of a St John's wort HMP. The extract shows an adsorption maxium at 340 nm. Dissolution of the film-coated tablets has been analysed in 500 ml 0.1 N HCl at 37°C at this wavelength. The 100% values have been analysed with an equivalent amount of extract in the dissolution medium.

broad efficacy, which cannot be attributed to clearly defined single substances. Therefore, the extract is defined as the active constituent which, irrespective of the preparation, consists of many known and unknown substances of different substance groups. Its quality – in accordance with Bauer and Tittel (1996) – should be defined by qualitative and quantitative test procedures as described before, by the manufacturing procedure and the starting drug. Therefore, the German 'Bundesinstitut für Arzneimittel und Medizinprodukte (BfArM)' in Berlin required the proof of the extract amount in the finished product by the batch-specific control. Hence, it appears, that in principle it is difficult to develop an appropriate dissolution concept for a herbal remedy, in which no clearly defined active constituent is known. Several questions arise: shall an extract dissolution curve be determined nonspecifically at an optimal wavelength, which is to be determined (Figure 9.8)? The maximum value would have to be determined batch-specifically from the employed extract. Shall the polar substances (e.g. flavonoid glycosides) be determined specifically by HPLC? Shall the slightly soluble substances be determined specifically by HPLC? If specific procedures are employed, the expenditure is very high, since the low concentrations in the preparation mostly do not permit direct measurements (this applies especially to the hypericins in St John's wort). Concerning the absorption of extracts and their components there are only few data and and no correlation studies on dissolution/absorption. Generally, the dosage of the extracts in solid dosage forms is very high in all St John's wort products. Thus, the release of an important part of the extract is guaranteed from the start.

Ideally, there is an *in vitro/in vivo* correlation for the dissolution. Because of the absence of *in vivo* data this correlation is not given for HMPs with their complex composition. Hamacher (1996) has postulated: 'The requirement of the release of the active constituent in single-dose solid dosage forms makes no sense as long as it is not known which definite constituents are responsible for the effect'.

Liquid dosage forms

To date only tinctures have become established as liquid preparations from St John's wort. Since the solubility has now become known of the various constituents of St John's wort in different ethanol–water mixtures, it is recommended that tinctures be prepared from the dried drug with 60–80% m/m ethanol. For homoeopathic mother tinctures (according to Homeopathic Pharmacopoeia (HAB1)) the freshly harvested, blossoming drugs are extracted with ethanol. To date there are no modern clinical trials available for liquid preparations. The first report on the current indications by Daniel (1974) is based, however, on the use of a liquid preparation (Hyperforat drops). Alcoholic extracts cannot be brought to solution in water (see also the passage on effervescent tablets) because of constituents which are difficult to dissolve. Therefore, tea preparation remains the only alcohol-free liquid preparation of St John's wort.

Stability

The stability testing of HMPs is carried out at present according to the guidelines of the ICH. At 25°C and 60% relative humidity the solid preparations proved stable if the package was adequate, as this must protect in particular from moisture absorption. The fluctuations observed regularly, especially in the determination of hypericins (by HPLC and also photometry) can be attributed to the analytical problems described above. A spread of ±20% around the initial value must, therefore, be accepted, so long as no clear trend towards breakdown kinetics is detectable. At 40°C and 70% relative humidity, however, the thermolability of hypericins, observed earlier by Niesel (1992) in tea preparations, begins to have an effect. A clear reduction in the hypericin content can be observed. In the past, hyperforin was regarded as extremely unstable. More recent investigations suggest that hyperforin is unstable in particular in lipophilic media, but that it can be kept reasonably stable in dried, alkanolic/aqueous extracts.

The stress conditions in the ICH guidelines do not generally prove suitable for herbal medicinal products. Natural substances are obviously not capable of surviving relatively extreme conditions, which only occur in nature in isolated cases. Accordingly nothing can generally be deduced from the data on stress conditions about the stability of a herbal medicinal product under more adequate conditions. Protection from long-term effects of heat and especially humidity should be taken into account in the storage and distribution of St John's wort preparations in climatic zones III and IV. Modern European packaging such as blisters made from aluminium and plastic foil, have not proved themselves in tropical regions. The traditional glass containers with safety closures, which also contain a drying agent, are preferable.

Tests on the stability of hypericin and pseudohypericin in analytical samples of a dry extract of Hyperici herba (Ze117) and standard solutions were done under different temperature and light conditions monitored by VIS spectroscopy and HPLC–VIS/DAD measurements by Wirz *et al.* (2001). All solutions were stable at −20°C in darkness over the investigated period of 140 days. Higher temperatures, light and the presence of pyridine turned out to accelerate the degradation of pseudohypericin, while exposure to light was most aggressive. Hypericin showed higher stability, light being the only factor investigated that decreased the concentration of hypericin. The instability in the presence of light was more pronounced in the extract solution both for hypericin and pseudohypericin (Figure 9.9). Under all the other storage conditions, the stability of pseudohypericin was better in the extract solutions than in the standard solutions. Cyclopseudohypericin was assumed to be one of the transformation products of pseudohypericin.

The transformation occurs as well in solid dosage forms under stress conditions. Cyclopseudohypericin and pseudohypericin differ in their absorption spectrum. At 555 nm cyclopseudohypericin reveals an absorption maximum, pseudohypericin does not. At 589 nm it is vice versa, pseudohypericin shows a maximum, which is missing in the cyclopseudohypericin

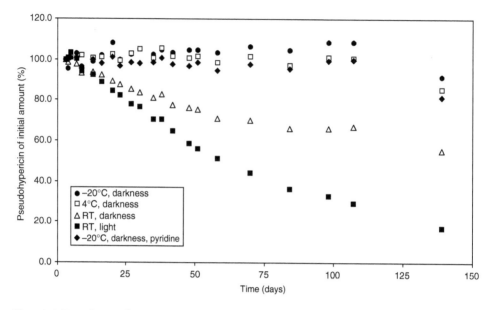

Figure 9.9 Degradation of pseudohypericin in the extract solutions under different conditions determined by HPLC. Pseudohypericin is not stable under room temperature. Light exposure influences the stability of pseudohypericin. For more details see Wirz *et al.* (2001).

spectrum. Growing amounts of cyclopseudohypericin would increase absorption at 555 nm. At the same time, absorption at 589 nm would decrease as a result of the transformation of pseudohypericin. Therefore, the ratio absorbance at 589 nm to absorbance at 555 nm is expected to decrease over time. It could be observed that the decrease of the ratio is nearly proportional to the degradation of pseudohypericin (Wirz *et al.* 2001). The result can be of interest in stability control to calculate total hypericin (including cyclopseudohypericin), because up to now cyclopseudohypericin is not available as a reference compound.

Bioavailability

The importance of an analytically measured bioavailability for HMPs is disputed. This is an acceptable comparison criterion among various products and preparations for synthetic drugs which dissolve with difficulty. As is so often the case, a proven particular instance has been postulated to a general criterion, which has now become the basis at least for almost all solid forms of drugs for regulatory purposes according to the 'Notice to applicants'. Bioavailability is defined as the extent to which, and the speed with which, an active substance is available to the receptor. This can never completely cover a complex extract, especially since in the case of St John's wort the proof that different biochemical regulation systems are affected by several substances, has recently become widespread. The bioavailability of a single constituent from a HMP is thus of very limited value. Not more information can be obtained from it, other than that the substance in question is absorbed and goes into the blood stream.

In the case of St John's wort, to study the bioavailability of hypericin and pseudohypericin is sensible from the point of view of drug safety. Analysis of the reports available on intoxication of animals, indicates dose-dependency of the phototoxic phenomena. In this respect, it is of interest whether blood levels which occur after administration of St John's wort, also lead to phototoxic reactions.

The absorption and elimination of hypericins in blood plasma has, therefore, been repeatedly investigated. The results obtained have differed quantitatively, but with a trend towards uniform data. Compared to pseudohypericin, hypericin exhibits a clearly higher half-life. The maximum serum concentration is, therefore, reached more swiftly for pseudohypericin than for hypericin. This can be attributed to the fact that the lag-time for hypericin at 1.9 h is clearly longer than that for pseudohypericin at 0.4 h (Kerb *et al*. 1996). The terminal elimination half life during steady state was analysed as 41.3 h (Kerb *et al*. 1996) and 41.7 h, respectively (Brockmöller *et al*. 1997) for hypericin and at 18.8 h and 22.8 h, respectively for pseudohypericin. On the other hand, hypericin is better absorbed than pseudohypericin; the C_{max} from a dosage of 100 μg is calculated as 1.0–1.1 μg/l for hypericin and as 0.7–0.9 μg/l for pseudohypericin. Accordingly a clearly higher AUC level was recorded for hypericin. Hypericin could not be detected in urine, either before or after hydrolysis (which was used to split off any metabolic products such as glucuronides) (Kerb *et al*. 1996).

Hypericin and pseudohypericin are apparently absorbed independently of the type of presentation. An extract suspended in an oily suspension and filled into soft gelatine capsules (Psychotonin® forte) resulted (although with a very small number of subjects) in a similar drug concentration–time profile to that of a methanolic extract processed into a sugar-coated tablet (LI 160). The kinetics, however, were swifter with a half-life of only 10–12 h (hypericin) and 2–2.5 h (pseudohypericin) and also a T_{max} of 4 and 2 h respectively after a single dose of ten capsules with a total content of 1.2 mg hypericin and 1.6 mg pseudohypericin (Sattler 1997). Compared to the data for solid dosage forms there is interestingly no indication of better availability from a lipophilic matrix. A direct comparison has not been made; however, the analytical differences to be expected when using various methods, ask for such a study.

The absorption of the poorly soluble hypericins obviously occurs without addition of solubilisers to the preparation, which has been demonstrated from the data for the extract Ze117 processed into a film-coated tablet (Boetcher *et al*. in press). The maximum plasma concentration of hypericin was achieved after 7.1 h (average of 12 test subjects); the half-life after administration of a dose of 250 mg of extract (108 μg hypericin) was 21.4 h, and after administration of 500 mg this was 24.6 h. Hypericin plasma concentration–time profiles at steady-state were rather flat with low peak-through fluctuations and maximum steady-state concentrations ranging between 1.4 and 4.8 μg/l. There was no undue accumulation, with adsorption and elimination characteristics unchanged over time. These figures concur very well with those from Brockmöller *et al*. (1997) for LI 160. This has also been confirmed in a direct comparison. The detailed evaluations produced linear pharmacokinetics for hypericin as regards absorption and elimination. Earlier announcements of non-linear pharmacokinetics in the low dose range, were later revised by the same authors.

Brockmöller *et al*. (1997) demonstrated in a parallel-design experiment, that even after high doses totalling 12 tablets (equivalent to 3600 mg of extract, three day's doses) with a total content of 11.25 mg hypericin, and also in steady-state after ingestion of 2 tablets 3 times daily (1800 mg extract with 2.18 mg hypericin and 3.44 mg pseudohypericin) for 14 days, sensitivity to UV radiation (A and B) was not, or only slightly, increased. Phototoxic phenomena did not occur at any of the sites irradiated. The average blood levels C_{trough} measured in steady-state were 30 μg/l for hypericin and 12 μg/l for pseudohypericin. These findings correlate with the fact that reports concerning skin reactions after taking St John's wort are very rare. These blood levels are clearly above the maximum levels achieved in those with a dosage of not more than 1 mg total hypericin daily, as proposed in the monograph of ESCOP, which has adopted the recommendations of Commission E. Accumulation of the hypericins was not observed.

The analysis of hypericins in plasma is not easy. In aqueous media there is a red shift of the maxima in ethanol at 560 and 590 nm, of 8 nm. Parallel to this there is a strong reduction in the molar absorption coefficients. The binding of hypericins to human serum albumin exhibits similar phenomena. A direct measurement in the aqueous system is, therefore, not possible at present. The hypericins have to be transferred to an organic phase. The detection sensitivity can be considerably increased using fluorescence detection. Even using this procedure, however, no direct measurement in aqueous systems is possible, since similar effects are observed in the fluorescence spectrum as in the UV/VIS spectrum. Sattler (1997) also observed varying binding behaviour to the proteins in serum when comparing hypericin with pseudohypericin. The analysis of most of the experiments mentioned above is based on a paper by Liebes *et al.* (1991), who developed a method using hypericin and synthetically manufactured desoxohypericin hexa-acetate as internal standards. According to this method the plasma is applied to sample preparation cartridges (separation columns) with 500 mg bonded silica gel (C18). The plasma proteins are washed out using a phosphate buffer, then the hypericins are extracted from the column with acetonitril or ethylacetate (solid phase extraction SPE). The analysis is made using an acetonitril/ammonium phosphate buffer. The measurement is taken with a fluorescence detector (470 nm ex/550 nm em). Pseudohypericin is not taken into account in this paper, since hypericin was central to the interest as a potential antiviral active substance. In a further development of the method (Kerb *et al.* 1996), pseudohypericin is included and dansylamide is used as internal standard. The plasma sample has been mixed with dimethylsulphoxide and a mixture of acetonitril and 2-butoxyethanol was added. The hypericins are then extracted with ethyl acetate at 37°C for 15 min. After centrifugation, the supernatant organic phase was removed and analysed. The yield of hypericin was then 78.6%, that of pseudohypericin 63.2%. The separation was carried out on isocratic RP-C18 with a THF–ACN phosphate buffer mixture in less than 10 min. Table 9.6 shows an overview about published chromatographic systems to analyse hypericins in plasma (Biber *et al.* 1998).

Pharmacokinetic data are also available for hyperforin. Contrary to expectations it was possible to detect hyperforin in the blood. Under conditions, to date not fully explained, hyperforin proved very susceptible to oxidation and for a long time was regarded as extremely unstable. The sensitivity is less pronounced in extracts than in lipophilic solutions in particular. Standard solutions in ethanol and methanol also proved quite stable.

The kinetics of hyperforin are similar to that of pseudohypericin. A retention time of approximately 1 h has been observed in surges. The elimination half life is approximately 9 h, C_{max} was reached after approximately 3 h. The steady-state simulation after regular intake of three tablets daily of the extract WS5572, which contains 5% hyperforin, give no indications of accumulation (Biber *et al.* 1998)

Special preparations

Herbal tea

The dried, homogeneously cut and mixed drug should preferably be filled into precisely dosed teabags for medicinal purposes. Adequate packaging of the bags provides protection from environmental influences such as moisture and light. Filling in single doses also prevents demixing of the tea drug. The risk of demixing is particularly great with St John's wort, since

Table 9.6 HPLC methods described in the literature for the quantification of hypericin (H) in plasma samples. Some of the methods also include the analysis of pseudohypericin (P)

Reference	Column[a]	Eluent[b] and flow	°C	Detection	H (min)	P (min)
Chi and Franklin (1999)	RP-8 (150 × 4.6 mm I.D., 5 μm) (Captial HPLC)	MeOH–0.03 M KH_2PO_4/K_2HPO_4 (pH 7) (7 : 3) Flow: 1 ml/min	60	Fluorimetric emission (excitation 315 nm, emission 519 nm)	3.8	—
Kerb et al. (1996)	LiChrospher 60 RP select B (250 × 4 mm I.D.) (Merck)	MeOH–THF–0.1 M NaH_2PO_4 (pH 4) (45 : 30 : 25) Flow: 0.8 ml/min	60	Fluorimetric emission (excitation 315 nm, emission 590 nm)	8.6	5.5
Liebes et al. (1991)	RP-phenyl (100 × 4.5 mm I.D., 4 μm) (Waters Associates)	A: ACN–0.1% $(NH_4)_3PO_4$/NaOH (pH 7) (3 : 7), B: ACN–H_2O (7 : 3) Elution profile: 0–15 min 100% A to 100% B, 15.01–19 min 100% B Flow: 1.2 ml/min	c	VIS absorption (590 nm)/ Fluorimetric emission (excitation 470 nm, emission >550 nm)	12.9	—
Stock (1992[d])	LiChrosorb 100 RP-18 (250 × 4 mm I.D., 5 μm) (Merck)	MeOH–ACN–H_3PO_4 85% (69.5 : 30 : 0.5) Flow: 1 ml/min	c	UV absorption (254 nm)/ Fluorimetric emission (excitation 580 nm, emission 600 nm)	13.9	3.4
Sattler (1997)	LiChrospher 100 RP-18 (250 × 4 mm I.D., 5 μm) (Merck)	A: ACN–MeOH– H_2O–H_3PO_4 85% (36 : 20 : 44 : 1) B: ACN–MeOH– H_3PO_4 85% (59 : 40 : 1) Elution profile: 0–1 min 100%A; 1–5 min 100%A to 100% B; 5–20 min 100%B. Flow:1.5 ml/min	RT	Fluorimetric emission (excitation 580 nm, emission 600 nm)	c	c

Notes
a Precolumns are not listed in this table.
b Wash out and reequilibration are not considered.
c Not mentioned in the reference.
d The method was also applied to analyse herbal drugs and herbal drug preparations of Hyperici herba (Table 9.1).

the hard, woody stalk parts are comparatively heavy and depending on the cut are also unwieldy. Blossoms and leaf parts on the other hand, are light and correspondingly easy to move. Dried leaves also powder easily. Hot water has proved an efficient solvent for natural substances in tea preparation. This also applies to St John's wort. Flavonoids in particular were extracted during 10 min when teabags, each of which contained 1.75 g of chopped St John's wort (Hyperici herba cum flore), were covered with water at 87°C (drinking water was boiled, then left to stand for 2 min). With an average of 15 mg rutin, 22.8 mg hyperoside, 4.5 mg quercitrin and 0.78 mg quercetin, a daily dosage of two teabags (total of 3.5 g of drug) achieved flavonoid dosages completely comparable with those produced in ready-to-use medicines (see Table 9.7). I3,II8-biapigenin and hyperforin, however, were not extracted with water. The amount of total hypericin in the lyophilisate of two tea bags was between 0.51 and 0.6 mg per daily dose and thus below the usual dosage in extracts, which is frequently 1 mg, but always in the region of the range used by the E Commission of 0.2–1.0 mg total hypericin (Engesser 1997). The extraction of hypericin from the tea infusion was incomplete, as was to be expected. Niesel (1992) measured 5.7% released hypericin and 22% pseudohypericin using HPLC, compared to the levels measured after analysis of the tea drug. After an extraction time of 10 min a balance obviously sets in, so that no increase in the yield can be achieved without changing the menstrum. Wirz (2001) reported similar results. Using TLC, Niesel showed that the plant acids also pass into the tea preparation. The TLC-fingerprint of a tea preparation did not differ in the range of the flavonoids (all spots between quercitrin and rutin) from the fingerprint of the drug. Boiling the St John's wort tea is not recommended due to the thermolability of hypericin.

Not much more can be said on the efficacy of the tea preparation. The range of applications is identical to that of extract preparations. In a comparative study of St John's wort tea and common yarrow tea, which is not considered to have any efficacy in the range of indications for St John's wort, but which tastes and looks similar, superiority of the St John's wort tea was ascertained for the indication depressed mood, from a visual analogue scale (Engesser 1997). According to evaluation of the scales, 13 out of 19 patients exhibited a better mood after use of St John's wort tea. The results were relativised, however, by the fact that the test subjects could

Table 9.7 Daily intake of St John's wort compounds with clinically approved extract products (reprint of Meier 2001) (Hyperforin was not included into this research)

Compounds	Overall	ZE117
Total hypericin (declared)	500–2700 µg	1000 µg
Total hypericin (analysed with photometry/DAC/ $E_{1\%/1\,cm} = 870$)	650–2800 µg	890–1170 µg
Hypericin HPLC	50–770 µg	150–300 µg
Pseudohypericin HPLC	180–1200 µg	400–600 µg
Total hypericin HPLC	230–2000 µg	550–925 µg
Rutin	10–30 mg	13–20 mg
Hyperosid	10–30 mg	10–15 mg
Quercitrin	2–4 mg	2–3 mg
Quercetin	1–5 mg	1–3 mg
Biapigenin	0.5–3 mg	0.5–1.2 mg
Amentoflavon	unknown	20–50 µg
Procyanidines	unknown	50–70 mg
Amino acids	unknown	30–60 µ mol
γ-amino butyric acid (GABA)	unknown	2–5 µ mol
Tryptophan	unknown	1 µ mol

not verify this result when asked; the personal evaluation and the measured result only agreed in 5 out of 19 cases. This shows that the relatively small differences measured are not clinically relevant. Further studies of tea preparations are not known to date. In the pharmacological test (displacement of 3H-flumazenil from benzodiazepine receptors) the lyophilisate of the tea preparations (the extract proportion was approximately 25%) was clearly less active than the ethanolic extract (50% m/m) Ze117. In rat brain homogenate the IC50 level of the lyophilisate was 548 ± 90 μg/ml compared to 21 ± 2 μg/ml for Ze117. This can be explained by the fact that the IC50 of Ze117 correlates with the content of amentoflavone. Similarly to biapigenin, however, the slightly soluble amentoflavone would not be present in the tea preparation.

Oleum hyperici

St John's wort oil has a special place among the preparations from *H. perforatum*. In folk medicine St John's wort is used chiefly as a topical agent for the treatment of wounds, and orally for dyspeptic symptoms. The monograph of the Commission E has confirmed this range of application and describes it as follows: 'Externally for the treatment and aftercare of injuries caused by sharp and blunt objects, myalgia and first degree burns, also internally for dyspeptic symptoms'. On the basis of the effects achieved with extracts, there is a trend now to use St John's wort for the indications 'to strengthen the nerves' or 'for nervous excitement' or 'to relax and for inner balance' (quotes from pack leaflets). No clinical proof of this is available, however. Furthermore, there are only a few indications of an antimicrobial activity *in vitro*.

Although St John's wort oil has a long history, the practise of maceration of fresh blossoms of St John's wort with olive oil for several weeks under the influence of sunlight, first appeared in the monograph in the supplementary volume of the sixth German Pharmacopoeia published in 1933. The monograph was not renewed later. The Swiss Pharmacopoeia has now published a new monograph (1999) for *Hyperici oleum*. Sunflower oil (*Helianthi oleum* Ph Eur) is used for the maceration. According to this, 20 g of fresh, blossoming St John's wort tips are covered with 80 g of sunflower oil as quickly as possible and subjected to fermentation while constantly shaking or stirring at a temperature of 15–30°C. After 50–80 days this is pressed, the reddened oil separated from the aqueous phase and filtered. The content of oil is determined by photometry using the absorption factor for hypericin of $A_{1\ cm}^{1\%} = 870$. The required content is laid down as not less than 1.0 mg/100 g. In the oil itself neither hypericin nor pseudohypericin could be detected. The hypericins were converted in the course of the extraction to oil hypericins as yet unknown. The reaction only takes place at room temperature if there is water present in the reaction mixture, which is why fresh plants are used. With dried drug, no corresponding discolouration takes place (Maisenbacher 1991). Experiments with peanut oil, sesame oil, Miglyol 812, isopropyl myristate and 2-octyldodecanol, showed that chiefly lipophilic substances are extracted, as well as quercetin and biapigenin and in particular hyperforin. Hyperforin was regarded until recently as the primary bactericidal substance and is for that reason desired in St John's wort oil. The stability of hyperforin in the oils has proved very limited, however. Only heat extraction with 2-octyldodecanol during storage under argon led to a stable system during the 250 days after manufacture (Maisenbacher and Kovar 1992). The heat extraction also had the effect of producing the red colouration even without water. For topical application the oil is used directly or processed into ointment bases; for oral ingestion the oil is filled into soft gelatine capsules with the addition of excipients.

Phytoequivalence/essential similarity

In the case of clearly defined, synthetic drugs, it is now customary to compare two products with the same active ingredient in terms of their bioavailability. Bioequivalence is considered using the serum level curves achieved and the area under the curve as a measure of a drug's absorption and distribution. Serum levels can be correlated with pharmacological and hence therapeutic activity. A judgement can then be made as to whether pharmacological and clinical data obtained for a comparator product can be extrapolated to the new preparation (with the same active ingredient – generic). A limiting factor is inter-individual differences.

In the case of sparingly soluble substances, dissolution can be used as a second criterion for comparison. Conversion of the dosage form of a drug into a soluble form in the gastrointestinal tract is a prerequisite for good bioavailability, on the basis that only dissolved molecules are absorbed. In the case of many drugs which are highly soluble in aqueous systems, although dissolution testing is required, the results are not very meaningful since the solubility of these substances is guaranteed.

In herbal medicines, the active ingredient consists of an extract with a complex composition, the individual constituents of which must determine efficacy, although their individual contribution to efficacy is not clear. The debate surrounding the active ingredients in St John's wort has been going on for years and more fuel is being added to it all the time. For many constituents and fractions of St John's wort, there is now evidence of pharmacological activity relating to the indication (Cott 1997, Simmen *et al*. 2001).

It has been shown recently that the interaction of various constituents may play a decisive role. In the forced swimming test, for example, fractions containing high amounts of hypericin and pseudohypericin were shown to be active but the pure compounds were inactive. The effects of hypericins could be won back by combining pure hypericins with another fraction, containing procyanidines (Butterweck *et al*. 1998).

However, such findings are few and far between, and there is still no equivalent to bioavailability which could be used to compare different herbal preparations from the same plant. This does not mean that herbal medicines cannot be compared. Various publications on this subject have appeared in the literature in Germany. Criteria other than bioavailability and dissolution must be used for assessment.

Initially, it was thought that standardisation (Hamacher 1996) would offer a way of enabling comparisons to be made. However, because an agreement could not be reached in most cases on which standardisation approach should be used, standardisation is now used primarily as a means of assuring the quality of the described extract.

Approaches to comparing preparations on a phytochemical level were taken further with an attempt to introduce the concept of 'phytoequivalence' (Uehlecke *et al*. 1994). Unfortunately, only theoretical examples were used. However, the fact that a means of making comparisons at this level did exist, particularly with fingerprint chromatography, was demonstrated at a later date on a number of examples, including St John's wort (Gaedcke 1995, Meier and Linnenbrink 1996). Qualitative differences between six important products on the German market could not be determined either by TLC or HPLC fingerprint methods with regard to flavonoids. Meanwhile, HMPs of St John's wort can be classified into hyperforin-rich and hyperforin-poor products as discussed before.

Phytochemical comparability is closely related to the preparation of the extract and the resulting characteristics. An additional level of comparison is therefore provided by the strength of the solvent used for extraction (Meier 2001) and the drug/extract ratio, which enables the drug equivalent used to be calculated.

A look at the current procedure for drawing up monographs (ESCOP, WHO, American Herbal Pharmacopoeia) shows that the relevant data are gathered together, evaluated and applied to the applications of the drug concerned. In general, different preparations of a similar character seem to be evaluated without any difference in efficacy being apparent. For example, evidence is provided of the efficacy of aqueous, methanolic/aqueous and ethanolic/aqueous valerian extracts. The starting position is similar for St John's wort: methanolic/aqueous and ethanolic/aqueous extracts have been tested using the same psychometric scales but hardly any differences were detected. In both cases, hydrophilic extraction seems to be important. For other plants, such as *Serenoa repens*, lipophilic extracts are required: ethanol, hexane and supercritical carbon dioxide are used as solvents, with no differences in the pharmacological profiles of the individual preparations being reported so far.

For preparations, the opposite approach must be taken: we have to ask whether the herbal medicine to be evaluated fulfils the requirements of the monograph. For St John's wort, the ESCOP monograph is currently the main relevant guideline.

In a comparative study, various factors must therefore be taken into consideration. For St John's wort, these are the starting material (drug quality), the extraction solvent, the manufacturing process, the extractives (expressed as the drug/extract ratio and dosage calculated from this), the hypericin and the hyperforin content as a measure of quantitative equivalence and the qualitative or semi-quantitative equivalence of the daily doses for known compounds (especially flavonoids and biapigenin) (Meier 2001).

The starting material for all preparations is the same: the drug is described in the Ph Eur monograph. Little is known about the extraction methods used, other than the fact that both maceration and percolation, as well as modified methods based on the counter-current principle can be assumed to produce qualitatively similar results.

The results of the analysis of clinically approved products which contains an ethanolic or methanolic dried extract equivalent at least to 2 g of drug standardised at least to 0.9 mg total hypericin per daily dose are listed in Table 9.7 (Meier 1999). The ranges are adequate and presented here as a guideline. In comparison to this, the values of the Ze117 extract, with which a large number of batches have been analysed, are given. It is remarkable, that the drug extract ratio is quite constant over all the products within a range of approximately 4–7 : 1. There is a trend to lower DER: the cultivation of *H. perforatum* leads to a less lignified herbal drug and therefore to higher amounts of extractives. Generally, Commission E and ESCOP monograph are still good guidelines as a base for new products.

The question of the dosage

One notes that in clinical studies using ethanol extracts, clearly smaller quantities, on average 500 mg, were used, than in the studies using methanol extracts with 900 and occasionally 1800 mg (Schulz and Hänsel 1999). From the results of the study, there is no indication that one extract is superior to the other with regard to one dosage over the other. Dose finding studies have not yet been carried out with St John's wort. Thus, the minimal dosage which guarantees a therapeutic effect is still not known today. Daily doses of 500 mg extract have the advantage that the therapy schedule of 2 × 1 dosing is easy to handle, which improves the compliance. There is a new trend to a once a day dosage of 500–650 mg of dried extract (ethanolic or methanolic). Therefore, compacted extracts will be filled into a gelatine capsule or an oblong tablet. A few reports of the phytotherapeutical practice approve the conception, but there is even more experience with the 2 × 1 and 3 × 1 daily dosage regimen. There is also a report that even lower doses of a HMP extracted from fresh plants show similar activity. However, the product fulfils the requirements of the ESCOP-Monograph concerning the daily dosage of total hypericin (1 mg/day) in the highest dose tested (Lenoir *et al.* 1999). Therefore, the theory, widely promulgated in recent years, that in phytotherapy, high dosages which lie above the usual administration are more effective, is to be questioned, especially using the example of St John's wort.

The large margin of therapeutical safety of plant extracts rather indicates that an increase in the dosage is not necessarily associated with an increase in the efficacy. There is a great need for research regarding this topic.

References

Bauer R. and Tittel G. (1996) Quality assessment of herbal preparations as a precondition of pharmacological and clinical studies. *Phytomedicine* 2, 193–6.

Biber A., Fischer H., Römer A. and Chatterjee S.S. (1998) Oral bioavailability of hyperforin from *Hypericum* extracts in rats and human volunteers. *Pharmacopsychiatry* 31, S36–43.

Blatter A. (2001) Optimierung der dünnschichtchromatographischen Identitätsprüfung von *Hyperici* herba Ph Eur und Vesuche zur Quantifizierung von Hypericin und Hyperforin. Diploma work, University of Basle, Departement of Pharmacy.

Bombardelli E. and Morazzoni P. (1997) Review: Seronoa repens. *Fitoterapia* 68, 99–126.

Brantner A., Kartnig Th. and Quehenberger F. (1994) Vergleichende phytochemische Untersuchungen an *Hypericum perforatum* L. und *Hypericum maculatum*. Crantz. *Sci. Pharm.* 62, 261–76.

Brockmöller J., Reum T., Bauer S., Kerb R., Hübner W.-D. and Roots I. (1997) Hypericin and pseudo-hypericin: Pharmacokinetics and effects on photosensitivity in humans. *Pharmcopsychiatry* 30, S94–101.

Brolis M., Gabetta B., Fuzzati N., Pace R., Panzeri F. and Peterlongo F. (1998) Identification by high-performance liquid chromatography-diode array detection-mass spectrometry and quantification by high-performance liquid chromatography-UV absorbance detection of active constituents of *Hypericum perforatum*. *J. Chromatogr.* A 825, 9–16.

Boetcher M., Mueck W., Haase C.G., Ochmann K., Rohde G., Unger S., Wensing G. and Hindmarch I. (2002) Pharmacodynamics, pharmacokinetics and tolerability of Remotiv® (*Hypericum* extract) in healthy human volunteers. *Pharmacopsychiatry* (in press).

Butterweck V. (1997) Beitrag zur Pharmakologie und Wirkstoff-Findung von *Hypericum perforatum* L. Thesis. Mathematisch-Naturwissenschaftliche Fakultät. Westfälische Wilhelms-Universität Münster.

Butterweck V., Petereit F., Winterhoff H. and Nahrstedt A. (1998) Solubilized hypericin and pseudo-hypericin from *Hypericum perforatum* L. exert antidepressant activity in the forced swimming test. *Planta medica* 64, 291–4.

Cameron D.W., Edmonds J.S. and Raverty W.D. (1976) Oxidation of emodine anthrone and stereochemistry of emodin bianthrone. *Aust. J. Chem.* 29, 1535–48.

Chi J.D. and Franklin M. (1999) Determination of hypericin in plasma by high-performance liquid chromatography. *J. Chromatogr.* B 724, 195–8.

Cott J.M. (1997) *In vitro* receptor binding and enzyme inhibition by *Hypericum perforatum* extract. *Pharmacopsychiatry* 30, S108–12.

Daniel K.W.O. (1974) Über die Behandlung psycho-somatischer Fehlhaltungen bzw. Störungen bei Kindern im Alter zwischen sechs und zwölf Jahren mit einem Vollextrakt aus *Hypericum perforatum*. *Physikalische Medizin und Rehabilitation* 15, 64–5.

Erdelmeier C.A.J. (1998) Hyperforin, possible the major non-nitrogenous secondary metabolite of *Hypericum perforatum* L. *Pharmacopsychiatry* 31, S2–6.

Engesser A.B. (1997) Pharmakologische und klinisch-pharmakologische Untersuchungen mit ausgewählten Teedrogen. Thesis. Universität of Basle.

Freytag W.E. (1984) Bestimmung von Hypericin und Nachweis von Pseudohypericin in *Hypericum perforatum* L. durch HPLC. *Dtsch. Apoth. Ztg.* 124, 2383–6.

Gaedcke F. (1995) Phytoäquivalenz. Was steckt dahinter? *Dtsch. Apoth. Ztg.* 135, 311–22.

Gaedcke F. (1997) Herstellung, Qualität, Analytik und Anwendung von Johanniskraut-Extrakten. *Zeitschrift für Arznei- und Gewürzpflanzen* 2, 63–72.

Gaedcke F. (1997) Johanniskraut und dessen Zubereitungen. Qualitätsbeurteilung über ein selektives, reproduzierbares HPLC-Verfahren. *Dtsch. Apoth. Ztg.* 137, 3753–7.

Gill M. and Giménez A. (1991) Austrovenetin, the principal pigment of the toadstool. *Dermocybe Austroveneta. Phytochem.* 30, 951–5.

Häberlein H., Tschiersch K.P., Stock S. and Hölzl J. (1992) Johanniskraut (*Hypericum perforatum* L.) Teil 1: Nachweis eines weiteren Naphthodianthrones. *Pharm. Ztg. Wiss.* 137, 169–73.

Hamacher H. (1996) Standardisierung komplexer Naturstoffgemische. Behördliche Zielvorgaben und ihre Realisierbarkeit. *Pharm. Ind.* 58, 339–46.

Hiermann A., Kartnig Th. and Azzam S. (1986) Ein Beitrag zur quantitativen Bestimmung der Procyanidine in *Crataegus*. *Scientia Pharmaceutica* 54, 331–7.

Hölzl J. and Ostrowski E. (1987) Johanniskraut (*Hypericum perforatum* L.) HPLC-Analyse der wichtigen Inhaltsstoffe und deren Variabilität in einer Population. *Dtsch. Apoth. Ztg.* 127, 1227–30.

Kapinus E.I., Falk H. and Tran, H.T.N. (1999) Spectroscopic investigation of the molecular structure of hypericin and its salts. *Monatshefte für Chemie* 130, 623–35.

Kerb R., Brockmöller J., Staffeldt B., Ploch M. and Roots I. (1996) Single-dose and steady-state pharmacokinetics of hypericin and pseudohypericin. *Antimicrob. Agents Chemother.* 40, 2087–93.

Klein-Bischoff U. and Klumpp U. (1993) Hypericin und Fluoreszenz, eine quantitative fluorimetrische Bestimmungsmethode. *Pharmazeutische Zeitung Wissenschaft* 138, 55–8.

Krämer W. and Wiartalla R. (1992) Bestimmung von Naphtodianthronen (Gesamthypericin) in Johanniskraut. *Pharmazeutische Zeitung Wissenschaft* 197, 202–7.

Lapke C. (2000) Freie Aminosäuren in Arzneipflanzen mit psychtroper Wirkung. Thesis. Freie Universität Berlin, Shaker-Verlag GmbH, Aachen.

Lenoir S., Degenring F.H. and Saller R. (1999) A double-blind randomised trial to investigate three different concentrations of a standardised fresh plant extract obtained from the shoot tips of *Hypericum perforatum* L. *Phytomedicine* 6, 141–6.

Liebes L., Mazur Y., Freeman D., Lavie D., Lavie G., Kudler N., Mendoza S., Levin B., Hochster H. and Meruelo D. (1991) A method for the quantitation of hypericin, an antiviral agent, in biological fluids by high-performance liquid chromatography. *Anal. Biochem.* 195, 77–85.

Maisenbacher P. (1991) Untersuchungen zur Analytik von Johanniskrautöl. Thesis. Fakultät für Chemie und Pharmazie. Eberhard-Karls-Universität Tübingen.

Maisenbacher P. and Kovar K.A. (1992) Analysis and stability of *Hyperici oleum*. *Planta Med.* 58, 351–4.

Meier B. (1999) The science behind *Hypericum*. *Adv. Therapy* 16, 135–47.

Meier B. (2001) Comparing phytopharmaceuticals: the example of St. John's wort. *Adv. Therapy* 18, 35–46.

Meier B. and Linnenbrink N. (1996) Status und Vergleichbarkeit pflanzlicher Arzneimittel. *Dtsch. Apoth. Ztg.* 136, 4205–20.

Melzer M., Fuhrken D. and Kolkmann, R. (1998) Hyperforin im Johanniskraut. Hauptwirkstoff oder nur Leitsubstanz? *Dtsch. Apoth. Ztg.* 138, 4754–60.

Micali G., Lanuzza F. and Currò P. (1996): High-performance liquid chromatographic determination of the biologically active principle hypericin in phytotherapeutic vegetable extracts and alcoholic beverages. *J. Chromatogr.* A 731, 336–9.

Morf S., Debrunner B., Meier B. and Kurth H. (1998) Automatische Probenvorbereitung von pflanzlichen Arzneimitteln. Ein Vergleich mit der konventionellen Methode am Beispiel Johanniskraut. *LaborPraxis* 22 (11), 56–62.

Mulinacci N., Bardazzi C., Romani A., Pinelli P., Vincieri F.F. and Costantini A. (1999) HPLC-DAD and TLC-densitometry for quantification of hypericin in *Hypericum perforatum* L. extracts. *Chromatographia* 49, 197–201.

Niesel S. (1992) Untersuchungen zum Freisetzungsverhalten und zur Stabilität ausgewählter wertbestimmender Pflanzeninhaltsstoffe unter besonderer Berücksichtigung moderner phytochemischer Analysenverfahren. Thesis. Freie Universität Berlin.

Orth, H.C.J. (1999) Isolierung, Stabilität und Stabilisierung von Hyperforin und Identifizierung seines Hauptabbauproduktes aus *Hypericum perforatum* L. Thesis. Eberhard-Karls-Universität Tübingen.

Ostrowski E. (1988) Untersuchungen zur Analytik, ^{14}C-Markierung und Pharmakokinetik phenolischer Inhaltsstoffe von *Hypericum perforatum* L. Thesis. Pharmazie und Lebensmittelchemie. Philipps-Universität Marburg.

Piperopoulos G., Lotz R., Wixforth A., Schmierer T. and Zeller K.P. (1997) Determination of naphtho-dianthrones in plant extracts from *Hypericum perforatum* L. by liquid chromatography-electrospray mass spectrometry. *J. Chromatogr.* B **695**, 309–16.

Porter L., Hrstich L. and Chan B (1986) The conversion of procyanidins and prodelphinidins to cyanidin and delphinidin. *Phytochemistry* **25**, 223–30.

Rehwald A. (1995) Analytical investigations of *Crataegus* species and *Passiflora incarnata* L. by HPLC. Swiss Federal Institute of Technology, ETH Zurich Thesis No. 10959. pp. 150–63.

Rocksloh K. (1999) Rezepturoptimierung von Johanniskrautextrakt-Dragéekernen unter Einbeziehung neuronaler Netze. Thesis. Eberhard-Karl-Universität Tübingen.

Rohr G. (1999) Analytical investigation on and isolation of procyandinins from *Crataegus* leaves and flowers, ETH Zürich Thesis No. 13020.

Sattler S. (1997) Naphthodianthrone aus *Hypericum perforatum* L.: Isolierung, Pharmakokinetik, Löslichkeitsverbeserung und Absorptionsstudien am Caco-2-Zellkulturmodell. Thesis. Philipps-University Marburg.

Schrader E. (2000) Equivalence of St. John's wort extract (Ze117) and fluoxetine: a randomized, controlled study in mild-moderate depression. *Int. Clin. Psychopharmacol.* **15**, 61–8.

Schrader E., Meier B. and Brattström A. (1998) *Hypericum* treatment of mild-moderate depression in a placebo-controlled study. A prospective, double-blind, randomized, placebo-controlled multicentre study. *Human Psychopharmacol.* **13**, 163–9.

Schütt H. (1996) Morphologische, phytochemische und botanische Untersuchungen zur Selektion hypericin-, pseudohypericin- und flavonoidreicher *Hypericum perforatum* L. Stämme. Thesis. Philipps-University, Marburg. Dissertationes Botanicae 263. Cramer, J. Berlin, Stuttgart.

Schütt H. and Hölzl J. (1994) Vergleichende Qualitätsuntersuchung von Johanniskraut-Fertigarzneimitteln unter Verwendung verschiedener quantitativer Bestimmungsmethoden. *Pharmazie* **49**, 206–9.

Schulz V. and Hänsel R. (1999) Rationale Phytotherapie – Ratgeber für die ärztliche Praxis. 4th Edition. Springer-Verlag, Berlin, Heidelberg, London. pp. 67–8.

Simmen, U., Higelin J., Berger-Büter K., Schaffner W. and Lundström K. (2001) Neurochemical studies with St. John's wort *in vitro. Pharmacopsychiatry* **34**, S137–42.

Stock S. (1992) ^{14}C-Markierung und Pharmakokinetik von Naphthodianthronen aus *Hypericum perforatum* L. sowie Untersuchungen zur Analytik von Oleum *Hyperici*. Thesis. Pharmazie und Lebensmittelchemie. Philipps-University Marburg.

Tateo F., Martello S., Lubian E. and Bononi M. (1998) Hypericin and hypericin-like substances: analytical problems. In *Food Flavors: Formation, Analysis and Packaging Influences.* E.T. Contis, C.T. Ho, C.J. Mussinan, T.H. Parliment, F. Shahidi and A.M. Spanier (Eds) Developments in food science Vol. 40. Elsevier (Amsterdam) 1998. pp 143–7.

Uehlecke, B., Frank, B and Reinhard E. (1994) Bewertung und Vergleichbarkeit von Phytopharmaka. Einführung des Begriffes 'Phytoäquivalenz'. *Dtsch. Apoth. Ztg.* **134**, 1772–4.

Wirz A. (2001) Analytical and phytochemical investigations on Hypericin and related compounds of *Hypericum perforatum*. Thesis No. 13553. Swiss Federal Institute of Technology, Zürich.

Wirz A., Meier B. and Sticher O. (2001) Stability of hypericin and pseudohypericin in extract solutions of *Hypericum perforatum* and in standard solutions. *Pharm. Ind.* **63**, 410–15.

Wirz A., Meier B. and Sticher O. (2001) Absorbance data of hypericin and pseudohypericin used as reference compounds for medicinal plant analysis. *Pharmazie* **56**, 52–7.

Woelk H. (2000) Comparision of St John's wort and imipramine for treating depression: randomised controlled trial. *Brit. Med. J.* **321**, 536–9.

Wurglics M., Westerhoff K., Kanzinger A., Wilke A., Baumeister A. and Schubert-Zsilavecz M. (2000) Johanniskrautextraktpräparate: Vergleich aufgrund der Hyperforin- und Hypericingehalte. *Dtsch. Apoth. Ztg.* **140**, 3904–9.

Zevakov V.A., Glyzin V.I., Shemeryankina T.B. and Patudin A.V. (1991) Quantitative evaluation of some species of St. John's wort for their level of hypericins by the HPLC method. *Chem. Nat. Compounds* **27**, 122.

10 The potential of *in vitro* cultures of *Hypericum perforatum* and of *Hypericum androsaemum* to produce interesting pharmaceutical compounds

Alberto C.P. Dias

Introduction

Medicinal plants represent a substantial part of the available genetic diversity in the world that has been endangered by the common practices of gathering seeds and biomass of wild species for the production of phytomedicines. A solution to this problem could be the utilisation of *in vitro* cultures of plants of interest for the mass production of a consistent and a high quality biomass (Murch *et al.* 2000c). Moreover, the utilisation of plant cells for the production of natural compounds of commercial interest has gained increasing attention over past decades (Alfermann and Petersen 1995, Stöckigt *et al.* 1995, Mülbach 1998, Fischer *et al.* 1999). Plant cell cultures are not constrained by environmental, ecological or climatic conditions and the biosynthesis of the desired compounds can be forced by appropriate culture conditions, use of elicitors and selection of high-yield producer strains, just to mention a few strategies.

Despite all the potential, the utilisation of plant cell cultures for the commercial large-scale production of phytopharmaceuticals is limited to a few examples like the production of shikonin, berberine or ginseng saponins (Mülbach 1998). However, the scientific community most probably has only a limited information of the industrial investigations since these are usually secretive. Nevertheless, *in vitro* cultures of plants are a powerful tool for studying the biosynthesis of secondary metabolites and the factors that influence it giving some possibilities of controlled production. This potential was nicely put in words by Zenk (1991) that labelled plant cell cultures as 'a pot of gold'.

Several *Hypericum* species have been used as medicinal plants around the world. *H. perforatum* (St John's wort) is one of the most known and used species. Its use has an anti-inflammatory and healing agent has been documented since the Middle Ages (Bombardelli and Morazzoni 1995). Nowadays, its use as a treatment for depression has become prominent and is one of the leaders of the medicinal plant market (Wills *et al.* 2000, Di Carlo *et al.* 2001). *H. androsaemum* (tutsan) plants have been largely used in Portugal by their hepatic protective and diuretic properties (Seabra and Vasconcelos 1992).

Several *in vitro* culture approaches have been carried out with *Hypericum* species. Data concerning *in vitro* cultures of *Hypericum* species found in literature is listed in Table 10.1. A major part of these studies have been performed with *H. perforatum*, which reflects the commercial importance of this species.

Phenolic production in *in vitro* cultures vs wild plants

A major problem faced by consumers and manufacturers of *H. perforatum* phytopharmaceuticals is the high discrepancies in the quality of the St John's wort products on the market (Murch *et al.* 2000d, Wills *et al.* 2000). This could be related to different ways of making medicinal preparations

Table 10.1 In vitro studies on Hypericum species

Species	Culture type	Type of study	References
H. perforatum	Shoot cultures	Hypericins accumulation	Mosen et al. 1993, Zdunek and Alfermann 1992, Kirakosyan et al. 2000a, Kirakosyan et al. 2000b, 2001
		In vitro plant multiplication, micropropagation and hypericin content	Cellorova et al. 1992, 1994, 1995, 1997, Cellarova and Kimakova 1999, Brutovska et al. 1994, 1998
	Shoot cultures	In vitro plant regeneration and micropropagation	Pretto and Santarém 2000 Murch et al. 2000a
	Shoot cultures	Melatonin and Serotonin biosynthesis	Murch et al. 2000b
	Calli cultures	Xanthones identification and accumulation	Dias et al. 2001
		Flavones identification and accumulation	Dias et al. 1998
Several Hypericum species, including H. perforatum and H. maculatum	Calli and cell cultures	Hypericins and flavonoids production	Kartnig and Brantner 1990, Kartnig et al. 1996, Rani et al. 2001
H. patulum	Suspended cells	Xanthones identification	Hshiguro et al. 1993, 1995a,b, 1996
H. erectum	Callus and shoot cultures	Procyanidins production and in vitro plant multiplication	Yazaki e Okuda 1990, 1994
H. canarensis	Shoot cultures	In vitro plant multiplication	Mederos et al. 1996, 1997
H. brasiliense	Calli and shoot cultures	In vitro plant multiplication	Cardoso and Oliveira 1996
H. androsaemum	Suspended cells	Xanthone biosynthesis	Schmidt and Beerhues 1997, Peters et al. 1998, Abd El-Mawla et al. 2001
	Calli and suspended cells	Xanthones identification and accumulation	Schmidt et al. 2000, Dias et al. 2000

but also is due to ecological, geographical and harvesting factors (Bütter *et al.* 1998, Kurth and Spreemann 1998). Due to uncontrolled harvesting as well as seasonal, climatic, and geographic restrictions the availability of *H. androsaemum* plants in nature, in Portugal, is strongly limited. To overcome these problems, *in vitro* cultures of both species could be an alternative choice if the active compounds are produced in such systems.

The pharmacological properties of both species have been attributed to phenolic compounds. Several *in vitro* cultures, namely *calli*, suspended cells and shoot cultures, were established from *H. perforatum* and from *H. androsaemum* and their phenolic composition were compared with those of the corresponding *in vivo* plants (Dias *et al.* 1999). The examination of the HPLC chromatograms clearly evidenced a different phenolic composition between the methanolic extracts of *in vivo* plants and those of *in vitro* cultures (Figures 10.1–10.3). Table 10.2 list the identified compounds in both *in vitro* cultures and wild plants extracts. Xanthones corresponds to the major peaks recorded in the chromatograms of the methanolic extracts of *calli* and suspended cells of both species (Figure 10.3). These compounds, not detected in the aerial part of *in vivo* plants, were also found in the extracts of the shoot cultures of *H. perforatum* and *H. androsaemum* (Figures 10.1–10.2). In terms of composition, the phenolic extracts of the shoot cultures were closer to those of the corresponding *in vivo* plants. Nevertheless, the global phenolic composition of *in vitro* cultures and of *in vivo* plants of *H. perforatum* and *H. androsaemum* is quite different, both in qualitative and quantitative terms (Figures 10.1–10.5). The existence of a modified secondary metabolism or even the production of new compounds by *in vitro* cultures is a common feature (Stöckigt *et al.* 1995, Mülbach 1998).

The differential phenolic accumulation observed between *in vitro* cultures and *in vivo* plants of *H. perforatum* and *H. androsaemum* will reflect differences in the biosynthetic pathways of these biologic systems. Figure 10.6 illustrates a very synthetic biosynthesis scheme of the major classes of phenolics found in *in vitro* cultures and in wild plants of both species. The general phenylpropanoid pathway is accordingly to that described elsewhere (Dixon and Paiva 1995, Dewick 1997, Dixon and Steele 1999). The biosynthesis of hypericins seems to involve the polyketide pathway, although few details are known (Dewick 1997, Nahrstdet and Butterweck 1997). No references were found in the literature concerning the biosynthesis of hyperforins. However, mevalonate and acetate should be the precursors of these compounds.

Flavonoids are one of the major class compounds accumulated in the aerial part of *in vivo* plants of *H. perforatum* and *H. androsaemum* (Figures 10.4, 10.5). However, in *in vitro* cultures of both species these phenolics were almost absent. In the opposite, xanthones are accumulated preferentially in *in vitro* cultures, namely in *calli*. These two types of compounds share common precursors (Figure 10.6). A key step in the flavonoid biosynthesis is the condensation of the 4-coumaroyl CoA with three molecules of malonyl-CoA, by the action of the chalcone synthase (CHS), resulting in the formation of the chalcone backbone (Dixon and Steele 1999). This molecule is then converted into flavanone, by the chalcone isomerase (CHI). This molecule could be converted to flavonols, like quercetin, after several enzymatic steps (Figure 10.6). A key step in the biosynthesis of xanthones is the sequential condensation of three molecules of malonyl-CoA with one of *m*-hydroxybenzoyl-CoA (or other benzoic acids), by the action of the benzophenone synthase (BFS), resulting in the formation of a benzophenone backbone (Beerhues 1996). A similar process was already described in *H. androsaemum* cell cultures (Peters *et al.* 1998). The mode of action of BFS is similar to other polyketide synthases, like the CHS (Schröder 1997). So we could speculate that under *in vivo* conditions the action of CHS is favoured originating the formation of flavonoids. In *in vitro* conditions the expression of BFS will be more active resulting in the change of the biosynthetic flow into the production of xanthones (Figure 10.6).

Another significant difference found was the differential accumulation of flavonoids observed in *in vitro* cultures and in *in vivo* plants of *H. perforatum* (Figures 10.2, 10.3 and 10.5; Table 10.2).

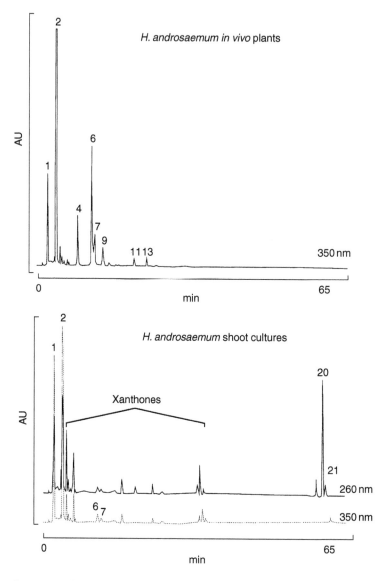

Figure 10.1 Chromatograms of methanolic extracts of *in vivo* plants and *in vitro* shoot cultures of *H. androsaemum*. Peak numbers correspond to compounds given in Table 10.2.

In vitro cultures accumulated flavones, wich were identified as being of the luteolin type (Dias *et al.* 1998, 1999), whereas wild plants produce mainly quercetin derivative flavonols. Apigenin, and occasionally luteolin, were the unique flavones detected in wild plants of *H. perforatum* (Dias *et al.* 1999). These two classes of flavonoids have a similar precursor flavanone molecule but have final divergent biosynthetic pathways (Figure 10.6). In *in vivo* conditions the enzymes flavanone 3-hydroxilase (F3H) and flavonol synthase (FLS) will be more active resulting into the formation of flavonols. *In vitro* conditions might have favoured the expression and/or activity of enzymes like the flavonoid 3′-hydroxilase and the flavone synthase (FNS) originating the accumulation of flavones of the luteolin type.

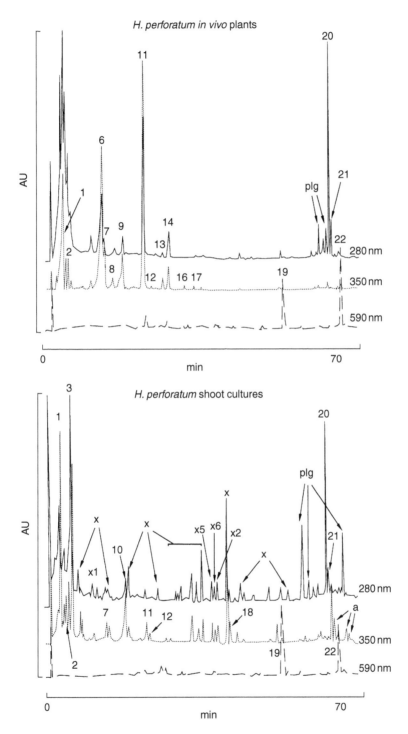

Figure 10.2 Chromatograms of methanolic extracts of *in vivo* plants and *in vitro* shoot cultures of *H. perforatum*. Peak numbers correspond to compounds given in Table 10.2.

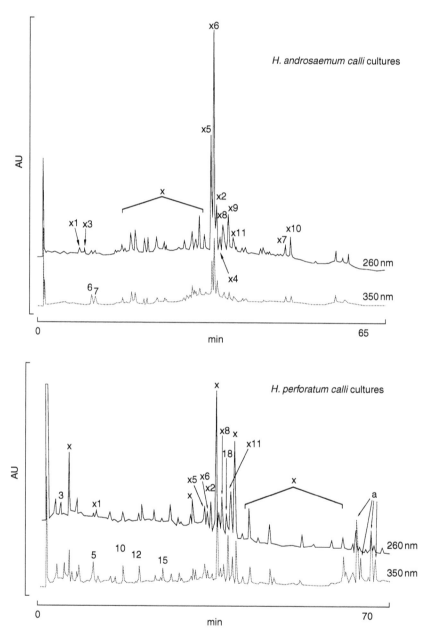

Figure 10.3 Chromatograms of methanolic extracts of *calli* cultures of *H. androsaemum* and *H. perforatum*. Peak numbers correspond to compounds given in Table 10.2.

An interesting fact was the accumulation of hyperforin and adhyperforin in shoots, but not in plants, of *H. androsaemum* (Figure 10.1 and 10.4). Till now, and to our knowledge, these compounds were only detected in *H. perforatum* plants. Recently, some phloroglucinol-type compounds were detected in plants of *H. androsaemum* (Valentão *et al.* unpublished results).

Table 10.2 Compounds identified in the methanolic extracts of *H. perforatum* and *H. androsaemum* biomass with their respective retention times (RT)[a]

Peak number	Compound	RT
1	3-caffeoylquinic acid	3.26
2	5-caffeoylquinic acid (chlorogenic acid)	4.26
3	Mangiferin	5.85
4	Quercetin 3-sulphate	8.71
5	Luteolin 5-glucoside[b]	10.21
6	Hyperoside (quercetin 3-galactoside)	12.39
7	Isoquercitrin (quercetin 3-glucoside)	13.15
8	Quercetin 3-xyloside	14.80
9	Quercitrin (quercetin 3-rhamnoside)	16.48
10	Luteolin 3'-glucoside[b]	18.98
11	Quercetin	20.85
12	Luteolin[b]	22.50
13	Kaempferol	24.68
14	Apigenin	25.70
15	Luteolin 5,3'-dimethylether[b]	26.34
16	I3,II8-biapigenin	29.27
17	Amentoflavone (I3', II8-biapigenin)	31.21
18	6-C-prenyl luteolin[b]	35.30
19	Pseudohypericin	48.91
20	Hyperforin	60.61
21	Adhyperforin	61.34
22	Hypericin	63.98
X1	1,3,5,6-tetrahydroxyxanthone[c]	11.05
X3	1,3,6,7-tetrahydroxyxanthone[c]	11.62
X5	1,3,6,7-tetrahydroxy-8-prenylxanthone[c]	33.18
X6	Toxyloxanthone B[c]	34.00
X2	1,3,5,6-tetrahydroxy-2-prenylxanthone[c]	34.58
X4	1,3,6,7-tetrahydroxy-2-prenylxanthone[c]	35.07
X8	1,3,7-trihydroxy-6-methoxy-8-prenylxanthone[c]	36.00
X9	1,3,6-trihydroxy-7-methoxy-8-prenylxanthone[c]	36.82
X11	Paxanthone[c]	38.27
X7	γ-mangostin[c]	48.07
X10	Garcinone B[c]	48.73

Note
(a) HPLC separation conditions as described elsewhere (Dias *et al*. 1998). The compounds indicated by (b) were identified from *H. perforatum calli* (Dias *et al*. 1998); the xanthones (c) were identified from *H. androsaemum* and *H. perforatum in vitro* cultures (Dias *et al*. 2000, 2001).

Although HPLC retention times were very close, none of these compounds were identified as hyperforin or adhyperforin. Similar to xanthones, *in vitro* conditions might have promoted the biosynthesis of hyperforin and adhyperforin whereas in *in vivo* environment the *H. androsaemum* plants do not have the capability to express this potential.

Hypericins and hyperforins production by *in vitro* cultures of *H. perforatum*

The utilisation of *H. androsaemum* and *H. perforatum in vitro* biomass as a direct substitute for the plants collected in nature will be inviable since their phenolic composition is quite different.

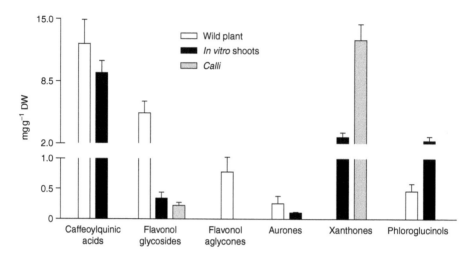

Figure 10.4 Specific accumulation of the major classes of phenolics detected in wild plants, and in 1-year-old *in vitro* cultures of *H. androsaemum*. Cultures were maintained in MS solid medium supplemented with 4.5 μM of IAA and 2.3 μM of KIN (*in vitro* shoots) or with 4.5 μM of NAA and 2.3 μM of KIN (*calli*). Results are mean of six independent replicates, except those of wild plants that are mean of 9 independent replicates collected in 1996; vertical bars indicate 95% confidence to average. In each class, all the values are statistically different ($P < 0.05$).

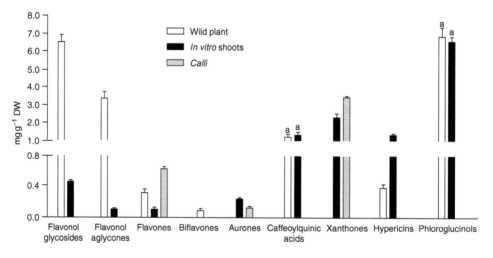

Figure 10.5 Specific accumulation of the major classes of phenolics detected in wild plants, and in 1-year-old *in vitro* cultures of *H. perforatum*. Cultures were maintained in MS solid medium supplemented with 4.5 μM of IAA and 2.3 μM of KIN (*in vitro* shoots) or with 4.5 μM of NAA and 2.3 μM of KIN (*calli*). Results are mean of six independent replicates, except those of wild plants that are mean of 12 independent replicates collected during 1996–9; vertical bars indicate 95% confidence to average. In each class, all the values are statistically different ($P < 0.05$), except those indicated with the same letter.

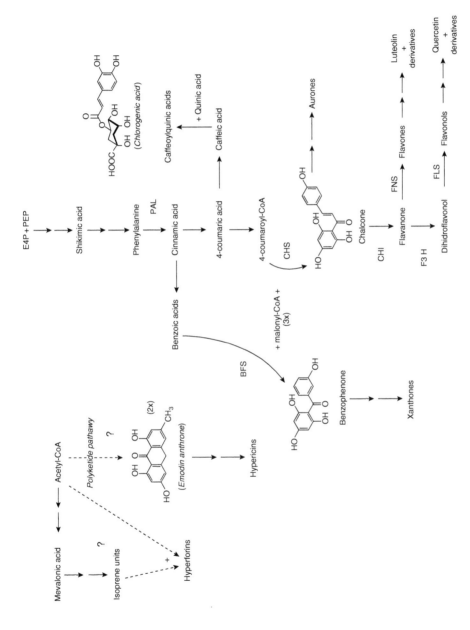

Figure 10.6 The biosynthesis pathway of the major classes of phenolics found in *in vitro* cultures and *in vivo* plants of *H. perforatum* and *H. androsaemum*. The scheme is based on the general phenolic pathways described in literature (see main text for references). Abbreviations: E4P, erythrose 4-P; PEP, phosphoenolpyruvate; PAL, phenylalanine ammonia-lyase; CHS, chalcone synthase; BFS, benzophenone synthase; CHI, chalcone isomerase; FNS, flavone synthase; F3H, flavanone 3-hydroxilase; FLS, flavonol synthase.

An alternative strategy is the utilisation of *in vitro* cultures for the production of valuable compounds or for obtaining a biomass containing selected bioactive compounds with increased therapeutic potential. *H. perforatum* has several compounds with noteworthy activity (Nahrstedt and Butterweck 1997, Greeson *et al.* 2001), being the hypericins and hyperforins the most interesting ones. Hypericin and its derivatives have antiviral activity against a broad spectrum of viruses, for example, against AIDS (Meruelo *et al.* 1988), and they are candidates for the photodynamic therapy of cancer (Anker *et al.* 1995). Hyperforin seems to be more responsible for the antidepressive activity of *H. perforatum* (Di Carlo *et al.* 2001, Greeson *et al.* 2001), in spite of other compounds like flavonoids and hypericins that could act synergistically for this activity (Wills *et al.* 2000).

Both hyperforins and hypericins were produced by shoot cultures of *H. perforatum* (Figures 10.2 and 10.5; Table 10.3). Several hormonal combinations, in solid or liquid MS medium, were tested (Dias and Ferreira, 2000). The highest specific production of total hypericins (1.24 ± 0.14 mg g^{-1} DW) and hyperforins (6.53 ± 0.56 mg g^{-1} DW) was observed in shoots maintained in solid medium, with the combination of $4.5\,\mu$M of IAA plus $2.3\,\mu$M of KIN (IK-shoots). Hyperforin was the major phloroglucinol detected in these shoots, although its accumulation was lower than that observed in *H. perforatum* plants (Table 10.3). Nevertheless, the overall accumulation of hyperforins was similar in wild plants and in IK-shoots due to the production of unidentified hyperforin-like compounds by *in vitro* cultures. Pseudohypericin and hypericin were the unique hypericins detected in IK-shoots (Table 10.3). The total amount of hypericins accumulated by IK-shoots of *H. perforatum* was higher than that found in the corresponding wild plants. This value is also higher compared to those reported, in the literature, for *in vitro* cultures of *H. perforatum* (Kartnig *et al.* 1996, Kirakosyan *et al.* 2000a,b, Kirakosyan *et al.* 2001). However, higher amounts could be achieved by medium optimisation and use of elicitors. Recently, the stimulation of the production of hypericins by shoot cultures of *H. perforatum* was achieved by using mannan (Kirakosyan *et al.* 2000b) and cork tissue (Kirakosyan *et al.* 2001). A negative aspect of the hypericins accumulation by IK-shoots of *H. perforatum* was the instability in their production. Total hypericins accumulation by these *in vitro* cultures showed a negative linear relationship with culture time (Dias *et al.* unpublished results). After 4 years of continuous cultivation, hypericins production by IK-shoots of *H. perforatum* decreased by 66% of its initial value. However, the accumulation of hyperforins remained similar during this time.

Table 10.3 Specific contents (mg g^{-1} DW) of hypericins and hyperforins in wild plants and in 1-year-old shoot cultures of *H. perforatum*[a]

Compound	Wild plants	In vitro *shoots*
Hypericins		
Pseudohypericin	0.11 ± 0.05	0.72 ± 0.08
Hypericin	0.27 ± 0.11	0.51 ± 0.06
Hyperforins		
Hyperforin	4.71 ± 1.34	3.04 ± 0.38
Adhyperforin	1.75 ± 0.72	0.32 ± 0.04
Other hyperforin-like compounds	0.38 ± 0.02	3.18 ± 0.47

Note

a Cultures were maintained in MS solid medium supplemented with $4.5\,\mu$M of IAA and $2.3\,\mu$M of KIN. Results are means $\pm$ SD of six independent replicates, except those of wild plants that are means of 12 independent replicates collected during 1996–9. All the values are statistically different ($P < 0.05$).

The instability of metabolites production by *in vitro* cultures is a well-known fact and a major drawback for their commercial utilisation (Mühlbach 1998).

The accumulation of hypericins and hyperforins was just observed in differentiated (shoot) cultures of *H. perforatum* (Figures 10.2 and 10.5). In plants, phenolic biosynthesis occurs mainly during cell differentiation and after leaf maturation (Matsuki 1996). Hypericin and related compounds are accumulated in special glands found in *H. perforatum* leaves (Cellarova *et al.* 1994). Moreover, the accumulation of hypericins in cell cultures was shown to be dependent on cellular and tissue differentiation (Kirakosyan *et al.* 2000b). It seems that organ differentiation is highly determinant for the production of hypericins and hyperforins by *in vitro* cultures of *H. perforatum*. Shoot cultures established and maintained in the same medium of *H. perforatum calli* cultures (MS medium supplemented with 4.5 µM of NAA plus 2.3 µM of KIN) accumulated these type compounds (Dias and Ferreira 2000). Notwithstanding the several hormonal variants tested and several *calli* and cells suspension lines of *H. perforatum* obtained, over an experimental period of 4 years, it was never detected hypericins or hyperforins in these undifferentiated cultures (Dias unpublished results). Nevertheless, the accumulation of hypericins in cell cultures of *H. perforatum* was already reported (Kartnig *et al.* 1996).

Xanthone biosynthesis and accumulation in *calli* and suspended cells of *H. androsaemum* and *H. perforatum*

Xanthones are compounds widely found in the Hypericaceae family that have interesting pharmacological properties. Xanthones exhibit strong and selective inhibition of monoamine oxidase A (Rocha *et al.* 1994) and some authors considered that they are involved in the antidepressant activity of *H. perforatum* (Bombardelli and Morazzoni 1995). Moreover, several xanthones have *in vitro* cytotoxicity and *in vivo* antitumour activity, as well as antibacterial and antifungic activities (Hostettmann and Hostettmann 1989, Hostettmann *et al.* 1995). The major phenolics produced by suspended cells and *calli* cultures of *H. androsaemum* and *H. perforatum* were xanthones (Figures 10.3–10.5). Several xanthones have been identified in cell cultures from other *Hypericum* species (see Table 10.1). It seems, therefore, that the accumulation of xanthones is characteristic of *in vitro* cultures of *Hypericum* sp. This could be exploited from biotechnological point of view.

Several 1,3,6,7 and 1,3,5,6-tetraoxygenated xanthones were identified and their accumulation was studied in *H. androsaemum* and *H. perforatum* cell cultures (Dias *et al.* 2000, 2001, Schmidt *et al.* 2000). Both *H. perforatum* and *H. androsaemum calli* and suspended cells accumulated common xanthones, it was shown that they have similar biosynthetic pathways (Figure 10.7). In both species, all correlations between individual xanthones connected by the arrows are positive except those involving cyclisation of the prenyl side chain. This could indicate that the formation of the pyran ring in xanthones is a key step in their biosynthesis.

The accumulation of xanthones in *H. androsaemum* and *H. perforatum* cell cultures was influenced greatly by the hormonal supplementation and by the culture type (*calli* or suspended cells) (Dias *et al.* 2000, 2001). In both species, the production of xanthones in *calli* was higher than that observed in suspended cells maintained in a similar medium (Figure 10.8). Cell suspensions of *H. androsaemum* and *H. perforatum* consisted mainly of isolated and small aggregates of cells in an undifferentiated state, whereas in *calli* a certain degree of differentiation occurred. This differentiation could have favoured xanthone production in *calli*. Lack of cellular differentiation is considered one of the major limitations in using cell suspensions for secondary metabolite production (Su 1995, Mülbach 1998).

Some hormonal effects on the xanthone production in *calli* of *H. androsaemum* seem to be associated with growth of cells. The increase in NAA concentration, in the range of 4.5–25 µM, in the

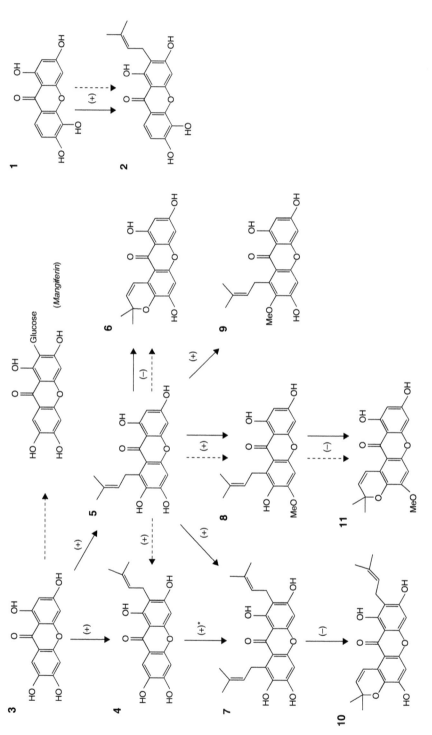

Figure 10.7 Proposed pathway of xanthone biosynthesis in *in vitro* cultures of *H. androsaemum* (- - - -) and *H. perforatum* based (——) on data published elsewhere (Dias *et al*. 2000, 2001). Significant (*P* < 0.05) positive (+) and negative (−) correlations between individual xanthones are indicated (+)*; positive correlation for 0.13 < *P* < 0.22. Compounds: (1) 1,3,5,6-tetrahydroxyxanthone; (2) 1,3,5,6-tetrahydroxy-2-prenylxanthone; (3) 1,3,6,7-tetrahydroxyxanthone; (4) 1,3,6,7-tetrahydroxy-2-prenylxanthone; (5) 1,3,6,7-tetrahydroxy-8-prenylxanthone; (6) toxyloxanthone B; (7) γ-mangostin; (8) 1,3,7-trihydroxy-6-methoxy-8-prenylxanthone; (9) 1,3,6-trihydroxy-7-methoxy-8-prenylxanthone; (10) garcinone B; (11) paxanthone.

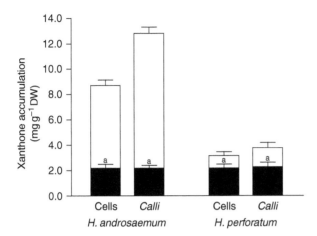

Figure 10.8 Specific xanthone accumulation in *calli* and suspended cells of *H. androsaemum* and of *H. perforatum* (□, 1,3,6,7 and ■, 1,3,5,6 oxygenation patterns). Cultures were maintained with MS medium supplemented with 4.5 μM of NAA plus 2.3 μM of KIN. Results are mean of six independent replicates, and vertical bars indicate 95% confidence to average. All the values are statistically different ($P < 0.05$) except those of xanthones with 1,3,5,6, oxygenation pattern signalised with the same letter.

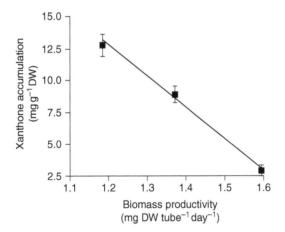

Figure 10.9 Relationship between specific xanthone accumulation and biomass productivity in *calli* of *H. androsaemum*. Cultures were maintained with MS medium supplemented with different molar concentrations of NAA, in the range of 4.5–22.5 μM, plus 2.3 μM of KIN. Results are mean of six independent replicates, and vertical bars indicate 95% confidence to average. The equation and the respective *r*-squared are as follows: $y = -2.4x + 41.7$; $r^2 = 0.999$; the regression model is very significant ($P < 0.001$).

presence of 2.3 μM of KIN, lead to an increase in biomass production and to a concomitant decrease in the total xanthone accumulation by *calli* (Dias *et al.* 2000). A tight negative linear relationship occurred between xanthone accumulation and biomass production by *calli* of *H. androsaemum* (Figure 10.9). The inverse relationship between secondary metabolism and growth is a common feature and sometimes the accumulation of particular compounds occurs only after cell growth stops (Su *et al.* 1995, Hagendoorn *et al.* 1997). The accumulation of xanthones in *H. perforatum calli* did not show any particular relation with biomass growth (Dias *et al.* 2001). Moreover, contrasting to that observed in *H. androsaemum*, the production of xanthones by *H. perforatum calli* was positively affected by the increase of NAA concentration, in the range studied.

Several hormonal combinations were experimented with *calli* and cell suspensions of *H. androsaemum* and *H. perforatum* (Dias *et al.* 2000, 2001). The highest production of xanthones in *H. androsaemum* (~1.3% DW) and *H. perforatum* (~0.7% DW) was observed in *calli* grown with the supplementation of 4.5 μM NAA plus 2.3 μM KIN and 4.5 μM 2,4-D plus 2.3 μM KIN, respectively. It was notorious that *H. androsaemum calli* and cell suspensions have the capability of accumulating higher levels of xanthones than the corresponding *in vitro* cultures of *H. perforatum* (Dias *et al.* 2000, 2001) (Figure 10.8). This fact was already pointed out by Schmidt and Beerhues (1997), they used cultured cells of *H. androsaemum* to study xanthone biosynthesis since they failed to get cultured cells of *H. perforatum* capable of accumulating these types of compounds.

Hypericum androsaemum suspended cells grown with the supplementation 4.5 μM NAA plus 2.3 μM KIN accumulated a lower xanthone amount (~0.9% DW) than that observed for *calli* maintained in similar conditons (Figure 10.8). *Calli* cultures are not a valid choice to produce metabolites at a commercial scale due to the lack of available technology and due to their low productivities. In this view, cell suspensions are the best option of choice. Nevertheless, higher contents in xanthones could be obtained with *H. androsaemum* cell suspensions (up to 2.2% DW) after yeast extract elicitation (Dias unpublished results). This content could be improved with further medium and elicitation process optimisation, which might render *H. androsaemum* suspended cells attractive for the commercial production of xanthones.

Conclusions and outlook

Both *H. androsaemum* and *H. perforatum in vitro* cultures have the potential to accumulate interesting pharmacological compounds. Namely, suspended cells of *H. androsaemum* have the potential to produce high amounts of xanthones. Moreover, the methanolic extract of suspended cells biomass of *H. androsaemum* proved to be a potent inhibitor of *Candida utillis* and *S. cerevisae* growth (Dias unpublished results). This could be related to the fact that *H. androsaemum* suspended cells accumulates mangostin, a powerful antifungic compound, and other related xanthones that might have similar activities (Figure 10.7). Therefore, these cells could be a good source of fungicide compounds and also a source of closely related xanthones that might allow correlation of structure and antifungic activity. Moreover, *H. androsaemum* and *H. perforatum* cell suspensions constitute a valuable tool to study xanthone biosynthesis and factors that control it.

Shoot cultures are the most interesting *in vitro* cultures of *H. perforatum*, in terms of production of valuable compounds, since they accumulate both hyperforins and hypericins. Moreover, ethanolic extracts of *H. perforatum* shoot cultures proved to have antiviral activity, *in vitro*, against African Swine Fever Virus (Dias *et al.* 1997). The activity of the ethanolic shoot cultures extracts was ten times more potent, in terms of biomass equivalents, than similar ethanolic extracts obtained from wild *H. perforatum* plants.

Unlike suspension cultures, shoot cultures are not a suitable system for the production of secondary metabolites at a commercial scale. Nevertheless, obtaining *H. perforatum* shoot cultures in liquid medium is possible and their phenolic composition is equivalent to those of shoot cultures maintained in solid medium with a similar hormonal composition (Dias and Ferreira 2000).

Recently, it was reported the invention of a new reactor, named *wave reactor*, that was claimed to be universal and specially indicated for the cultivation of organogenic cultures like shoot cultures (Lettenbauer *et al.* 1999). It might be possible that shoot cultures in liquid medium could be a potential alternative for large-scale production of medicinal plant phytochemicals. Nevertheless, in a personal view, it is unlikely that *H. perforatum* shoot cultures could be a viable substitute for plant biomass in the near future. However, *H. perforatum* shoot cultures constitute an important model in the study of biosynthesis of hypericins and hyperforins and the factors that control their production. Few details are known concerning the biochemical pathways involved in the production of both hypericins and hyperforins (Dewick 1997, Briskin 2000). Shoot cultures of *H. perforatum* could constitute *the pot of gold* to elucidate enzymes and precursors of these compounds.

At present, most medicinal plants have been studied mainly at the phytochemical and pharmacological level (Briskin 2000). Studies concerning plant physiology, biosynthesis of specific phytochemicals and the factors (biotic and abiotic) that regulate their production are limited. The use of *H. perforatum* and *H. androsaemum in vitro* cultures to elucidate these mechanisms will be of great importance to improve maximum growth of plants on the field with the best phytochemical composition. The utilisation of shoot cultures as a micropropagation system will allow the improvement of these species by selection of elite individuals and enable genetic engineering for improved medicinal content (Murch 2000a,c). Another approach, could be the utilization of *in vitro* cultures for the study of defense mechanisms of *H. perforatum* against *Colletotrichum gloeosporioides* (anthracnose), a fungus that is responsible for heavy losses in cultivated plants. Similar strategies have been done with success in other plants like mango (Jayasankar and Litz 1998) and beans (Bolwell *et al.* 1998).

References

Abd El-Mawla, A.M., Schmidt, W. and Beerhues, L. (2001) Cinnamic acid is a precursor of benzoic acids in cell cultures of *Hypericum androsaemum* L. but not in cell cultures of *Centaurium erythraea* RAFN. *Planta* 212, 288–93.

Alfermann, A.W. and Petersen, M. (1995) Natural product formation by plant cell biotechnology. *Plant Cell, Tissue and Organ Culture* 43, 199–205.

Anker, L., Gopalakrishna, R., Jones, K.D., Law, R.S. and Coudwell, W.T. (1995) Hypericin in adjuvant brain tumor therapy. *Drugs Future* 20, 511–17.

Beerhues, L. (1996) Benzophenone synthase from cultured cells of *Centaurium erythraea*. *FEBS Lett.* 383, 264–6.

Bolwell, G.P., Davies, D.R., Gerrish, C., Auh, C.K. and Murphy, T.M. (1998) Comparative biochemistry of the oxidative burst produced by rose and french bean cells reveals two distinct mechanisms. *Plant Physiol.* 116, 1379–85.

Bombardelli, E. and Morazzoni, P. (1995) *Hypericum perforatum*. *Fitoterapia* 66, 43–68.

Briskin, D.P. (2000) Medicinal plants and phytomedicines. Linking plant biochemistry and physiology to human health. *Plant Physiol.* 124, 507–14.

Brutovska, R., Cellarova, E. and Dolezel, J. (1998) Cytogenetic variability of *in vitro* regenerated *Hypericum perforatum* L. plants and their seed progenies. *Plant Sci.* 133, 221–9.

Brutovska, R., Cellarova, E., Davey, M.R., Power, J.B. and Lowe, K.C. (1994) Stimulation of multiple shoot regeneration from seedling leaves of *Hypericum perforatum* L. by pluronic F-68, *Acta Biotechnol.* 14, 347–53.

Büter, B., Orlacchio, C., Soldati, A. and Berger, K. (1998) Significance of genetic and environmental aspects in the field cultivation of *Hypericum perforatum*. *Planta Med.* 64, 431–7.

Cardoso, M.A. and Oliveira, D.E. (1996) Tissue culture of *Hypericum brasiliense* Choisy: shoot multiplication and callus induction. *Plant Cell, Tissue and Organ Culture* 44, 91–4.

Cellarova, E. and Kimakova, K. (1999) Morphoregulatory effect of plant growth regulators on *Hypericum perforatum* L. seedlings. *Acta Biotechnol.* 19, 163–9.

Cellarova, E., Brutovska, R., Daxnerova, Z., Brunakova, K. and Weigel, R.C. (1997) Correlation between hypericin content and the ploidy somaclones of *Hypericum perforatum* L.. *Acta Biotechnol.* 17, 83–90.

Cellarova, E., Daxnerova, Z., Kimakova, K. and Haluskova, J. (1994) The variability of the hypericin content in the regenerants of *Hypericum perforatum*. *Acta Biotechnol.* 14, 267–74.

Cellarova, E., Kimakova, K. and Brutovska, R. (1992) Multiple shoot formation and phenotypic changes of Ro regenerants in *Hypericum perforatum*. *Acta Biotechnol.* 12, 445–52.

Cellarova, E., Kimakova, K., Daxnerova, Z. and Martonfi, P. (1995) *Hypericum perforatum* (St. John's wort): *In vitro* culture and the production of hypericin and other secondary metabolites. In *Biotechnology in Agriculture and Forestry*, vol. 33, Y.P.S. Bajaj (Ed.), Springer-Verlag, Berlin, pp. 261–75.

Dewick, P.M. (1997). *Medicinal Natural Products.* Wiley, West Sussex, UK, pp. 56–60.

Di Carlo G., Borrelli, F., Ernst, E. and Izzo, A.A. (2001) St John's wort: Prozac from the plant kingdom. *Trends Pharmacol. Sci.* 22, 292–7.

Dias, A.C.P. and Ferreira, M.F. (2000) Influence of the medium composition on the accumulation of hypericins and hyperforins in shoot cultures of *Hypericum perforatum*. In *Natural Products Research in the New Millennium*, Abstracts book of the 48th GA meeting, P1D/13. September 3–7, Zurich.

Dias, A.C.P., Seabra, R.M., Andrade, P.B. and Ferreira, M.F. (1999) The development and evaluation of an HPLC-DAD method for the analysis of the phenolic fractions from in vivo and in vitro biomass of Hypericum species. *J. Liquid Chromatogr. Related Technol.*, 22, 215–27.

Dias, A.C.P., Seabra, R.M., Andrade, P.B., Ferreres, F. and Ferreira, M.F. (2001) Xanthone production in *calli* and suspended cells of *Hypericum perforatum*. *J. Plant Physiol.* 158, 821–7.

Dias, A.C.P., Seabra, R.M., Andrade, P.B., Ferreres, F. and Ferreira, M.F. (2000) Xanthone biosynthesis and accumulation in *calli* and suspended cells of *Hypericum androsaemum*. *Plant Sci.* 158, 93–101.

Dias, A.C.P., Sinogas, C., Antas, T. and Ferreira, M.F. (1997) Antiviral activity of ethanolic fractions of *in vitro* shoot cultures and *in vivo* plants of *Hypericum perforatum*. In *Abstracts Book of the International Symposium Bioassay Methods in Natural Product Research and Drug Development*. 24–27 August, Uppsala.

Dias, A.C.P., Tomás-Barberán, F., Ferreira, M.F. and Ferreres, F. (1998) Unusual flavonoids produced by *callus* of *Hypericum perforatum*. *Phytochemistry* 48, 1165–8.

Dixon, R.A. and Paiva, N.L. (1995) Stress-induced phenylpropanoid metabolism. *Plant Cell* 7, 1085–97.

Dixon, R.A. and Steele, C.L. (1999) Flavonoids and isoflavonoids – a gold mine for metabolic engineering. *Trends Plant Sci.* 4, 394–400.

Fischer, R., Emans, N., Schuster, F., Hellwig, S. and Drossard, J. (1999) Towards molecular farming in the future: using plant-cell-suspension cultures as bioreactors. *Biotechnol. Appl. Biochem.* 30, 109–12.

Greeson, M.J., Sanford, B. and Monti, D.A. (2001) St John's wort (*Hypericum perforatum*): a review of the current pharmacological, toxicological, and clinical literature. *Psychopharmacology* 153, 402–14.

Hagendoorn, M.J.M., Jamar, D.C.L., Meykamp, B. and van der Plas, L.H.W. (1997) Cell division versus secondary metabolite production in *Morinda citrifolia* cell suspensions. *J. Plant Physiol.* 150, 325–30.

Hostettmann, K. and Hostettmann, M. (1989) Xanthones. In *Methods in Plant Biochemistry, Plant Phenolics*, vol.1, J.B. Harborne (Ed.), Academic Press, New York, pp. 493–508.

Hostettmann, K., Marston, A. and Wolfender, J.-L. (1995) Strategy in the search for new biologically active plant constituents. In *Phytochemistry of Plants used in Traditional Medicine*, K. Hostettmann, A. Marstrom, M. Maillard and M. Hamburger (Eds), Clarendon Press, Oxford, pp.17–45.

Hshiguro, K., Fukumoto, H., Nakajima, M. and Isoi, K. (1993) Xanthones in cell suspension cultures of *Hypericum patulum*. *Phytochemistry* 33, 839–40.

Hshiguro, K., Fukumoto, H., Suitani, A., Nakajima, M. and Isoi, K. (1996) Prenylated xanthones from cell suspension cultures of *Hypericum patulum*. *Phytochemistry* 42, 435–7.

Hshiguro, K., Nakajima, M., Fukumoto, H. and Isoi, K. (1995a) Co-occurence of prenylated xanthones and their cyclization products in cell suspension cultures of *Hypericum patulum*. *Phytochemistry* 38, 867–9.

Hshiguro, K., Nakajima, M., Fukumoto, H. and Isoi, K. (1995b) A xanthone substituted with an irregular monoterpene in cell suspension cultures of *Hypericum patulum*. *Phytochemistry* 39, 903–6.

Jayasankar, S. and Litz, R.E. (1998) Characterization of embryogenic mango cultures selected for resistance to *Colletotrichum gloeosporioides* culture filtrate and phytotoxin. *Theoret. Appl. Genetics* 96, 823–31.

Kartnig, T. and Brantner, A. (1990) Secondary constituents in cell cultures of *Hypericum perforatum* and *Hypericum maculatum. Planta Med.* 56, 634.

Kartnig, T., Göbel, I. and Heydel, B. (1996) Production of hypericin, pseudohypericin and flavonoids in cell cultures of various *Hypericum* species and their chemotypes. *Planta Med.* 62, 51–3.

Kirakosyan, A., Hayashi, H., Inoue, K., Charchoglyan, A. and Vardapetyan, H. (2000a) Stimulation of the production of hypericins by mannan in *Hypericum perforatum* shoot cultures. *Phytochemistry* 53, 345–8.

Kirakosyan, A., Vardapetyan, R.R. and Charchoglyan, A. (2000b) The content of hypericin and pseudohypericin in cell cultures of *Hypericum perforatum. Russ. J. Plant Physiol.* 47, 270–3.

Kirakosyan, A., Vardapetyan, R.R., Charchoglyan, A.G., Yamamoto, H., Hayashi, H. and Inoue, K. (2001) The effect of cork pieces on pseudohypericin production in cells of *Hypericum perforatum* shoots. *Russ. J. Plant Physiol.* 48, 815–8.

Kurth, H. and Spreemann, R. (1998) Phytochemical characterization of various St. John's wort extracts. *Adv. Therapy* 15, 117–28.

Lettenbauer, C., Eibl, R., Mallol, A., Oksman-Caldentey, K.M. and Eibl, D. (1999) The wave reactor – a universal reactor for mass propagation of hairy root cultures. In *EURESCO Conferences: Plant Cell Biology and Biotechnology Applications: Signal Recognition, Transduction Mechanisms and Gene Regulation*, Rolduc, The Netherlands.

Matsuki, M. (1996) Regulation of plant phenolic synthesis: from biochemistry to ecology and evolution. *Aust. J. Bot.* 44, 613–34.

Mederos, S., Andrés, S.L. and Luis, J.G. (1997) Rosmanol controls explants browning of *Hypericum canariensis* L. during the *in vitro* establishment of shoots. *Acta Soc. Bot. Pol.* 66, 347–9.

Mederos, S., Lopez-Bazzocchi, I., Ravelo, A. and Gonzalez, A. (1996) Hypericin from clonal propagation of *Hypericum canariensis* L. *Plant Tissue Culture* 6, 7–13.

Meruelo, D., Lavie, G. and Lavie, D. (1988) Therapereutic agents with dramatic antiretroviral activity and little toxicity at effective doses: aromatic polycyclic diones hypericin and pseudohypericin. *Procedings of the Natural Academy of Sciences USA* 85, 5230–4.

Mosen, W., Zdunek, K., Woerdenbag, H.J., Pras, N. and Alfermann, A.W. (1993) Production of natural products by shoot organ cultures cultivated in vitro. *Pharmaceutical World-Sci.* 15, Suppl. H10.

Mühlbach, H.P. (1998) Use of plant cell cultures in biotechnology. In *Biotechnology Annual Review*, M.R. El-Gewely (Ed.), Elsevier Science, Amsterdam, pp. 113–76.

Murch, S.J., Choffe, K.L., Victor, J.M.R., Slimmon, T.Y., KrishnaRaj, S. and Saxena, P.K. (2000a) Thidiazuron-induced plant regeneration from hypocotyl cultures of St. John's wort (*Hypericum perforatum* L. cv "Anthos"). *Plant Cell Rep.* 19, 576–81.

Murch, S.J., KrishnaRaj, S. and Saxena, P.K. (2000b) Tryptophan is a precusor for melatonin and serotonin biosynthesis in *in vitro* regenerated St. John's wort (*Hypericum perforatum* L. cv "Anthos"). *Plant Cell Rep.* 19, 698–704.

Murch, S.J., KrishnaRaj, S. and Saxena, P.K. (2000c) Phytopharmaceuticals: mass-production, standardization, and conservation. *Sci. Rev. Altern. Med.* 4, 39–43.

Murch, S.J., KrishnaRaj, S. and Saxena, P.K. (2000d) Phytopharmaceuticals: problems, limitations, and solutions. *Sci. Rev. Altern. Med.* 4, 33–7.

Nahrstedt, A. and Butterweck V. (1997) Biologically active and other chemical constituents of the herb of *Hypericum perforatum* L.. *Pharmacopsychiatry* 30 (suppl.), 129–34.

Peters, S., Schmidt, W. and Beerhues, L. (1998) Regioselective oxidative phenol couplings of 2,3',4,6-tetrahydrosybenzophenone in cell cultures of *Centaurium erythraea* RAFN and *Hypericum androsaemum* L. *Planta* 204, 64–9.

Pretto, F.R. and Santarém, E.R. (2000) *Callus* formation and plant regeneration from *Hypericum perforatum. Plant Cell, Tissue and Organ Culture* 62, 107–13.

Rani, N.S., Balaji, K. and Ciddi, V. (2001) Production of hypericin from tissue culture of *Hypericum perforatum. Indian J. Pharmaceut. Sci.* 63, 431–2.

Rocha, L., Marston, A., Kaplan, M.A.C., Stoeckli-Evans, H., Thull, U. and Testa, B. (1994) An antifungal γ-pyrone and xanthones with monoamine oxidase inhibitory activity from *Hypericum brasiliense*. *Phytochemistry* 36, 1381–5.

Schmidt, W., Abd El-Mawla, A.M.A., Wolfender, J.L., Hostettmann, K. and Beerhues, L. (2000) Xanthones in cell cultures of *Hypericum androsaemum*. *Planta Med.* 66, 380–2.

Schmidt, W. and Beerhues, L. (1997) Alternative pathways of xanthone biosynthesis in cell cultures of *Hypericum androsaemum* L.. *FEBS Lett.* 420, 143–6.

Schröder, J. (1997) A family of plant-specific polyketide synthase: facts and predictions. *Trends Plant Sci.* 2, 373–8.

Seabra, R.M. and Vasconcelos, M.H. (1992) Análise de amostras de hipericão existentes no mercado. *Revista Portuguesa de Farmácia* 17, 15–20.

Stöckigt, J., Obitz, P., Falkenhagen, H., Lutterbach, R. and Endreß, S. (1995) Natural products and enzymes from plant cell cultures. *Plant Cell, Tissue and Organ Culture* 43, 97–109.

Su, W.W. (1995) Bioprocessing technology for plant cell suspension cultures. *Appl. Biochem. Biotechnol.* 50, 189–229.

Wills, R.B.H, Bone, K. and Morgan, M. (2000) Herbal products: active constituents, modes of action and quality control. *Nutr. Res. Rev.* 13, 47–77.

Yazaki, K. and Okuda, T. (1990) Procyanidins in callus and multiple shoot cultures of *Hypericum erectum*. *Planta Med.* 56, 490–1.

Yazaki, K. and Okuda, T. (1994) *Hypericum erectum* Thunb. (St. John's wort): in vitro culture and the production of procyanidins. In *Biotechnology in Agriculture and Forestry*, vol. 26, Y.P.S. Bajaj (Ed.), Springer-Verlag, Berlin, pp. 167–78.

Zdunek, K. and Alfermann, A.W. (1992) Initiation of shoot organ cultures of *Hypericum perforatum* and formation of hypericin derivatives. *Planta Med.* 58, A621–2.

Zenk, M.H. (1991) Chasing the enzymes of secondary metabolism: plant cell cultures as a pot of gold. *Phytochemistry* 30, 3861–3.

11 The clinical pharmacology of *Hypericum perforatum*

E. Ernst and A.A. Izzo

Preclinical pharmacology

Mechanism of action

The antidepressant effect of *Hypericum perforatum* (St John's wort) extracts is well documented in animal studies. Inhibition of monoamine oxidase (MAO) by hypericin was believed to be the primary mode of action of the antidepressant effect of St John's wort. However, this initial assumption has not been confirmed in several subsequent studies. Indeed, *Hypericum* extracts are only weak inhibitors of MAO-A and MAO-B activity, and the extract constituents, including hypericin, exert MAO inhibition only at concentrations ($EC_{50} > 10\ \mu M$) higher than those found clinically (Di Carlo *et al.* 2001, Nathan 1999, Vitiello 1999, Barnes *et al.* 2001).

Studies using rat striatum synaptosomal preparation, intact rat astrocytes or microdialysis in the rat nucleus accumbens and striatum suggest that *Hypericum* extracts inhibit the uptake of several brain neurotransmitters and that the phloroglucinol derivative is one of the major constituents required for antidepressant activity (Muller *et al.* 1998, Neary and Bu 1999, Di Matteo *et al.* 2000). St John's wort (hyperforin) possesses a pharmacological profile which is unique amongst all other antidepressants: it inhibits the uptake of 5-HT, noradrenaline and dopamine with the same potency and it also inhibits the uptake of glutamate and GABA. By inhibiting the major route of terminating the action of neurotransmitters, hyperforin produces an increase of neurotransmitters into the synaptic cleft. The reuptake action of hyperforin is not associated with specific binding to the different transporter molecules on presynaptic nerves, but with a completely novel mechanism related to sodium conductive pathways (Singer *et al.* 1999, Wonneman *et al.* 2000).

At a receptor level, chronic treatment with *Hypericum* extracts down {4970}regulates β_1-adrenoceptors and upregulates post-synaptic 5-HT$_{1A}$ receptors (Teufel-Mayer 1997). However, in contrast to most antidepressants, which downregulate 5-HT$_{2A}$ receptors after chronic administration, *Hypericum* extract upregulates 5-HT$_{2A}$ receptors (Nathan 1999).

Clinical pharmacology

Studies conducted with *Hypericum* extract fall into six main categories:

- Pharmacokinetic studies
- Pharmacodynamic studies
- Clinical studies on the antidepressant efficacy
- Clinical studies investigating the possible phototoxic effect
- Clinical studies on herb/drug interactions, and
- Case reports of toxic effects, including herb/drug interactions.

Pharmacokinetics

Hypericum contains a number of active constituents: in addition to flavonoids (quercetin, hyperoside and rutin), tanning agents and procyanidines, *Hypericum* extracts also contain biapigenin and the phloroglucine derivative hyperforin, which is now believed to be the main active ingredient. Levels of hyperforin, pseudohypericin and hypericin have been measured in four pharmacokinetic studies (Staffeldt 1994, Kerb 1996, Biber 1998, Agrosi *et al.* 2000). Key data of two of these studies (Kerb 1996, Biber 1998) are summarised in Tables 11.1 and 11.2.

Plasma levels of hyperforin have been studied in six healthy volunteers after oral administration of film coated tablets containing 300 mg of hypericum extract containing 14.8 mg hyperforin (Biber 1998). Absorption of hyperforin takes place after a lag time of approx 1 h; the maximum

Table 11.1 Pharmacokinetic parameters of hypericin and pseudohypericin determined in the single-dose study after administration of 300, 600 or 1200 mg of hypericum extract (LI160) in healthy volunteers ($n = 12$). Data from Kerb *et al.* 1996

Parameter	Hypericin			Pseudohypericin		
	300 mg LI 160	900 mg LI 160	1800 mg LI 160	300 mg LI 160	900 mg LI 160	1800 mg LI 160
t_{lag} (h)	2.1	1.9	1.9	0.5	0.4	0.4
C_{max} (μg/l)	1.3	7.2	16.6	3.4	12.1	29.7
t_{max} (h)	5.5	6.0	5.7	3.0	3.0	3.0
C_{mean} (μg/l)	0.9	4.4	9.4	1.9	5.8	14.0
AUC (h μg/l^{-1})	41.4	198	494	45	140	285
$T_{1/2ka}$ (h)	1.4	0.6	0.8	0.9	1.3	1.4
$T_{1/2\alpha}$ (h)	1.9	6	5.8	1.2	1.4	1.5
$T_{1/2\beta}$ (h)	24.5	43.1	48.2	18.2	24.8	19.5
Cl/F (ml/min)	101	63.3	51.0	195	188	185
V_{ss}/F (l)	111	69.6	73.3	117	61	50.0

Notes
t_{lag} = lag time of absorption; C_{max} = maximum plasma concentration; t_{max} = time to achieve C_{max}; C_{mean} = mean concentration in plasma (AUC$_{0-12}$/12); AUC = area under the curve; $T_{1/2ka}$ = Median half-life of absorption; $T_{1/2\alpha}$ = Median half-life of distribution; $T_{1/2\beta}$ = Median half-life of excretion; Cl/F = clearance/systemic availability; V_{ss}/F = volume of distribution at steady state/systemic availability.

Table 11.2 Pharmacokinetic parameters of hyperforin after oral administration of 300, 600 and 1200 mg *Hypericum* extract W5572 (5% hyperforin) to six volunteers. Data from Biber 1998

Parameter	300 mg WS5572	600 mg WS5572	900 mg WS5572
C_{max} (μg/l)	153.15	301.8	437.3
T_{max} (h)	3.58	3.5	2.83
AUC (μgh/l)	1335.9	2214.6	3377.9
$T_{1/2\alpha}$ (h)	3.14	2.55	2.52
$T_{1/2\beta}$ (h)	9.46	8.52	9.65
MRT (h)	12.12	11	12.57
Cl (ml/min)	199.3	238.2	340.3

Notes
C_{max} = maximum plasma concentration; T_{max} = time to achieve C_{max}; AUC = area under the curve; $T_{1/2\alpha}$ = Median half-life of distribution; $T_{1/2\beta}$ = Median half-life of excretion; MRT = Mean residence time; Cl = clearance (dose/AUC).

plasma levels of approximately 150 ng/ml (280 nM) are reached 3.5 h after administration. Hyperforin pharmacokinetics were linear up to 600 mg of the extract. Increasing the doses to 90 or 1200 mg of extract resulted in lower C_{max} and AUC values than those expected from linear extrapolation of data from lower doses. Plasma concentration curves in volunteers fitted well in an open two-compartment model. The elimination half-life and retention time in plasma are long (9 and 12 h, respectively).

In a repeated dose study, no accumulation of hyperforin in plasma was observed. Using the observed AUC values from the repeated dose study, the estimated steady-state plasma concentrations of hyperforin after 3 × 300 mg/day of the extract (i.e. after normal therapeutic dose regimen) is approximately 100 ng/ml (180 nM) with no remarkable fluctuation in plasma levels. Hyperforin does not accumulate in the circulation even after administration of 900 mg of extract given once daily for 8 days (Biber 1998).

Staffeldt *et al.* studied in 12 healthy male subjects the pharmacokinetics of hypericin (H) and pseudohypericin (PH) after administration of St John's wort (LI 160, 300, 900 or 1800 mg corresponding to 250, 750 or 1500 µg hypericin and 526, 1578 or 3156 pseudohypericin) (Staffeldt 1994). The median maximal plasma levels were 1.5, 4.1, and 14.2 µg/l for H and 2.7, 11.7, and 30.6 µg/l for PH, respectively, for the three doses given above. The median elimination half-life times of H were 24.8–26.5 h, and varied for PH from 16.3 to 36.0 h. Ranging between 2.0 and 2.6 h, the median lag-time of absorption was remarkably prolonged for H when compared to PH (0.3–1.1 h). The areas under the curves (AUC) showed a nonlinear increase with raising dose; during long-term dosing (3 × 300 mg/day), a steady-state was reached after 4 days. Mean maximal plasma level during the steady-state treatment was 8.5 µg/l for H and 5.8 µg/l for PH, while mean trough levels were 2.3 µg/l for H and 3.7 µg/l for PH. In spite of their structural similarities there are substantial pharmacokinetic differences between H and PH.

Single-dose and steady-state pharmacokinetics of H and PH were studied in 13 healthy volunteers by administration of St John's wort (LI 160 300, 900 or 1800 mg) (Kerb 1996). Oral administration resulted in a median peak levels in plasma (C_{max}) of 1.3, 7.2 and 16.6 µg/l for H and 3.4, 12.1 and 29.7 µg/ml for PH, respectively. The C_{max} and the area under the curve values for the lowest dose were disproportionately lower than those for the higher doses. A lag time of 1.9 h for H was remarkably longer than the 0.4-h lag time for PH. Median half-lives for absorption, distribution and elimination after *Hypericum* 900 mg were 0.6, 60 and 43.1 h for H and 1.3, 1.4 and 24.8 h for PH. Fourteen-day treatment with 300 mg *Hypericum* three times a day resulted in median steady-state through level of 7.9 µg/l for H and 4.8 µg/l for PH after 7 and 4 days, respectively.

Kinetic parameters after intravenous administration of *Hypericum* extract in two subjects corresponded to those estimated after oral dosage. Both H and PH were initially distributed into a central volume of 4.2 and 5.0 l, respectively. The mean distribution volumes at steady-state were 19.7 l for H and 39.3 l for PH, and the mean total clearance rates were 9.2 ml/min for H and 43.3 ml/min for PH. The systemic availability of H and PH from LI 160 was roughly estimated to be 14 and 21%, respectively (Kerb 1996).

The oral absorption of both hyperforin and hypericin after oral administration of soft (or hard) gelatin capsules (containing *Hypericum* alcoholic extract 300 mg, 5% hyperforin and 0.3% hypericin) was studied in an open, single dose, two-way, randomised, cross-over study involving 12 healthy subjects (Agrosi *et al.* 2000). Peak plasma concentration (C_{max}) of hyperforin was 168 µg/l for the soft gelatin formulation and 84.25 for the hard gelatin capsule. The T_{max} values for hyperforin were 2.5 h for the soft gelatin compared to 3.08 h for the hard formulation, whereas the total AUC was respectively 1483 h µg/l and 584 h µg/l. Bioassay of hypericin confirmed the higher absorption of soft gelatin capsules compared with the hard gelatin capsule.

Pharmacodynamics

The therapeutic effects of *Hypericum* extracts for the treatment of mild to moderate depression cannot, of course, be demonstrated in healthy subjects. However, some understanding of the mechanism of action of the active ingredient has been gained from studies using quantitative EEG analysis and measurements of cortical-evoked potentials to model effects on cognitive function and sleep quality. Five such studies have been published (Johnson *et al.* 1992, Johnson *et al.* 1994, Schulz *et al.* 1994, Schellenberg *et al.* 1998, Dimpfel *et al.* 1999).

In a pilot study with a cross-over design including 12 volunteers (Johnson *et al.* 1992), the effects of *Hypericum* extract at 3×300 mg daily were compared to those of placebo on a range of parameters: (i) quantitative EEG (resting and under conditions of increased attentiveness); (ii) visual evoked potentials (VEPs) and auditory evoked potentials (AEPs); (iii) objective psychometric tests (Brickenkamp's d_2 test for alertness and the Mackworth Clock test for attentiveness); (iv) subjective assessment (von Zerssen's Scale). The effects of *Hypericum* on the EEG profile were broadly similar to those of other mood-elevating treatments. They consisted of reduced alpha-activity and correspondingly increased beta- (particularly $beta_1$-) and theta-activities. These changes are indicative of a tension-relieving action and of increased alertness. This conclusion is also supported by the reductions in latent times observed for both VEPs and AEPs under active treatment. In addition, significantly improved performance in the Mackworth Clock test was observed in the active treatment group.

The second study compared *Hypericum* extract at the same dose level as before with a standard mood-elevating therapy (the tricyclic antidepressant maprotiline) in 24 volunteers (Johnson *et al.* 1994). The effects of treatment were evaluated using quantitative analysis of resting EEG and measurement of VEPs and AEPs, as in the first study. Treatment with *Hypericum* extract was again shown to result in enhanced activation in the beta- ($beta_2$-) and theta-regions of the EEG, further supporting its action as a stress- or tension-relieving but not as a sedative agent. Except for opposing effects on theta-activity, the resting EEG profile was similar with *Hypericum* compared to maprotiline. As in the pilot study, considerable reductions in the latent times for both VEPs and AEPs were observed under treatment with *Hypericum*, providing further support for a positive effect on attentiveness.

The third study investigated the effects of 4 weeks' therapy with *Hypericum* extract (3×300 mg daily) vs placebo on sleep in older volunteers using a cross-over design (Schulz *et al.* 1994). Evaluation was by continuous recording of EEG, electro-oculogram (EOG), electromyogram (EMG), ECG and nasal respiration during 8-h sleep periods in the laboratory. The results showed no significant effect of treatment on sleep onset, total sleep duration or sleep efficiency. An increase in the proportion of time spent in deep sleep (non-REM stages 3 and 4 or slow wave sleep) was, however, noted under active treatment, together with a slight reduction in REM sleep latency. This pattern of action is distinct from that of synthetic treatments for low mood disorders, such as tricyclic antidepressants and monoamine oxidase inhibitors, which characteristically cause a prolongation in REM latency and a suppression of REM sleep. The slight decrease in total sleep time observed in this study under active treatment suggests an overall stimulatory rather than a sedative effect.

A double-blind randomised, placebo-controlled parallel-group trial was performed to evaluate the central pharmacodynamic effects of two *Hypericum* extracts with different contents of hyperforin (0.5 and 5.0%) but identical hypericin content (Schellenberg *et al.* 1998). The quantitative EEG results of the placebo group on days 1 and 8 showed no significant changes with regard to their physiological daily rhythm. In both verum groups (0.5% and 5.0% hyperforin content), reproducible central pharmacodynamic effects were apparent in comparison to placebo, in particular with the extract containing 5.0% hyperforin. A peak pharmacodynamic efficacy was observed between 4 and 8 h post administration. The extract containing 5.0% hyperforin showed a marked tendency to produce higher increases in EEG baseline power performances than the one containing 0.5% hyperforin. This suggests that extracts rich in hyperforin have a shielding effect on the central nervous system.

In a single blind study, Dimpfel *et al.* compared the effects of two commercial extracts of St John's wort (Texx 300 and Jarsin 300) with those of placebo in 35 healthy young volunteers (Dimpfel *et al.* 1999). Measurements were carried out on the first day and after repetitive drug (t.i.d) over 3 weeks. Texx 300 produced an increase of α_1 (serotoninergic, 8% increase), α_2 (dopaminergic, 32% increase) and β (14% increase) frequencies in comparison with placebo. Jarsin 300 produced less increase within the α_2 frequencies. In addition, β increases were observed after Jarsin 300 following repetitive administration. Slight differences could be detected between the two preparations with respect to their maximum effects, but in general the profile of action was similar for both drugs becoming more clear after the repetitive dosing. However, the effects of Texx 300 occurred a little bit earlier possibly due to a better intestinal absorption.

Clinical studies on the antidepressant efficacy

The effectiveness of St John's wort in depression has been investigated in eleven (Ernst 1995, Linde *et al.* 1996, Volz 1997, Friede and Wustenberg 1998, Mulrow *et al.* 1998, Kim 1999, Stevinson 1999, Gaster 2000, Linde 2000, Volz and Laux 2000, Williams 2000) systematic reviews/meta-analyses (total number of trials covered 29; see Table 11.3). Mainly due to slight differences in the inclusion criteria (e.g. restriction to trials with a minimum of 6 weeks observation or with a minimum quality score) the respective study collections differed to a considerable amount (Linde *et al.* 2001). However, the conclusions were very similar. St John's wort has been shown to be superior to placebo in mild to moderate depression. There is growing evidence that St John's wort is as effective as other antidepressants for mild to moderate depression and causes fewer adverse effects but further trials are still needed to establish long-term effectiveness and safety.

Clinical trials investigating the possible phototoxic effect

Increased sensitivity to sunlight associated with the naphthodianthrone derivatives hypericin and pseudohypericin is a potential risk of *Hypericum* extracts. The phenomenon of phototoxicity following the ingestion of large amounts of St John's wort has been reported in grazing animals. Hypericin is believed to be the photosensitising agent present in St John's wort (Vandenboagaerde *et al.* 1998). Hypericin induces phototoxic effects in cultures of human keratinocytes (Bernd *et al.* 1999) and, in the presence of light, induces changes in calf lens proteins that theoretically could lead to the formation of cataracts (Schey 2000). However, the blood concentrations obtained during antidepressive therapy with St John's wort are too low to cause phototoxic reactions. It has been estimated that it would require a dose of St John's wort 30–50 times greater than the recommended daily dose to lead to severe phototoxic reactions in humans, and that by shielding the patient from ultraviolet light for a week, serious complications would be avoided (Di Carlo *et al.* 2001).

The possible clinical effects of treatment with *Hypericum* extracts (or hypericin) on sensitivity to ultraviolet light were investigated in five studies (Brockmoller *et al.* 1997, Schempp *et al.* 1999, Gulick *et al.* 2000, Schempp *et al.* 2000, Barnes 2001). Overall, no or mild phototoxic

Table 11.3 Systematic reviews/meta-analyses of clinical trials of St John's wort. Data from Linde 2001

Author/year	Indication	Comparisons	Studies	Features[a]	Results	Author's conclusion
Ernst (1995)	Depression	Placebo and antidepressants	11 RCT	y/y/y/y/n	Most of 8 placebo-controlled trials positive. Three trials against standard medication with similar effects	St John's wort is superior to placebo and seems equally effective as standard medication
Volz (1997)	Depression	Placebo and antidepressants	11 RCT/CCT	p/p/n/n/n	Most placebo-controlled trials positive; similarly effective as (not adequately dosed) antidepressants	A therapy using St John's wort for mild and moderate depression can be attempted. Further studies needed
Linde et al. (1996) and Linde (2000)	Depression	Placebo and antidepressants	27RCT	y/y/y/y/y	Treatment response: RR 2.47 (95% CI 1.69–3.61) vs placebo and 1.01 (0.87–1.16) vs antidepressants	St John's wort more effective than placebo. Inadequate evidence to assess equivalence with antidepressants
Friede (1998)	Anxiety in depressed patients	Placebo, amitriptyline	8 RCT	?/y/y/y/n	Trials collectively show reduction of anxiety symptoms over placebo. Only 1 trial vs amitriptyline	St John's wort is effective for depressed patients with anxiety
Kim (1999)	Depression	Placebo and antidepressants	6 RCT	p/y/y/y/y	Treatment response: RR 1.48 (95% CI 1.2–2.8) vs placebo and 0.98 (0.67–1.28) vs antidepressants	St John's wort more effective than placebo and similarly effective as low dose antidepressant; quality problems

Study	Condition	Comparison	Number of trials	Features[a]	Results	Conclusion
Stevinson (1999)	Depression	Placebo and antidepressants	6 RCT	y/y/y/y/n	Only trials published after Linde 1996; trials show effects better than placebo/ similar to antidepressants	Data confirm findings of earlier trials, but still insufficient evidence to assess equivalence with antidepressants
Mulrow et al. (1998) and Williams (2000)	Depression	Placebo and antidepressants	14 RCT	y/y/n/y/y	Treatment response: RR 1.9 (95% CI 1.2–2.8) vs placebo and 1.2 (1.0–1.4) vs antidepressants	Data suggest that St John's wort is superior to placebo, insufficient evidence regarding equivalence with antidepressants
Gaster (2000)	Depression	Placebo and antidepressants	8 RCT	p/y/p/y/n	4 placebo-controlled trials with positive results, in 4 trials	Data suggest that St John's wort is superior to placebo, insufficient evidence re equivalence with antidepressants
Volz (2000)	Mild to moderate depression	Fluoxetine	17 + 9 CCT	n/y/n/n/y	No direct comparison of Hypericum and fluoxetine available. Mean depression score (HAMD) reduction in hypericum trials 53%, in fluoxetine trials 55%	Response rates are similar; findings difficult to interpret because of the indirect comparison

Notes
a Features: 1 = comprehensive search, 2 = explicit inclusion criteria, 3 = formal quality assessment, 4 = summary of results for each included study, 5 = meta-analysis; y = yes, n = no, p = partly, ? = unclear review on all pharmacologic treatments for the respective condition.
RCT = randomised controlled trials.
CCT = non-randomised controlled trials.

effects were observed after *Hypericum* extract; however, treatment with hypericin resulted in moderate or severe phototoxicity.

A placebo-controlled randomised clinical trial with monitoring of hypericin and pseudohypericin plasma concentration was performed to evaluate the increase in dermal photosensitivity in humans after application of high dose of *Hypericum* extract. The study was divided into a single dose and a multiple dose part (Brockmoller *et al.* 1997). In the single dose study (placebo-controlled double-blind crossover design) 13 healthy male volunteers of various skin types received placebo or a single dose of *Hypericum* extract of either 900, 1800 or 3600 mg at day zero. Individual doses were separated by a washout phase of at least 14 days. The results of the study showed no significant effects of treatment at any dose level on the sensitivity of subjects overall either to UVA or UVB. In the multiple dose study (open design) 50 healthy volunteers received *Hypericum* at a dose of 1800 mg daily for 15 days. Increased (approx 20%) sensitivity to UVA at day 15 was recorded. This effect was significantly greater (approximately 30% reduction) in subjects classified as having light-sensitive skin than in those of less sensitive skin type (approximately 10% reduction). The corresponding investigation of effects following irradiation with UVB showed a slight (not significant) increased sensitivity (Brockmoller *et al.* 1997).

In another single-dose study (Schempp 1999), administration of St John's wort 1800 mg to 12 healthy volunteers resulted in a mean serum total hypericin concentration of 43 ng/ml and a mean skin blister fluid concentration of 5.3 ng/ml. After administration of St John's wort 300 mg three-times daily for 7 days to achieve steady-state concentrations, the mean serum total hypericin concentration was 12.5 ng/ml and the mean skin blister fluid concentration was 2.8 ng/ml; these concentrations are below those estimated to be phototoxic (>100 ng/ml).

A study reported that HIV-positive patients treated with oral hypericin 0.05 mg/kg for 28 days developed mild symptoms of photosensitivity on exposure to sunlight, and that two patients developed intolerable symptoms of photosensitivity when the dose was increased to 0.16 mg/kg (Barnes 2001).

In a dose-escalating study involving 30 HIV-infected patients treated with oral (0.5 mg/kg daily) or intravenous hypericin (starting dosage: 0.25 mg kg twice or three times weekly), 16 patients discontinued treatment before completing 8 weeks of therapy because of moderate or severe phototoxicity; severe cutaneous phototoxicity was observed in 11 of 23 evaluable patients (Gulick *et al.* 2000).

In a study investigating the photosensitising capacity of topical St John's wort, volunteers ($n = 8$ for each preparation) applied *Hypericum* oil (containing hypericin 110 μg/ml) or *Hypericum* ointment (containing hypericin 30 μg/ml) to their forearms before exposure to solar-simulated radiation. Visual assessment detected no change in erythema after application of either preparation, although evaluation of skin erythema using a more sensitive photometric measurement revealed an increase in erythema index after treatment with *Hypericum* oil (Schempp *et al.* 2000).

Clinical studies on herb/drug interactions

Strong support for the hepatic enzyme-inducing properties of St John's wort comes from recent preclinical studies. Although not all investigations yielded the same results, most agreed that *Hypericum* extracts activate enzymes of the CYP P450, namely CYP3A4, which is possibly the most important of the CYP P450 family of hepatic enzymes involved in the metabolism of many common drugs (Durr *et al.* 2000, Di Carlo *et al.* 2001). Other experimental evidence indicates that St John's wort can induce intestinal P-glycoprotein (which is involved in the elimination of drugs) in rats (Durr *et al.* 2000).

Five clinical studies have investigated the enzyme-inducing properties of St John's wort using either internal (6β-hydroxyxostisol/cortisol ratio) or external probe stubstrates (dextromethorphan, alprazolam, caffeine) (Kerb 1997, Ereshefsky 2000, Gewertz 2000, Markowitz 2000, Roby 2000). Although a clinical study did not yield the same results, four clinical studies showed an increase or a trend to increase the metabolic capacity of cytochrome enzyme (Table 11.4).

In another clinical study, the administration of St John's wort extract (300 mg three-times daily) to 8 healthy male volunteers during 14 days resulted in a 1.4-fold increased expression of duodenal P-glycoprotein, supporting the importance of intestinal P-glycoprotein (in addition to hepatic CYP3A4) for St John's wort/drug interactions (Durr *et al.* 2000).

Other clinical trials (Herberg 1994, Donath *et al.* 1999, Johne *et al.* 1999, Piscitelli 2000, Roots *et al.* 2000, Schmidt *et al.* 2000) have shown that St John's wort lowered the plasma level of the cardioactive drug digoxin, the anticoagulant drug phenprocoumon and the protease inhibitor (anti HIV) indinavir, but it did not have negative additive effects on alcohol-induced changes in vigilance (Table 11.5).

Case reports

Case reports of adverse events

There are several case reports which have highlighted the possibility of mania-induction after taking St John's wort. None of these cases involved rechallenge with St John's wort and in all cases there were other pharmacological factors and/or underlying illness that could have been responsible for or contributed to the precipitation of mania.

A case of hypomania was reported in a woman with panic disorders and unipolar major depression who had discontinued sertraline treatment one week before starting St John's wort tincture (Schneck 1998). O'Breasil and Argouarch reported two cases of hypomania in individuals with no history of bipolar disorders. A man who had received electroconvulsive therapy and who had previously taken various antidepressants, experienced a hypomanic episode 6 weeks after starting St John's wort (dosage not stated). Another man with symptoms of post-traumatic stress disorder was diagnosed with an acute manic episode after 3 months of self-treatment with St John's wort (dosage not stated) (O'Breasil and Argouarch 1998). Nierenberg *et al.* reported two cases of mania in patient with bipolar depression who began self-treatment with standardised St John's wort extract 9000 mg/day (Nierenberg 1999).

Finally, a detailed case report highlighted the possibility of neuropathy related to the possible phototoxic effects of St John's wort. This is probably caused by demyelinisation of cutaneous axons by photoactivated hypericin. A woman developed stinging pains in areas exposed to the sun (face and hands) 4 weeks after starting treatment with St John's wort 500 mg/day (extract and hypericin content not stated); her symptoms improved 3 weeks after stopping St John's wort and disappeared over the next 2 months (Bove 1998).

Case reports of herb–drug interaction

Multiple case reports (Rey 1998, Gordon 1998, Bon 1999, Khawaja 1999, Lanz *et al.* 1999, Nebel *et al.* 1999, Barbanel 2000, Breidenbach 2000, Breidenbach 2000, Roots *et al.* 2000, Ruschitzka *et al.* 2000, Yue 2000, Izzo and Ernst 2001) of drug interactions with St John's wort have been reported (Table 11.6). Case reports or case series indicate that St John's wort lowered plasma levels of cyclosporin (with rejection episodes), theophilline, warfarin (with decreased

Table 11.4 Clinical studies dealing with the effect of St John's wort on hepatic cytochrome enzymes. Data from Izzo and Ernst (2001)

St John's wort dosage and duration of the treatment	Probe dosage and duration of the treatment	Study design	Sample size and description	Clinical result of interaction	Possible mechanism	References
300 mg three-times daily for two-weeks	6β-hydroxycortisol, D-glucaric acid and cortisol	Before-after comparison	27 men and 23 women (21–35 years old)	Increased (41%) urinary excretion of 6β-hydroxycortisol; no changes in urinary excretion of cortisol	Hepatic enzyme induction	Kerb 1997
(standardised to 0.3% hypericin) 300 mg 3 times daily for 14 days	6β-hydroxycortisol/ cortisol ratio	Open label, before-after comparison	4 men and 9 women (18–45 years old)	Increased (114%) urinary excretion of 6β-hydroxycortisol/ cortisol ratio	Hepatic enzyme induction	Roby 2000
(Solaray®, standardised to 0.3% hypericin) 300 mg 3 times daily for 3 days	Alprazolam (1 mg in 3 subjects or 2 mg in 4 subjects) and destromethorphan (30 mg) before and after St John's wort treatment (day 3)	Before-after comparison	3 women and 4 men (24–32 years old)	Trend to decrease (48%) plasma concentration of alprazolam 2 mg/kg	Hepatic enzyme induction	Markowitz 2000
300 mg 3 times daily for 8 days	Destromethorphan 30 mg before and after St John's wort treatment (day 8)	Before-after comparison	16 subjects (sex and age not reported)	Trend to increase the metabolism of destromethorphan	Hepatic enzyme induction	Ereshefsky 2000
300 mg 3 times daily for 8 days	Caffeine 200 mg before and after St John's wort treatment (day 8)	Before-after comparison	16 subjects and urine (sex and age not reported)	No changes in plasma and urine caffeine metabolite levels	Not applicable	Gewertz 2000

Table 11.5 Herb–drug interactions: clinical trials on human volunteers. Data from Izzo and Ernst (2001)

St John's wort dosage and duration of the treatment	Co-medication dosage and duration of the treatment	Study design	Sample size and description	Clinical result of interaction	Possible mechanism	References
(LI160) 300 mg 3 times daily for seven days	Alcohol individual dose to achieve a 0.45–0.8 mg/ml blood concentration at day 7 (concomitantly to St John's wort)	Randomized, placebo-controlled double blind, cross-over trial	16 men and 16 women (25–40 years old)	No changes in cognitive capacities	Not applicable	Schmidt et al. 2000
(Aristofora®) 3 capsules daily for 9 days; last day 6 capsules along with alcohol (each capsule containing 0.25 mg hypericin)	Alcohol Individual doses to achieve a 0.05% blood concentration at days 15 (concomitantly to St John's wort)	Placebo-controlled, three-way, cross-over trial (one group received a mixture of valerian and St John's wort)	12 women and 6 men (mean age 45.6 ± 11.2)	St John's wort did not decrease alcohol-induced changes in vigilance (either alone or in combination with valerian)	Not applicable	Herberg 1994
(LI160) 900 mg daily for 14 days	Amitriptyline 75 mg twice daily for 14 days along with St John's wort	Open-label study	12 depressed patients (age and sex not reported)	Decreased plasma concentrations of amitriptyline (21.7%) and of its metabolite nortryptiline (40.6%)	Induction of hepatic enzymes	Roots et al. 2000
(LI160) 300 mg three times daily for 10 days	Digoxin 15 days treatment; Days 1–4 administered alone; days 5–15 along with St John's wort	Single blind, placebo-controlled trial with parallel groups	12 women and 13 men (12 placebo and 13 treated) (22–32 years old)	Decreased plasma concentration of digoxin trough levels (33.3%), AUC (25%) and C_{max} (26%)	Induction of the intestinal P-glycoprotein	Johne et al. 1999
(Preparation standardised to 0.3% hypericin) 300 mg 3 times daily for 16 days	Indinavir After achieving the steady state, a single dose of 800 mg (before and after St John's wort treatment)	Open-label, before–after comparison	6 men and 2 women (29–50 years old)	Decreased (57%) plasma concentration (AUC) of indinavir	Hepatic enzyme induction	Piscitelli 2000
(LI160) 300 mg daily for 11 days	Phenprocoumon 12 mg, single dose before and after St John's wort (or placebo) (day 11)	Randomized, single blind, placebo-controlled, cross-over trial (2-week washout period)	10 healthy men (18–50 years old)	Decreased (17.4%) plasma concentration of phenprocoumon (AUC of the free phenprocoumon)	Hepatic enzyme induction	Donath et al. 1999

Table 11.6 Case reports and case series of possible interactions between St John's wort and prescribed drugs. Data from Izzo and Ernst (2001)

St John's wort dosage and duration of the treatment	Patient's age and sex	Patient's diagnosis	Prescribed drug dosage and duration of the treatment	Other drugs used concomitantly	Clinical result of interaction	Possible mechanism	References
300 mg two times daily[b]	61-year-old female	Heart transplant	Cyclosporin[b]	None mentioned	Lowering of blood cyclosporin; rejection episode	Hepatic enzyme induction of intestinal P-glycoprotein	Bon 1999
300 mg three times daily[b]	54-year-old female	Lung fibrosis	Cyclosporin[b]	Prednisolone	Lowering of blood cyclosporin levels	Hepatic enzyme induction of intestinal P-glycoprotein	Bon 1999
[b]	30 patients[a]	Kidney transplants	Cyclosporin[b]	Other unreported drugs	Lowering of blood cyclosporin (47%)	Hepatic enzyme induction of intestinal P-glycoprotein	Breidenbach 2000
[b]	10 patients[a]	Liver transplants	Cyclosporin[b]	None mentioned	Lowering of blood cyclosporin (49%); rejection episode in one patient	Hepatic enzyme induction of intestinal P-glycoprotein	Breidenbach 2000
300 to 900 mg daily (or St John's wort tea)[b]	5 patients[a]	Kidney transplant	Cyclosporin[b]	None mentioned	Lowering of blood cyclosporin levels	Hepatic enzyme induction of intestinal P-glycoprotein	Roots et al. 2000
[b]	Woman in her mid-twenties	None reported	Cyclosporin[b]	None mentioned	Fall in cyclosporin blood levels (75%)	Hepatic enzyme induction of intestinal P-glycoprotein	Rey 1998
(LI160) 300 mg three times daily for three weeks	61-year-old[a]	Heart transplant with mild depression	Cyclosporin 125 mg twice daily for 11 months	Azathioprine, corticosteroids	Lowering of plasma cyclosporin to 95 μg/l; rejection episodes	Hepatic enzyme induction of intestinal P-glycoprotein	Ruschitzka et al. 2000
(LI160) 300 mg three times daily for three weeks	63 year-old[a]	Heart transplant	Cyclosporin 125 mg twice daily for 20 months	Azathioprine corticosteroids	Lowering of blood cyclosporin to 87 μl/L; rejection episode	Hepatic enzyme induction of intestinal P-glycoprotein	Ruschitzka et al. 2000

b	39-year-old female	Depression and migraine	Loperamide[b]	Valerian	Brief episode of acute delirium (disoriented, agitated, a confused state)	Potentiation of MAO inhibition	Khawaja 1999
b	Female[a]	None reported	Oral contraceptive[b]	None mentioned	Changed menstrual bleeding	Hepatic enzyme induction	Yue 2000
b	8 females aged 23–31 years	None reported	Oral contraceptive for a long time[b]	None mentioned	Intermenstrual bleeding	Hepatic enzyme induction	Yue 2000
b	Female[a]	None reported	Oral contraceptive ethinylestradiol 0.03 mg/desogestrel 0.15 mg	None mentioned	Intermenstrual (breakthrough) bleeding	Hepatic enzyme induction	Bon 1999
b	Female[a]	None reported	Oral contraceptive ethinylestradiol 0.02 mg/desogestrel 0.15 mg	None mentioned	Intermenstrual (breakthrough) bleeding	Hepatic enzyme induction	Bon 1999
b	44-year-old female	No pathology reported	Oral contraceptive ethinylestradiol 0.03 g/desogestrel 0.15 mg	None mentioned	Intermenstrual (breakthrough) bleeding	Hepatic enzyme induction	Bon 1999
300 mg three times daily for 3 days	84-year-old female	Depression and anxiety	Nefazodone 100 mg twice daily[b]	None	Nausea, vomiting, headache	Synergistic serotonin uptake inhibition	Lantz et al. 1999
b	78-year-old female	Depression	Sertraline 50 mg daily[b]	Calcium carbonate and conjugate estrogens	Dizziness, nausea, vomiting, headache	Synergistic serotonin uptake inhibition	Lantz et al. 1999
b	64-year-old male	Depression	Sertraline 75 mg/day[b]	None mentioned	Nausea, epigastric pain, and anxiety	Synergistic serotonin uptake inhibition	Lantz et al. 1999
300 mg twice daily for 2 days	82-year-old male	Depression, status post left cerebrovascular accident	Sertraline 50 mg/day[b]	Aspirin and multivitamins	Nausea, vomiting, anxiety, and confusion	Synergistic serotonin uptake inhibition	Lantz et al. 1999
300 mg three times daily for 2 days	79-year-old male with insulin-dependent diabetes	Depression and diabetes	Sertraline 50 mg/day[b]	Insulin	Nausea, anxiety, and feelings of restlessness and irritability	Synergistic serotonin uptake inhibition	Lantz et al. 1999

(Continued)

Table 11.6 (Continued)

St John's wort dosage and duration of the treatment	Patient's age and sex	Patient's diagnosis	Prescribed drug dosage and duration of the treatment	Other drugs used concomitantly	Clinical result of interaction	Possible mechanism	References
Dosage unclear; five weeks treatment	28-year-old male	Depression	Sertraline 50 mg daily for 5 weeks	Testosterone (after post-orchidectomy)	Maniac episode	Synergistic serotonin uptake inhibition	Barbanel 2000
600 mg/day for 10 days	50-year-old female	Asthma and depression	Paroxetine 40 mg daily for 8 months. Replacing paroxetine with St John's wort for 10 days. After this period, an acute dose of 20 mg	No other tranquillisers	Nausea, weakness, fatigue, groggy and lethargic state	Synergistic serotonin uptake inhibition	Gordon 1998
(0.3% hypericin) 300 mg daily for two months	42-year-old white female	None reported	Theophylline 300 mg bid for several months followed by one acute dose of 80 mg bid	Furosemide, potassium, morphine, zolpidem, valproic acid, ibuprofen, amitriptyline, albuterol, prednisone, zafirlukast, triamcinolone	Decreased theophylline levels	Hepatic enzyme induction	Nebel et al. 1999
b	75-year-old female	Polymorbid	Phenprocoumon[b]	None mentioned	Increased 'Quick-Wert' test (indicating decreased anticoagulant effect)	Hepatic enzyme induction	Bon 1999

b	79-year-old female	None reported	Warfarin 2.5 years[b]	None mentioned	Decreased INR (from 2.5–3.8 to 1.7)	Hepatic enzyme induction	Yue 2000
b	65-year-old female	None reported	Warfarin 2.5 years[b]	None mentioned	Decreased INR (from 2.4–3.6 to 2.0–2.1)	Hepatic enzyme induction	Yue 2000
b	76-year-old male	None reported	Warfarin 10 days[b]	None mentioned	Decreased INR (from 2.6 to 1.1)	Hepatic enzyme induction	Yue 2000
b	61-year-old female	None reported	Warfarin many years[b]	None mentioned	Decreased INR (INR before treatment not available; INR after 1.2)	Hepatic enzyme induction	Yue 2000
b	84-year-old female	None reported	Warfarin more than six months[b]	None mentioned	Decreased INR (from 2.9–3.6 to 1.5)	Hepatic enzyme induction	Yue 2000
b	56-year-old female	None reported	Warfarin[b]	None mentioned	Decreased INR (from 2.6 to 1.5)	Hepatic enzyme induction	Yue 2000
b	85-year-old female	None reported	Warfarin long time[b]	None mentioned	Decreased INR (from 2.1–4.1 to 1.5)	Hepatic enzyme induction	Yue 2000

Notes

a sex (and/or age) not reported.

b dose (and/or duration of the treatment) not reported.

INR = International Normalised Ratio.

LI160 = *Hypericum* extract standardised to 0.3% hypericin.

anticoagulant effect); in addition, breakthrough bleeding has been described when St John's wort is given concomitantly to oral contraceptives. Induction of hepatic cytochrome enzymes (with increased metabolism of drugs) and/or induction of intestinal P-glycoprotein (with increased elimination of drugs) could explain the mechanism of the observed interactions. Finally, the symptoms of a central serotonergic syndrome (mental status changes, tremor, autonomic instability, gastrointestinal upsets, headache, myalgia and restlessness) described when St John's wort is given in parallel with other 5-HT reuptake inhibitors (e.g. the antidepressant paroxetine, sertraline and nefazodone) might be due to an additive effect on brain 5-HT reuptake.

References

Agrosi, M., Mischiatte, S., Harrasser, PC. and Savio, D. (2000) Oral bioavailability of active principles from herbal products in humans. A study of *Hypericum perforatum* extract using the soft gelatin capsule technology. *Phytomed.* 7, 455–62.

Barbanel, D.M. (2000) Mania in a patient receiving testosterone replacement post-orchidectomy taking St John's wort and sertraline. *J. Psychopharm.* 14, 84–6.

Barnes, J. (2001) Citation from: Pitisuttithum, P., Migasena, S., Suntharasamaie, P. International conference on AIDS 1996;11:285. *J. Pharm. Pharmacol.* 53, 583–90.

Barnes, J., Anderson, L.A. and Phillipson, J.D. (2001) St John's wort (*Hypericum perforatum* L.): a review of its chemistry, pharmacology and clinical properties. *J. Pharm. Pharmacol.* 53, 583–90.

Bernd, A., Simon, S., Ramirez Bosca, A. and Kippenberg, S. *et al.* (1999) Phototoxic effects of Hypericum extract in cultures of hyman deratinocytes compared with those of psoralen. *Photochem Photobiol.* 69, 218–21.

Biber, A., Fisher, H., Romer, A. and Chatterjee, S.S. (1998) Oral bioavailability of hyperforin from Hypericum extracts in rats and human volunteers. *Pharmacopsychology* 31, 36–43.

Bon, S. (1999) Johanniskraut Ein Enzymindiktor. *Schweitzer Apothekerzeitung* 16, 535–6.

Bove, G.M. (1998) Acute neuropathy after exposure to sun in a patient treated with St John's Wort. *Lancet* 352, 1121–2.

Breidenbach, T. (2000) Profound drop of cyclosporin A whole blood trough levels caused by St John's wort (*Hypericum perforatum*). *Transplantation* 69, 2229–32.

Breidenbach, T. (2000) Drug interaction of St John's wort with cyclosporin. *Lancet* 355, 1912.

Brockmoller, J., Reum, T., Bauer, S., Kerb, R., Hubner, W.D. and Roots, E. (1997) Hypericin and pseudohypericin: pharmacokinetics and effects on photosensitivity in humans. *Pharmacopsychology* 30, 94–101.

Di Carlo, G., Borrelli, F., Ernst, E. and Izzo, A.A. (2001) St John's wort: prozac from the plant kingdom. *Trends Pharmacol Sci.* 22, 292–7.

Di Matteo, V., Di Giovanni, G., Di Mascio, M. and Esposito, E. (2000) Effect of acute administration of Hypericum perforatum-CO_2 extract on dopamine and serotonin release in the rat central nervous system. *Pharmacopsychology* 33, 14–18.

Dimpfel, W., Todorova, A. and Vonderheid-Guth, B. (1999) Pharmacodynamic properties of St John's wort. A single blind neurophysiological study in healthy subjects comparing two commercial preparations. *Eur. J. Med. Res.* 4, 303–12.

Donath, F., Roots, I., Langheinrich, M. and Hubmer, W. (1999) Interactions of St John's wort extract with phenprocoumon. *Eur. J. Clin. Pharmacol.* 55, A22-A22.

Durr, D., Stieger, B., Kullak-Ublick, G.A., Rentsch, K.M. and Steinert, H.C., Meier, P.J. *et al.* (2000) St John's wort induces intestinal P-glycoprotein/MDR1 and intestinal and hepatic CYP3A. *Clin. Pharmacol. Ther.* 68, 598–604.

Ernst, E. (1995) St John's wort; an anti-depressant? A systematic, criteria-based review. *Phytomed.* 2, 67–71.

Ereshefsky, B. (2000) Determination of SJW differential metabolism at CYP2D6 and CYP3A4, using dextromethorphan probe methodology. Presented at the 39th Annual Meeting of the New Clinical Drug Evaluation Program. *Meeting of the NCDEU*, 130–1.

Friede, M. and Wustenberg, P. (1998) Johanniskraut zur Therapie von Angstsyndromen bei depressiven Verstimmunger. *Ztsch Phytother.* **19**, 309–18.

Gewertz, N. (2000) Determination of differential effects of St John's wort on the CYP1A2 and NAT2 metabolic pathways using caffeine probe methodology (abstract). Proceedings 39th Annual meeting, New Clinical Drug Evaluation Unit, June. *Meeting of the NCDEU*,131-.

Gordon, J.B. (1998) SSRIs and St John's wort possible toxicity. *Am. Fam. Physician* **57**, 950.

Gulick, R.M., McAuliffe V., Holden-Wiltse, J., Crumpacker, C., Liebes, L. and Stein, D.S. *et al.* (2000) Phase I studies of hypericin, the active compound in St John's wort, as an antiretroviral agent in HIV-infected adults. *Ann. Intern. Med.* **130**, 510–4.

Gaster, B. and Hoiroyd, J. (2000) St John's wort for depression. *Arch. Intern. Med.* **160**, 152–6.

Herberg, K.W. (1994) Alternative zu synthetischen Psychopharmaka? *Therapiewoche* **44**, 704–13.

Izzo, A.A. and Ernst, E. (2001) Interactions between herbal medicines and prescribed drugs: a systematic review. *Drugs.* **15**, 2163–75.

Johne, A., Brockmöller, J., Bauer, S., Maurer, A., Langheinrich, M. and Roots, I. (1999) Pharmacokinetic interaction of digoxin with an herbal extract from St John's wort (*Hypericum perforatum*). *Clin. Pharmacol. Ther.* **66**, 338–45.

Johnson, D., Siebenhuner, G., Hofer, E., Sauerwein-Giese, E. and Frauendorf, A. (1992) The effects of St John's wort extract upon central nervous system activity. *Neurologie/Psychiatrie* **6**, 436–44.

Johnson, D., Ksciuk, H., Woelk, H., Sauerwein-Giese, E. and Frauendorf, A. (1994) Effects of Hypericum extract LI 160 compared with maprotiline on resting EEG and evoked potentials in 24 volunteers. *J. Geriatr. Psychiatr. Neurol.* **7**, S44–S46.

Kerb, R. (1997) Urinary 6-(-hydroxycortisol excretion rate is affected by treatment with hypericum extract (abstract). *Eur. J. Clin. Pharmacol.* **52**(suppl), A186.

Kerb, R., Brockmoller, J., Staffeldt, B., Ploch, M. and Roots, I. (1996) Single-dose and steady-state pharmacokinetics of hypericin and pseudohypericin. *Antimicr. Agents Chem.* **40**, 2087–93.

Khawaja, I.S. (1999) Herbal medicines as a factor in delirium. *Psychiatric Services* **50**, 969–70.

Kim, H.L. S.J. (1999) St John's wort for depression. A meta-analysis of well-defined clinical trials. *J. Nervous Mental Dis.* **187**, 532–8.

Lantz, M.S., Buchalter, E. and Giambanco, V. (1999) St John's wort and antidepressant drug interactions in the elderly. *J. Geriatric. Psych Neurol.* **12**, 7–10.

Linde, K. (2000) St John's wort for depression. *The Cochrane Library* **4**, 1–17.

Linde, K., Ramirez, G., Mulrow, C.D., Pauls, A., Weidenhammeer, W. and Melchart, D. (1996) St John's wort for depression – an overview and meta-analysis of randomised clinical trials. *BMJ* **313**, 253–8.

Linde, K., Ter Riet, G., Hondras, M., Vickers, A., Saller, R. and Melchart, D. (2001) Systematic reviews of complementary therapies – an annotated bibliography. Part 2: herbal medicine. *Complementar. Alt. Med.* **1**, 5-.

Markowitz, J.S. (2000) Effect of St John's wort (*Hypericum perforatum*) on cytochrome P-450 2D6 and 3A4 activity in healthy volunteers. *Life Sci.* **66**, 133–9.

Muller, W.E., Singer, A., Wonnemann, M., Hafner, U., Rulli, M. and Schafer, C. (1998) Hyperforin represents the neurotransmitter reuptake inhibiting constituent of Hypericum extract. *Pharmacopsychology* **31**, 16–21.

Mulrow, C.D., Williams, J.W. and Trivedi, M. *et al.* (1998) Treatment of depression – newer pharmacotherapies. *Psychopharmacol. Bull.* **34**, 409–15.

Nathan, P.J. (1999) The experimental and clinical pharmacology of St John's wort (*Hypericum perforatum* L.). *Mol. Psychiat.* **4**, 333–8.

Neary, J.T. and Bu, Y. (1999) Hypericum LI 160 inhibits uptake of serotonin and norepinephrine in astrocytes. *Brain Res.* **816**, 358–63.

Nierenberg, A.A. (1999) Mania associated with St John's wort. *Biol. Psychiat.* **46**, 1707–8.

Nebel, A., Schneider, B.J., Baker, R.K. and Kroll, D.J. (1999) Potential metabolic interaction between St John's Wort and theophylline. *Ann. Pharmacother.* **33**, 502.

O'Breasil, A.M. and Argouarch, S. (1998) Hypomania and St John's wort. *Can. J. Psychiat.* **43**, 746–7.

Piscitelli, S.C. (2000) Indinavir concentrations and St John's wort. *Lancet* **355**, 547–8.

Rey, J.M. (1998) Hypericum perforatum (St John's Wort) in depression: pest or blessing. *MJA* **169**, 583–6.

Roby, C.A. (2000) St John's wort: effect on CYP3A4 activity. *Clin. Pharmacol. Ther.* 67, 451–7.

Roots, I., Johne, A. and Maurer, A. *et al.* (2000) Arzneimittel interaktionen von hypericum extract. *Proc. Germ. Soc. Pharmacol.*

Roots, I., Johne, A. and Schmider, I. *et al.* (2000) Interaction of a herbal extract from St John's wort with amitriptyline and its metabolites (abstract). *Clin. Pharmacol. Ther.* 67, 159-.

Ruschitzka, F., Meier, P.J., Turina, M., Lüscher, T.F. and Noll, G. (2000) Acute heart transplant rejection due to St John's Wort. *Lancet* 355, 548–9.

Schey, K.L. (2000) Photooxidation of lens alpha-crystallin by hypericin (the active ingredient in St John's wort). *Photochem. Photobiol.* 72, 200–3.

Schellenberg, R., Sauer, S. and Dimpfel, W. (1998) Pharmacodynamic effects of two different hypericum extracts in healthy volunteers measured by quantitative EEG. *Pharmacopsych.* 31, 44–53.

Schempp, C., Winghofer, B., Langheinrich, M., Schopt, E. and Simon, J.C. (1999) Hypericin levels in human serum and interstitial skin blister fluid after oral single-dose and steady-state administration of *Hypericum perforatum* extract (St John's wort). *S. Pharmacol. Appl. Skin. Physiol.* 12, 299–304.

Schempp, C.M., Ludtke, R., Winghofer, B. and Simon, J.C. (2000) Effect of topical application of *Hypericum perforatum* extract (St John's wort) on skin sensitivity to solar simulated radiation. *Photodermatol. Photoimmunol. Photomed.* 16, 125–8.

Schmidt, U., Harrer, G., Huhn, U., Berger-Deinert, W. and Luther, D. (2000) Wechselwirkungen von hypericum-extrakt mit alkohol. *Nervenheilkunde* 12, 314–9.

Schneck, C. (1998) St John's wort and hypomania. *J. Clin. Psychiatry* 59, 689-.

Schulz, H. and Jobert, M. (1994) Effects of Hypericum extract on the sleep EEG in older volunteers. *J. Geriatr. Psychiatr. Neurol.* 7, S39–S43.

Singer, A., Wonneman, M. and Muller, W.E. (1999) Hyperforin, a major antidepressant constituent of St John's wort, inhibits serotonin uptake by elevating free intracellular Na^+. *J Pharmacol Exp. Ther.* 290, 1363–8.

Staffeldt, B., Kerb, R., Brockmoller, J., Ploch, M. and Roots, I. (1994) Levels of hypericin and pseudohypericin after oral intake of the *Hypericum perforatum* extract LI 160 in healthy volunteers. *J. Geriatr. Psych. Neurol.* 7, S47–S53.

Stevinson, C. (1999) Hypericum for depression: an update of the clinical evidence. *Eur Neuropsychopharmacol*, 9, 501–5.

Teufel-Mayer, T. and Gleitz, J., (1997) Effect of long-term administration of hypericum extracts on the affinity and density of the central serotonergic 5-HT_1 and 5-HT_2 receptors. *Pharmacopsychology* 30, 113–16.

Vandenboagaerde, A.L., Kamahubwa, A., Delaey, E., Himpens, B.E., Merlevede, W.J. and De Witte, P.A. (1998) Photocytotoxic effect of pseudohypericin versus hypericin. *Photochem. Photobiol.* B. 45, 87–94.

Vitiello, B. (1999) *Hypericum perforatum* extracts as potential antidepressants. *J. Pharm. Pharmacol.* 51, 513–17.

Volz, H-P. (1997) Controlled clinical trials of hypericum extracts in depressed patients – an overview. *Pharmacopsychiatry* 30(suppl), 72–6.

Volz, H.P. and Laux, P. (2000) Potential treatment for subthreshold and mild depression: a comparison of St John's wort extracts and fluoxetine. *Comprehensive Psych.* 41, 133–7.

Williams, J.W. (2000) A systematic review of newer pharmacotherapies for depression in adults: evidence report summary. *Ann. Int. Med.* 132, 743–56.

Wonneman, M., Singer, A. and Muller, W.E. (2000) Inhibition of synaptosomal uptake of ^{3}H-L-glutamate and ^{3}H-GABA by hyperforin, a major constituent of St John's wort: the role of amiloride sensitive sodium conductive pathways. *Neuropsychopharmacology* 23, 188–96.

Yue, Q.Y.B. (2000) Safety of St John's Wort (*Hypericum perforatum*). *Lancet* 355, 576–7.

12 Hypericin as a potential antitumour agent

(The work of Professor Kyung-Tae Lee's group in Seoul)

William E. Court

St John's wort (*Hypericum perforatum* L.) has acquired considerable popularity amongst users of herbal and homoeopathic medicines, being valued in particular as an antidepressant and also used in wound healing treatments. In Britain it is claimed that some two million people have employed the plant to treat moderate or mild depression. In 1984, Muldner and Zoller conducted a clinical trial using a standardised extract and showed effectiveness against anxiety symptoms within 4–6 weeks and Linde *et al.* (1996), in a review of the use of St John's wort for depression, noted that the plant was as effective as many common antidepressants. More recently, in Germany, workers considering 300 persons taking either the antidepressant imipramine or extract of St John's wort concluded that the two drugs were 'equivalent' although St John's wort yielded fewer side effects (*British Medical Journal* (2000) **321**, 536). Nevertheless, some researchers are concerned that St John's wort may increase sensitivity to sunlight and may also reduce the plasma levels of simultaneously administered drugs by induction of metabolism (Baede-Van Duk *et al.* 2000).

Other workers have sought different aspects of the pharmacology of St John's wort. Professor Kyung-Tae Lee's team in Seoul, Korea systematically investigated the antitumour potential of the naphthodianthrone pigment hypericin (MW 504.5), a photoactive polycyclic anthrone isolated from *H. triquetrifolium* Turra and other *Hypericum* spp. and a compound regarded by many as the most important in the plant extract (Figure 12.1).

In vitro this pigment has been shown to oxidise lipids, amino acids and proteins and to disrupt the normal function of cellular membranes and demonstrates antiviral (Meruelo and Lavie 1988) and antineoplastic properties (Zhang *et al.* 1996) in some cell types as well. Such manifestations have been positively linked to protein tyrosine kinase (PTK) activity (Agostinis *et al.* 1995, Kil *et al.* 1996) and protein kinase C (PKC) activity (Takahashi and Nakanishi 1989), suggesting that inhibition of PKC is involved in these processes.

Phosphorylation of the tyrosine residues of proteins is regarded as an important biochemical reaction that mediates a wide variety of cellular signals (Hunter and Cooper 1985, Yarden and Ullrich 1988) including control of the cell cycle and cell differentiation. Many cellular plasma membrane receptors, such as receptors for epidermal growth factor (EGF), platelet-derived growth factor (PDGF) and insulin, possess an integral, intracellular, tyrosine kinase moiety that is activated upon the binding of its specific ligand to the extracellular region of the receptor (Yarden and Ullrich 1988, Ullrich and Schlessinger 1990). The occurrence and involvement of unusually high PTK activity in a wide variety of growth related disease states has been established (Bishop 1987).

Hyperproliferation of cells producing nonmalignant growths is also frequently related to enhanced PTK activity. For example, Ross (1986) cites the enhanced PTK activity of the PDGF receptor which results from its exposure to sustained levels of PDGF encountered in atherosclerosis

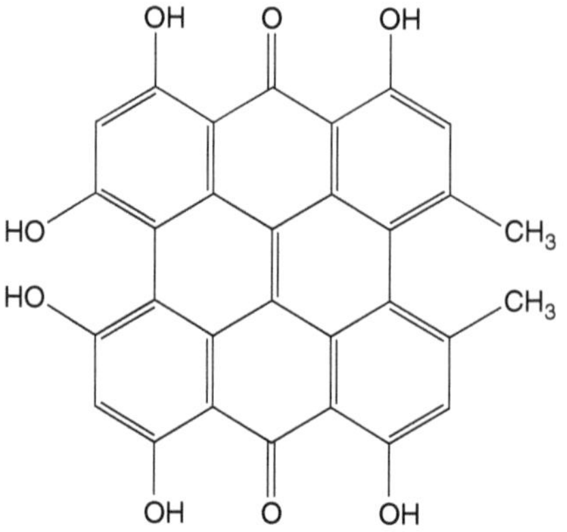

Figure 12.1 The chemical structure of hypericin.

and restenosis. Ross (1986), Bishop (1987) and Ullrich and Schlessinger (1990) observed that the enhanced PTK activity of oncogene products or the overexpression of their normal counterparts was essential for their transforming activity. The overexpression of PTK oncoproteins, yielding enhanced kinase activity, can also alter the developmental pattern of cell types into which they have been introduced (Filvaroff *et al.* 1990).

The realisation of the occurrence of PTKs in a wide range of disease states led to the idea of PTK blockers used to oppose the hyperproliferative conditions resulting from enhanced PTK activity (Yaish *et al.* 1988, Levizki 1990). Naturally occurring compounds such as erbstatin (Umezawa and Imoto 1991), herbimycin A (Murakami *et al.* 1988, Fukazawa *et al.* 1991), staurosporin (Secrist *et al.* 1990) and flavonoids (Akiyama *et al.* 1987) as well as chemically synthesised compounds such as thiazolidine-diones (Geissler *et al.* 1990) and tyrphostins (Lyall *et al.* 1989, Levizki *et al.* 1991) were found capable of inhibiting the growth of cultured cells by their specific interference with PTK activities.

Lee's team decided to investigate the similar potential of hypericin and they reported (Kil *et al.* 1996) the potent antitumour activity *in vitro* of hypericin against several types of tumour cells although there was no demonstrable cytotoxicity on normal cells such as *Macaccus rheus* monkey kidney cells (MA-104) and primary cultured rat hepatocytes up to 500 μM concentration. The method used was the MTT (3-(4,5-dimethylthiazol-2-yl)-2,5-diphenyl tetrazolium bromide) colorimetric assay technique described earlier by Denizot and Lang (1986) and Baek *et al.* (1996) and applied after a 2 day hypericin treatment. Results summarised in Table 12.1 indicate significant specificity in inhibiting P388 leukaemia and L-1210 leukaemia cells. In a similar manner gastrointestinal tumour cells (SNU-1 and SNU-C4) were also inhibited. It was concluded that, because of this antiproliferative effect on tumour cells and lack of toxicity on normal cells *in vitro*, hypericin had potential as a chemotherapeutic agent.

Lee's team also investigated the method of growth inhibition by hypericin and, in particular, the effect of hypericin on the activity of the EGF receptor in A-431 human epidermoid carcinoma cell membranes. Hypericin inhibited the autophosphorylation of the EGF receptor which occurs under physiological conditions and the tyrosine phosphorylation of RR-SRC peptide (Arg-Arg-Leu-Ile-Glu-Asp-Ala-Glu-Tyr-Ala-Ala-Arg-Gly) catalysed by an EGF receptor. They also noted that A-431 cells treated with hypericin demonstrated, by western immunoblot analysis, inhibition of the tyrosine phosphorylation of EGF-dependent endogenous EGF

Table 12.1 The sensitivity of cultured cells to hypericin (from Kil *et al.* 1996)

Cell line	Origin	Source	$IC_{50}^{*}(\mu M)$
P388	Leukaemia	Mouse	0.15×10^{-3}
L-1210	Leukaemia	Mouse	0.38
SNU-C4	Colon cancer	Human	0.14
SNU-1	Stomach cancer	Human	0.34
MH1C1	Moris hepatoma	Rat	2.0
HepG2	Hepatoma	Human	112
A-431	Epidermoid carcinoma	Human	1.7
MA-104	Kidney cell	Monkey	>500
Hepatocyte**	Liver	Rat	>500

Notes

  * IC_{50} values cited represent the means of three independent experiments and are defined as the drug concentration that resulted in a 50% decrease in cell numbers as compared with the control cultures in the absence of inhibitor.
  ** Hepatocytes were primary cultured cells isolated by the two-step collagenase method.

receptors. In addition, hypericin inhibited the T cell PTK, P56^{lck} in a dose-dependent fashion with an $IC_{50} = 5\ \mu M$. The tyrosine phosphorylation on RR-SRC peptide and the EGF-induced receptor autophosphorylation, either *in vitro* or in intact cells, was inhibited by hypericin at the same concentration as that in A431 cell proliferation. It was concluded that hypericin directly inhibited EGF receptor and P56$^{lck\ PTK}$ activity *in vitro* and could mediate such action *in vivo*.

In a further communication Kim *et al.* (1998) observed that the development of a malignant phenotype could be regarded as a defect in the control of the differentiation process in which the neoplastic cells exhibit a change in the tightly controlled homeostatic balance between proliferation and maturation that occurs in normal cells. The cellular mechanisms controlling these phenomena are crucial in the understanding of the onset of malignancy and may prove important in the invention of novel approaches to combating cancers.

Cell proliferation and differentiation can be achieved via cAMP-dependent protein kinase A (PKA) or PKC activities. When cells were treated with phorbol ester (phorbol 12-myristate 13-acetate (TPA)), a compound initiating PKC activity, cells from the human histocytic lymphoma U-937 cell lines commenced proliferating and differentiating to a monocyte/macrophage-like cell (Way *et al.* 1992). Other workers have suggested that PKA is also a key compound in the differentiation and proliferation processes (Cho-Chung and Clair 1993).

Differentiation therapy for leukaemia, using differentiation-inducing compounds and combinations of various inducers, has been demonstrated as a new approach in the treatment of human leukaemia as the therapeutic use of retinoic acid caused a complete remission of acute promyelocytic leukaemia (Chomienne *et al.* 1996). Kim *et al.* (1998) studied the antiproliferative and differentiation-inducing activity of hypericin in U-937 human histocytic lymphoma cell lines. Leukaemia cells are blocked at certain stages in the maturation processes and display a highly proliferative capacity. The potential value of differentiation inducers as therapeutic agents depends on their ability to overcome the maturation blockade. Kil *et al.* (1996) showed that hypericin produced a selective cytotoxicity on leukaemia cells and PTK inhibitory activity and their further study (Kim *et al.* 1998) revealed that hypericin has a potent differentiation-inducing activity in the myelogenous leukaemia-derived U-937 cells. Hypericin produced the expression of differentiation markers such as NBT (nitroblue-tetrazolium) reducing ability, phagocytic activity, cell size and the appearance of AS-D esterase activity in U-937 cells. It was noted that compound NA-382, a PKC inhibitor, reduced the differentiation induced by hypericin treatment but the compound H-89, a PKA inhibitor, yielded no such effect. Although the precise mechanism of the differentiation induction by hypericin in U-937 cells is not as yet understood, it is obvious that PKC rather than PKA fulfils an important role in the modulation of differentiation. Previous work indicated that tyrosine phosphorylation induced by EGF was significantly inhibited by hypericin in a dose-dependent manner (Agostinis *et al.* 1995, Kil *et al.* 1996). Thus, it was suggested that if the target of the action by hypericin was limited to tyrosine kinase or a closely related reaction, then the probable common step in differentiation was possibly related to tyrosine residues in cellular proteins. Kondo *et al.* (1989) noted that genistein and herbimycin A, inhibitors of the protein phosphorylation activity of PTK *in vitro*, also induced differentiation of several cells under specific conditions. Consequently, it seems probable that specific inhibition of protein phosphorylation at tyrosine residues makes the cells, directly or indirectly, less proliferative and more susceptible to physiological conditions favouring differentiation. Kim *et al.* (1998) showed that hypericin-induced differentiation of U-937 cells was inhibited by the PKC inhibitor NA-382 and, therefore, they suggested that differentiation by hypericin was related to PKC activity and was stimulated by inhibition of tyrosine phosphorylation in response to hypericin-mediated signalling.

In a further paper Lee *et al.* (1999) noted that leukaemia cells were blocked at some stages of the maturation process and demonstrated a highly proliferative capacity. The potential value of differentiation inducers as therapeutic agents resides in their ability to overcome the maturation blockades. The human promyelocytic leukaemia-derived cell line HL-60 can be induced to differentiate morphologically and functionally into mature granulocytes *in vitro* by various substances including DMSO (dimethylsulfoxide) (Collins *et al.* 1978) and ATRA (all *trans*-retinoic acid) (Breitman *et al.* 1980). In addition, such cells have been reported to differentiate to monocyte/macrophage by the action of TPA (12-O-tetradecanoylphorbol-13-acetate) (Rovera *et al.* 1979) and active vitamin D_3 ($1\alpha,25(OH)_2D_3$) (Miyaura *et al.* 1981). Therefore, differentiation therapy for leukaemia, employing differentiation-inducing compounds or combinations of such compounds, has been shown to be a new approach to the treatment of some human leukaemias since ATRA produced a complete remission of acute promyelocytic leukaemia (Chomienne *et al.* 1996). However, the differentiation inducers ATRA and TPA have been shown to induce apoptosis in leukaemia cells, apoptosis being the self-controlled cell death characterised by nucleosomal fragmentation and by a number of morphological changes that differ from those of necrosis. The induction of apoptosis in proliferating tumour cells may therefore be useful in anticancer treatment.

Working with human promyelocytic leukaemia-derived HL-60 cells, Lee *et al.* (1999) demonstrated a potent ability of hypericin to induce differentiation and apoptosis. Hypericin produced the expression of differentiation markers such as NBT-reducing ability, phagocytic activity, cell size and the appearance of naphthol AS-D esterase activity indicating that hypericin induced differentiation to monocytes in HL-60 cells. It was also shown that, at the same concentration range as for induction of differentiation, apoptotic DNA fragmentation occurred in HL-60 cells. The mechanism for the induction of differentiation and apoptosis by hypericin in HL-60 cells is not as yet clearly understood although hypericin is known as a potent inhibitor of of PKC (Takahashi and Nakanishi 1989) and of PTK (Agostinis *et al.* 1995, Kil *et al.* 1996). Lee *et al.* (1999) pointed out that genistein and herbimycin A, known inhibitors of the protein phosphorylation action of PTK *in vitro*, can induce the differentiation of several cells under specific conditions. On the other hand, they noted that the apoptosis-inducing activity of some chemical compounds apparently varied with both the cell type and the cytotoxic insult, for example, TPA-induced apoptosis, but prevented apoptotic death. H-7, a known PKC inhibitor, produced apoptosis in HL-60 cells but also prevented apoptotic cell death in mouse thymocytes exposed to corticosteroids. However, in Lee's team's work employing HL-60 cells, hypericin induced both differentiation and apoptosis and they speculated that hypericin-induced differentiation may operate by inhibition of PTK activity of cells and may cause apoptosis via PKC inhibition in quite different ways from ATRA and active vitamin D_3.

During the maturation process for myeloid cells, apoptosis happens as a normal physiological phenomenon. Mature granulocytes and erythrocytes are known to undergo spontaneous apoptosis at the end of their lifespan and therefore it has been suggested that apoptosis induced during maturation of myeloid and erythroid cells modulates the cell number prior to the migration of mature cells from the bone marrow into the peripheral circulation.

As hypericin may promote the physiological processes of the cells including differentiation, maturation and controlled cell death, it may be useful for the treatment of patients suffering from leukaemia. Logically, before any further progress can be made, this potential value must be adequately tested clinically in a wide range of subjects using carefully standardised preparations under controlled conditions.

References

Agostinis, P., Vandenbogaerde, A., Donella-deana, A., Pinna, L.A., Lee, K.T., Goris, J., Merlevede, W., Vandenheede, J.R. and De Witte, P. (1995) Photosensitized inhibition of growth factor regulated protein kinase by hypericin. *Biochem. Pharmacol.* **49**, 1615–22.

Akiyama, T., Ishida, J., Nakagawa, S., Ogawara, H., Watanabe, S.I., Itoh, N., Shibuya, M. and Fukami, Y. (1987) Genistein, a specific inhibitor of tyrosine-specific protein kinases. *J. Biol. Chem.* **262**, 5592–5.

Baede-Van Duk, P.A., Galen, E. Van and Lekkerkerker, J.F.F. (2000) Combinations of *Hypericum perforatum* (St. John's wort) with other pharmaceuticals is risky. *Nederlands Tijdschrift voor Geneskunde* **144**, 811–2.

Baek, S.H., Han, D.S., Yook, C.N., Kim, Y.C. and Kwak, J.S. (1996) Synthesis and antitumor activity of Cannabigerol. *Arch. Pharm. Res.* **19**, 228–30.

Bishop, J.M. (1987) The molecular genetics of cancer. *Science* **235**, 305–11.

Breitman, T.R., Selonick, S.E. and Collins, S.J. (1980) Induction of differentiation of the human promyelocytic cell line by retinoic acid. *Proc. Natl. Acad. Sci. USA* **77**, 2936–40.

Cho-Chung, Y.S. and Clair, T. (1993) The regulatory subunit of cAMP-dependent protein kinase as a target for chemotherapy of cancer and other dysfunctional-related diseases. *Pharmacol. Ther.* **60**, 265–88.

Chomienne, C., Fenaux, P. and Degos, L. (1996) Retinoid differentiation therapy in promyelocytic leukemia. *FASEB. J.* **10**, 1025–30.

Collins, S.J., Ruscetti, F.W., Gallagher, R.E. and Gallo, R.C. (1978) Terminal differentiation of human promyelocytic leukemia cells induced by dimethylsulfoxide and other polar compounds. *Proc. Natl. Acad. Sci. USA* **75**, 2458–62.

Denizot, F. and Lang, R. (1986) Rapid colorimetric assay for cell growth and survival: modifications to the tetrazolium dye procedure giving improved sensitivity and reliability. *J. Immun. Methods* **89**, 271–7.

Hunter, T. and Cooper, J.A. (1985) Protein tyrosine kinase. *Annu. Rev. Biochem.* **54**, 897–930.

Filvaroff, E., Stern, D.F. and Dotto, G.P. (1990) Tyrosine phosphorylation is an early and specific event involved in primary keratinocyte differentiation. *Mol. Cell Biol.* **10**, 1164–73.

Fukazawa, H., Li, P.-M., Yamamoto, C., Murakami, Y., Mizuno, S. and Uehara, Y. (1991) Specific inhibition of cytoplasmic protein kinases by herbimycin A *in vitro*. *Biochem. Pharmacol.* **42**, 1661–7.

Geissler, J.F., Traxler, P., Regenass, U., Murray, B.J., Roesel, J.L., Meyer, T., McGlynn, E., Storni, A. and Lydon, N.B. (1990) Thiazolidine-diones: biochemical and biological activity of a novel class of tyrosine protein kinase inhibitors. *J. Biol. Chem.* **265**, 22255–61.

Kil, K.-S., Yum, Y.-N, Seo, S.-H. and Lee, K.-T. (1996) Antitumor activities of hypericin as a protein tyrosine kinase blocker. *Arch. Pharm. Res.* **19**, 490–6.

Kim, J.-I., Park, J.-H., Park, H.-J., Choi, S.-K. and Lee, K.-T. (1998) Induction of differentiation of the human histocytic lymphoma cell line U-937 by hypericin. *Arch. Pharm. Res.* 21, 41–5.

Kondo, K., Watanabe, T., Sasaki, H., Uehara, Y. and Oishi, M. (1989) Induction of *in vitro* differentiation of mouse embryonal carcinoma (F9) and erythroleukemia (MEL) cells by herbimycin A, an inhibitor of protein phosphorylation. *Mol. Cell Biol.* 7, 285–93.

Lee, K.T., Kim, J.-I., Rho, Y.-S., Chang, S.-G., Jung, J.-C., Park, J.-H., Park, H.-J. and Miyamoto, K.-I. (1999) Hypericin induces both differentiation and apoptosis in human promyelocytic leukemia HL-60 cells. *Biol. Pharm. Bull.* 22, 1271–4.

Levizki, A. (1990) Tyrphostins: potential antiproliferative agents and novel molecular tools. *Biochem. Pharmacol.* 40, 313–19.

Levizki, A., Grazit, A., Osherov, N., Posner, I. and Gilon, C. (1991) Inhibition of protein-tyrosine kinases by tyrphostins. *Meth. Enzymol.* 201, 347–61.

Linde, K., Ramirez, G., Mulrow, C.D., Pauls, A., Weidenhammer, W. and Melchart, D. (1996) St. John's wort for depression – an overview and meta-analysis of randomised clinical trials. *Brit. Med. J.* 313, 253–8.

Lyall, R.M., Zilberstein, A., Gazit, A., Gilon, C., Levizki, A. and Schlessinger, J. (1989) Tyrphostins inhibit epidermal growth factor (EGF)-receptor tyrosine kinase activity in living cells and EGF-stimulated cell proliferation. *J. Biol. Chem.* 264, 14503–9.

Meruelo, D. and Lavie, G. (1988) Therapeutic agents with dramatic antiretroviral activity and little toxicity at effective doses: aromatic polycyclic diones hypericin and pseudohypericin. *Proc. Natl. Acad. Sci. USA* 85, 5230–4.

Muldner, H. and Zoller, M. (1984) Clinical trial of *Hypericum perforatum* extract as an anti-anxiety agent. *Arzneim. Forsch.* 34, II (8), 918.

Murakami, Y., Mizuno, S., Hori, M. and Uehara, Y. (1988) Reversal of transformed phenotypes by herbimycin A in src oncogene expresses rat fibroblasts. *Cancer Res.* 48, 1587–90.

Ross, R. (1986) The pathogenic of atherosclerosis: an update. *New Engl. J. Med.* 314, 488–500.

Rovera, G., Santoli, K. and Damsky, C. (1979) Human promyelocytic leukemia cell in culture differentiates into macrophage-like cells when treated with phorbol diesters. *Proc. Natl. Acad. Sci. USA* 76, 2779–83.

Secrist, J.P., Sehgal, I., Poeis, G. and Abraham, R.T. (1990) Preferential inhibition of the platelet-derived growth factor receptor tyrosine kinases by staurosporine. *J.Biol. Chem.* 265, 20394–400.

Takahashi, I. and Nakanishi, S. (1989) Hypericin and pseudohypericin specialty inhibit protein kinase C: possible relation to their antiretroviral activity. *Biochem. Biophys. Res. Commun.* 165, 1207–12.

Ullrich, A. and Schlessinger, J. (1990) Signal transduction by receptor with tyrosine kinase activity. *Cell* 61, 203–12.

Umezawa, K. and Imoto, M. (1991) Use of erbstatin as protein-tyrosine inhibitor. *Methods Enzymol.* 201, 379–80.

Way, D.K., Messer, B.R., Garris, T.O., Qin, W., Cook, P.P. and Parker, P.P. (1992) Modulation of protein kinase C by phorbol esters in monoblast U937 cell. *Cancer Res.* 52, 5604–9.

Yaish, P., Gazit, A., Gilon, C. and Levitski, A. (1988) Blocking of EGF dependent cell proliferation by EGF receptor kinase inhibitors. *Science* 242, 933–5.

Yarden Y. and Ullrich, A. (1988) Growth factor receptor tyrosine kinases. *Annu. Rev. Biochem.* 57, 443–78.

Zhang, W., Hinton, D.R., Surnock, A.A. and Couldwell, W.T. (1996) Malignant glioma sensitivity to radiotherapy, high-dose tamoxifen and hypericin: corroborating clinical response *in vitro*: case report. *Neurosurgery* 38, 587–91.

13 Neuropsychopharmacological studies on Indian *Hypericum perforatum* Linn

Vikas Kumar, P.N. Singh and S.K. Bhattacharya

Introduction

Hypericum species were known to ancient communities as useful medicinal plants. The use of *Hypericum perforatum* (HP), in particular, as a remedy was described and recommended throughout the Middle Ages. HP Linn is a perennial plant belonging to the Guttiferae family. It is commonly known as St John's wort. Some taxonomists classify the genus *Hypericum* in a separate family, the Hypericaceae. The genus *Hypericum* encompasses approximately 400 species, of which ten morphologically and chemically distinct species grow in central Europe (Hoelzl 1993). HP is distributed in Europe, Asia, North Africa and North America. Indian HP (IHp) is a rhizomatus perennial herb growing up to a height of 3 feet, distributed in the western Himalayas at altitudes of 3000–10,500 feet. HP contains numerous compounds with documented biological activity. Most researchers consider its effects to be due to variety of constituents rather than any single component. Constituents that have stimulated the most interest include the napthodianthrones, hypericin and pseudohypericin, a broad range of flavonoids, including quercetin, quercetrin, amentoflavone and hypericin, the phloroglucinols hyperforin and adhyperforin, essential oils and xanthones (Benigni *et al*. 1971, Upton 1997).

HP flowers at the time of the summer solstice and in medieval Europe it was considered to have powerful magical properties that enabled it to repel evil. Medicinally, it was used to treat emotional and nervous complaints (Andrew 1996). In the Nineteenth century the herb fell into disuse, but recent research has brought it back into prominence as an extremely valuable remedy for nervous problems. In the European folk medicine tradition, *Hypericum* was used as an antiphlogistic to treat bronchial and urogentital tract inflammations, hemorrhoids, traumas, burns, scalds and ulcers (Bombardelli and Morazzoni 1995). In Russia, *Hypericum* was used in gastroenteritis, rheumatism, boils, hemorrhoids, coughs, excessive bleedings, wounds and ulcers (Hutchins 1991). Native Americans used indigenous species of *Hypericum* root internally to treat consumption and fevers, and externally for snakebite. The aerial parts were used to treat contusions, burns and ulcers (Frichsen Brown 1989).

HP has been widely researched for its antidepressant effects (Muldner and Zoller 1984, Hahn 1992, Hubner *et al*. 1994, Muller and Rossol 1994, Sommer and Harrer 1994, Vorbach *et al*. 1994, Ernst 1995, Butterweck *et al*. 1996, 1997, De Smet and Nolen 1996, Bhattacharya *et al*. 1998, Chatterjee *et al*. 1998a, 1998b, Kumar *et al*. 1999, Gaster and Holroyd 2000, Challem 2001, Gruenwald 2001). The findings clearly show that *Hypericum* extract is clinically effective as an antidepressant drug and works by biochemical mechanisms of action similar to that of the tricyclic antidepressants or the specific serotonin reuptake inhibitors. Accordingly, these findings had an important impact on the recognition of the potential use of HP as an antidepressant drug in many countries all over the world. One advantage of HP is that it did not induce cardiac and anticholinergic side effects as commonly seen with trycyclic antidepressants and MAO inhibitors (David 2000).

As far as IHp is concerned, to date there are no reports on its neuropsychopharmacological profile. Therefore, it was thought worthwhile to evaluate its neuropsychopharmacological actions. The focus of this study was on evaluation of IHp extract on various neuropsychopharmacological activities and elucidation of their mechanisms of action with the help of neurochemical and receptor binding techniques (Kumar *et al.* 1999, 2000a, 2000b, 2000c, 2001a, 2001b, 2001c, 2002).

Materials and methods

Animals

Adult Charles Foster albino rats (150 ±10 g) and Wistar mice (22 ± 5 g), of either sex, were obtained from the Central Animal House of the Institute (Institute of Medical Sciences, Banaras Hindu University, Varanasi) and were randomly distributed into different experimental groups. The rats were housed in groups of six in polypropylene cages at an ambient temperature of 25°C ± 1°C and 45–55% RH, with a 12:12 h light/dark cycle. Animals were provided with commercial food pellets (Brooke Bond-Lipton, India) and water *ad libitum* unless stated otherwise. Experiments were conducted between 09.00 and 14.00 h. Animals were acclimatised for at least 1 week before using them for the experiments and exposed only once to every experiment. 'Principles of laboratory animal care' (NIH publication number 85-23, revised 1985) guidelines were followed.

Drug treatments

The plant IHp was collected during August from the Company Garden, Saharanpur, India. A specimen of the plant is preserved with Indian Herbs, Saharanpur. 50% ethanolic extract (yield 26.75% w/w, standardized for 4.5–5% hyperforin, HPLC) of the dried leaves, flowers and stem of the plant, was orally administered as 0.3% carboxymethyl cellulose (CMC) suspension, in the doses of 100 and 200 mg/kg, once daily for three consecutive days. Pilot studies indicated that single dose administration of IHp had little to no acute behavioural effects, hence the extract of IHp was administered orally at two different dose levels once daily for three consecutive days. Control rats were treated with equal volume of vehicle (0.3% CMC suspension). Standard drugs were used in each set of experiments (mentioned below) accordingly and were administered intraperitoneally to rodents 30 min before experiments for comparison. Experiments were conducted on day 3, 1 h after the last drug administration. In case of antistress activity, the test drugs and vehicle were administered for 14 days, once daily, 1 h before the induction of stress. Experiments were conducted on day 14, 1 h after the last stress procedure and 2 h after drug or vehicle administration.

Drugs and chemicals

The following drugs and chemicals were used and all the reagents and chemicals used were of analytical grade.

- Imipramine (Sun Pharma, India) (15 mg/kg, i.p.) was used as the standard antidepressant agent.
- Lorazepam (Cipla Ltd, India) (0.5 mg/kg, i.p.) was used as the standard anxiolytic agent.
- Piracetam (UCB Pharma Ltd, India) (500 mg/kg, i.p.) was used as the standard nootropic agent.
- Indomethacin (IDPL, India) (20 mg/kg, i.p.) was used as the standard anti-inflammatory agent whereas pentazocine (Ranbaxy, India) (10 mg/kg, i.p.) and aspirin (Astra-IDL Ltd, India) (25 mg/kg, i.p.) were used as standard analgesic agents.

13 Neuropsychopharmacological studies on Indian *Hypericum perforatum* Linn

Vikas Kumar, P.N. Singh and S.K. Bhattacharya

Introduction

Hypericum species were known to ancient communities as useful medicinal plants. The use of *Hypericum perforatum* (HP), in particular, as a remedy was described and recommended throughout the Middle Ages. HP Linn is a perennial plant belonging to the Guttiferae family. It is commonly known as St John's wort. Some taxonomists classify the genus *Hypericum* in a separate family, the Hypericaceae. The genus *Hypericum* encompasses approximately 400 species, of which ten morphologically and chemically distinct species grow in central Europe (Hoelzl 1993). HP is distributed in Europe, Asia, North Africa and North America. Indian HP (IHp) is a rhizomatus perennial herb growing up to a height of 3 feet, distributed in the western Himalayas at altitudes of 3000–10,500 feet. HP contains numerous compounds with documented biological activity. Most researchers consider its effects to be due to variety of constituents rather than any single component. Constituents that have stimulated the most interest include the napthodianthrones, hypericin and pseudohypericin, a broad range of flavonoids, including quercetin, quercetrin, amentoflavone and hypericin, the phloroglucinols hyperforin and adhyperforin, essential oils and xanthones (Benigni *et al*. 1971, Upton 1997).

HP flowers at the time of the summer solstice and in medieval Europe it was considered to have powerful magical properties that enabled it to repel evil. Medicinally, it was used to treat emotional and nervous complaints (Andrew 1996). In the Nineteenth century the herb fell into disuse, but recent research has brought it back into prominence as an extremely valuable remedy for nervous problems. In the European folk medicine tradition, *Hypericum* was used as an antiphlogistic to treat bronchial and urogential tract inflammations, hemorrhoids, traumas, burns, scalds and ulcers (Bombardelli and Morazzoni 1995). In Russia, *Hypericum* was used in gastroenteritis, rheumatism, boils, hemorrhoids, coughs, excessive bleedings, wounds and ulcers (Hutchins 1991). Native Americans used indigenous species of *Hypericum* root internally to treat consumption and fevers, and externally for snakebite. The aerial parts were used to treat contusions, burns and ulcers (Frichsen Brown 1989).

HP has been widely researched for its antidepressant effects (Muldner and Zoller 1984, Hahn 1992, Hubner *et al*. 1994, Muller and Rossol 1994, Sommer and Harrer 1994, Vorbach *et al*. 1994, Ernst 1995, Butterweck *et al*. 1996, 1997, De Smet and Nolen 1996, Bhattacharya *et al*. 1998, Chatterjee *et al*. 1998a, 1998b, Kumar *et al*. 1999, Gaster and Holroyd 2000, Challem 2001, Gruenwald 2001). The findings clearly show that *Hypericum* extract is clinically effective as an antidepressant drug and works by biochemical mechanisms of action similar to that of the tricyclic antidepressants or the specific serotonin reuptake inhibitors. Accordingly, these findings had an important impact on the recognition of the potential use of HP as an antidepressant drug in many countries all over the world. One advantage of HP is that it did not induce cardiac and anticholinergic side effects as commonly seen with trycyclic antidepressants and MAO inhibitors (David 2000).

- Panax ginseng (PG) (B.E. Ltd, India) (100 mg/kg, p.o.) was used as the standard adaptogenic agent for comparison.
- Dopamine (DA), norepinephrine (NE), serotonin (5-HT), methylhydroxyphenyl glycol (MHPG), dihydroxyphenyl acetic acid (DOPAC) and 5-hydroxy indole acetic acid (5-HIAA) were procured from Sigma, St Louis, USA.
- *Copper reagent.*   50 ml of 2% sodium carbonate dissolved in 0.1% N sodium hydroxide + 1 ml of 2% sodium potassium tartarate + 1 ml of 1% copper sulphate.
- *Hydrogen peroxide 7.5 nM.*   1.043 ml of 30% w/w hydrogen peroxide was made upto 100 ml with sodium chloride + EDTA solution (9 g of NaCl + 29.22 mg of EDTA was dissolved in 1 l to make sodium chloride + EDTA solution).
- *Potassium phosphate buffer (65 nM, pH 7.8).*   Potassium dihydrogen phosphate (KH_2PO_4, 2.2 g) and dipotassium phosphate (K_2PO_4, 11 g) were dissolved in 250 ml and 1 l distilled water respectively and then mixed together. The pH was adjusted to 7.8 with KH_2PO_4.
- *Sodium carbonate buffer (0.05 M, pH 10.2).*   Sodium carbonate (Na_2CO_3, 5.3 g) and sodium bicarbonate ($NaHCO_3$, 4.2 g) were dissolved separately in distilled water (1 l) which served as stock solution. Buffer was prepared by mixing Na_2CO_3 (64 ml) and $NaHCO_3$ (70 ml). The pH of the buffer was adjusted to 10.2 by using the above stock solution accordingly.
- *Sucrose solution.*   Sucrose (10.95 g) was dissolved in distilled water (100 ml).
- *Tris HCl buffer (40 mM, pH 7.4).*   Tris (4.8 g) was dissolved in distilled water (1 l) and the pH was adjusted to 7.4 with dilute HCl.

Methods

General neuropharmacological screening

(1) *Potentiation of pentobarbital-induced sleeping time*: Pentobarbital (40 mg/kg, i.p.) was administered to control and drug treated animals. Onset of sleep (loss of righting reflex) was noted and duration of sleep was measured as the period between the loss of righting reflex and its return (Ojima *et al.* 1995). IHp (100 and 200 mg/kg, p.o.) and diazepam (2 mg/kg, i.p.) were administered 45 min and 30 min prior to pentobarbital injection, respectively.

(2) *Locomotor activity*: The spontaneous locomotor activity was assessed with the help of photoactometer (Techno, India) (Ramanathan *et al.* 1999). Each animal was observed for a period of 10 min in a square closed field arena (30 × 30 × 30 cm) equipped with six photocells in the outer wall. Interruptions of photocell beams (locomotor activity) were recorded by means of a six digit counter. The rats were observed for locomotor activity in the apparatus, one at a time.

(3) *Effect on muscle grip performance of mice*: Effect on motor co-ordination was examined on rota-rod apparatus (Techno). Each animal was placed on a rotating rod (20 rpm) in a pre-test session and only those animals, which stayed on the rod for not less than 3 min, were selected for the test session. The test session was performed on the same day as the pre-test session. Fall-off time (when the mouse falls from the rotating rod) for each animal was noted before and after drug administration (Kulkarni and Joseph 1998). IHp (100 and 200 mg/kg, p.o.) and diazepam (2 mg/kg, i.p.) were administered 45 and 30 min before test session, respectively.

(4) *Maximal electroshock (MES) seizures in rats*: According to this method, the supramaximal electroshock (150 mA) was given through a pair of corneal electrodes for 0.2 s duration using a Techno convulsiometer. The hind limb extensor response was taken as the positive end point (Mitra 1990). Albino rats were prescreened and only those showing positive hind limb tonic extensor response were used after an interval of at least 48 h. IHp (100 and 200 mg/kg, p.o.) and

phenobarbitone (2.5 mg/kg, i.p.), was administered 45 and 30 min prior to MES challenge, respectively.

(5) *Pentylenetetrazole (PTZ) induced convulsions in mice*: The mice were challenged with pentylenetetrazole (80 mg/kg, i.p.). The number of mice, which exhibited seizures, the latency to first convulsion and percent lethality were recorded (Rudzik *et al.* 1973). IHp (100 and 200 mg/kg, p.o.) and diazepam (10 mg/kg, i.p.), were administered 45 min and 30 min prior to PTZ challenge, respectively.

Antidepressant activity

(1) *Behavioural despair test (Willner 1984)*. The rat was placed in a cylinder (45×20 cm) containing 38 cm water ($25 \pm 2°C$), so that the rat could not touch the bottom of the cylinder with its hind limb or tail, or climb over the edge of the chamber. Two swim sessions were conducted, an initial 15 min pretest, followed by a 5 min test 24 h later. Drugs were administered after pretest. The period of immobility (remained floating in water without struggling and making only those movements necessary to keep its head above water) during 5 min test period was noted.

(2) *Learned helplessness test (Sherman et al. 1979)*. This model is based on the assumption that, exposure to uncontrollable stress associated with repeated experiences of failure to escape from the stress produces a helpless situation, which results in performance deficits in subsequent learning tasks. A typical experiment involves two parts:

a   *Inescapable shock pretreatment.* Electric foot shocks were delivered in $20 \times 10 \times 10$ cm chamber with plexiglass walls and cover. The floor was made of steel grids to deliver electric shock. A constant current shocker was used to deliver 60 scrambled, randomised inescapable shocks (15 s duration, 0.8 mA, every min) to grid floor. Control rats were placed for 1 h in identical chambers but no shocks were administered. Inescapable shock pretreatment was performed in the morning.

b   *Conditioned avoidance training.* In order to evaluate escape and avoidance performance, avoidance training was initiated 48 h after inescapable shock pretreatment in the jumping box (Techno, India). The jumping box was divided into two equal chambers ($27 \times 29 \times 25$ cm) by a plexiglass partition with a gate providing access to the adjacent compartment through a 14×17 cm space. Animals were placed singly in one of the chambers of jumping box and were allowed to habituate to the test environment for 5 min (for the first session only) and then were subjected to 30 avoidance trails (inter-trial intervals being 30 s). During the first 3 s of each trial, a light signal (conditioned stimulus) was presented, allowing the animals to avoid shocks. If a response did not occur within this period, a 0.8 mA shock (3 s duration) (unconditioned stimulus) was applied via the grid floor. In case no escape response occurred within this period, shock and light conditioned stimulus were terminated. Avoidance sessions were performed for 3 consecutive days (day 3, 4 and 5) in the morning, and the number of escape failures, referred to as no crossing response during shock delivery, was recorded.

(3) *Tail suspension test (Chermat et al. 1986)*. A mouse was hung on a wire in an upside down posture so that its nostrils just touched the water surface in a container. After initial vigorous movements, the mouse assumed an immobile posture and the period of immobility during a 5 min observation period was noted. This test is a reliable and rapid screening method for antidepressants, including those involving the serotonergic system (Bhattacharya *et al.* 1999).

(4) *Reserpine induced hypothermia (Askew 1963)*. On the day before testing, rats were dosed with 2 mg/kg reserpine (Sigma, USA) subcutaneously. Rats had free access to food and water. Eighteen hours after reserpine administration, the animals were placed into individual cages. The initial rectal temperature was determined by insertion of an electric thermometer (telethermometer) to a constant depth of 5 cm. Following administration of the IHp extract, the rectal temperature was measured again at 60 min interval for 7 h.

(5) *5-Hydroxytryptophan (5-HTP) induced head twitches in mice*: Mice were treated with 5-HTP (100 mg/kg, i.p.) and the number of head twitches displayed by each mouse was counted by the staggering method using three 2 min periods (19–21 min), (23–25 min) and (27–29 min) after 5-HTP administration. The effect of IHp on 5-HTP induced head twitches was investigated (Mitra 1990).

(6) *L-dopa induced hyperactivity and aggressive behaviour in mice (Mitra 1990)*. Mice were treated with L-dopa (100 mg/kg, i.p.). Stages of activity and aggressive behaviour were recorded by a scoring system at every 10 min for 30 min after L-dopa administration by the 'blind method'. The different parameters of observation were, piloerection, salivation, increase in motor activity, irritability, reactivity, jumping, squeaking and aggressive fighting. The scores were graded in the following manner:

0   No effect.
1   Piloerection, slight salivation, slight increase in motor activty.
2   Piloerection, salivation, marked increase in motor activty and irritability.
3   Piloerection, profuse salivation, marked increase in motor activty, reactivity, jumping, squeaking and aggressive fighting.

Anxiolytic activity

(1) *Open-field test (Bronstein 1972)*: The open-field apparatus was made of plywood and consisted of squares (61 × 61 cm). The entire apparatus was painted black except for 6 mm thick white lines, which divided the floor onto 16 squares. Open-field was lighted by a 40 W bulb focusing onto the field from a height of about 100 cm. The entire room, except the open-field, was kept dark during the experiment. Each animal was centrally placed in the test apparatus for 5 min and the following behavioural aspects were noted:

- *Ambulation*: this was measured in terms of the number of squares crossed by the animal;
- *Rearings*: number of times the animal stood on its hind limbs;
- *Self groomings*: number of times the animal groomed facial region, and licked/washed/ scratched various parts of its body;
- *Activity in center*: number of central squares crossed by the animal; and,
- *Fecal droppings*: number of fecal droppings excreted during the period.

(2) *Elevated plus-maze test (Pellow and File 1986)*: The maze had two opposite arms, 50 × 10 cm, crossed with two enclosed arms of the same dimension but having 40 cm high walls. The arms were connected with a central square, 10 × 10 cm, giving the apparatus shape of a plus sign. The maze was kept in a dimly lit room and elevated 50 cm above the floor. Naive rats were placed individually in the center of the maze, facing an enclosed arm. Thereafter, the number of entries and time spent on the open and enclosed arms were recorded during the next 5 min. An arm entry was defined when all four paws of the rat were in the arm. A neutral 'blind' observer made observations.

(3) *Elevated zero-maze test (Shepherd et al. 1994)*: The maze comprised a black perspex annular platform (105 cm in diameter, 10 cm width) elevated to 65 cm above the ground level, divided

equally into four quadrants. The two opposite quadrants were enclosed by a black perspex wall (27 cm high) on both the inner and outer edges of the platform, while the remaining two opposite quadrants were surrounded by perspex 'lip' (1 cm high) which served as a tactile guide to animals on these open areas. The apparatus was illuminated by dim white light arranged in such a manner as to provide similar lux levels in open and enclosed quadrants. Rats were placed on one of the enclosed quadrants for a 5 min test period. The maze was cleaned with 5% ethanol/water solution and dried thoroughly between test sessions. During the 5 min test period, time spent on open arms, number of 'head dips' over the edges of platform, and number of 'stretched attend postures' from closed to open quadrants were recorded. Animals were scored as being in the open area when all four paws were in the open quadrants and in the enclosed area only when all four paws had passed the open-closed divide.

(4) *Social interaction test (File and Hyde 1978)*: The rats were first housed individually for 5 days before testing. The apparatus used for the test was a wooden box (60 × 60 × 35 cm) with a solid floor and was placed in a dimly lit room. On day 6, the rats were placed individually in the box and given two 7.5 min familiarisation sessions at 2 h interval. On day 7, rats were paired on weight and sex basis and placed in the box for 7.5 min. During this time total time spent by the rat pair in 'social interaction', including sniffing, following, grooming, kicking, boxing, biting and crawling under or over the partner, was recorded by a neutral 'blind' observer.

(5) *Novelty induced suppressed feeding latency (Bodnoff et al. 1988) test*: The test apparatus was a wooden box (60 × 60 × 35 cm) with a solid floor placed in a dimly lit room. The floor of the wooden box was covered with 2 cm layer of wooden chips, and laboratory chow pellets were evenly placed on the floor. A similar arrangement was made in the home cages of the rats. Food was removed from the home cage 48 h prior to testing, but water was provided *ad libitum*. Naive rats were placed individually in the test chamber and the latency to begin eating (defined as chewing of the pellet and not merely sniffing or playing with it), was recorded. If the rat had not eaten within 300 s, the test was terminated and latency score 300 s was assigned. A neutral 'blind' observer made observations.

Nootropic activity

(1) *Transfer latency in elevated plus maze*: This test was used to assess the retention of learning and memory (Itoh *et al.* 1990). The plus-maze consisted of two opposite open arms, 50 × 10 cm, crossed with two enclosed arms, of the same dimensions with walls 40 cm high. The arms were connected with a central square (10 × 10 cm) to give the apparatus a plus sign appearance. The maze was kept in a dimly lit room elevated 50 cm above floor level. On day 1, a rat was individually placed on the far end of one of the open arms, facing away from the center, and the time taken by the animal to enter one of the closed arms (transfer latency day 1) was recorded with the help of a stop watch. The rat was left in the enclosed arm for 10–15 s and returned to its home cage. On day 2, the procedure was repeated and the day 2 transfer latency was recorded. Similarly after an interval of 1 week, on day 9, the transfer latency was again recorded.

(2) *Passive avoidance test*: This test uses normal behaviour of rats and was developed by King and Glasser (1970). The step through passive avoidance behaviour was evaluated by using the light-dark apparatus, which has two walls of wood and the remaining two walls of transparent plexiglass. It was divided into two equal compartments (30 × 25 × 30 cm) by a plexiglass with a 10 × 10 cm opening in the center. A guillotine door between the two compartments controlled the opening. The light compartment was painted white and a 15 W lamp illuminated it. The interior of the dark chamber was painted black and had a ceiling. Each compartment had

a copper grid floor. To ensure electrical separation, there was a 1.5-cm gap between the two floors in the light-dark box, at the opening between the two chambers. On day 1, a rat was placed in the white box and the time taken to enter into the dark box was noted. As soon as the rat entered the dark box, the guillotine door was closed and foot electric shock (0.5 mA, 3 s) was delivered. The rat was then replaced in its home cage. On the following day (24 h retention interval) each rat was again placed in the white box and was given a 5 min inhibition period. Latency to step through to the dark chamber was recorded. Electric shock was not delivered on day 2. If the animal remained in the white box for a 5 min test period, the maximum score of 300 s was assigned (Ramanathan 1997). On day 9 (after a gap of one week), latency to step through was again recorded to test the retention of the passive avoidance learning.

(3) *Active avoidance test*: Active avoidance learning acquisition and its retention were tested by the method of Spignoli *et al.* (1986a). The apparatus used was the conventional shuttle avoidance box (Techno, India) which consisted of two grid-floor compartments ($29 \times 29 \times 25$ cm each) separated by a plexiglass transparent partition with a single opening (14×17 cm), and buzzer. The rats were placed individually on the right compartment of a shuttle box and allowed to adapt for 15 s. Thereafter, the rats were exposed to a 15 s acoustic buzzer stimulus (conditioned stimulus, CS) followed by both the acoustic stimulus and electric shock (unconditioned stimulus, UCS; 1.5 mA, 50 Hz) through the grid floor of the right for 30 s. Jumping to the unelectrified adjacent (safe) left compartment during CS was designated as conditioned response (CR1), while jumping to the safe left chamber during the initial 15 s adaptation period was designated as anticipatory conditioned response (CR2). The number of trials required by the animal to reach the criterion of two consecutive correct responses represents the learning rate. A 60 min inter-trial interval period was maintained. For statistical analysis, rats not reaching criterion within eight trials were arbitrarily assigned a score of 9. All the rats were subjected to this training schedule and were retested 24 h later and at day 9 (after a gap of 1 week) for retention of the learned task. Besides CR1, CR2 and trial scores, total time taken and the total number of shocks received to reach criterion were also recorded.

The effect of IHp extract was tested against three amnesic models listed below on the learning and memory parameters described earlier. The behaviour experimental procedures and treatment protocols were the same as mentioned earlier.

(4) *Scopolamine induced amnesia*: This test was used to assess the memory functions (Itoh *et al.* 1990). Scopolamine hydrobromide (1 mg/kg, s.c.) was administered immediately after the learning trial on day 1.

(5) *Sodium nitrite induced amnesia*: Sodium nitrite (25 mg/kg, s.c.) was administered immediately after the learning trial on day 1 (Satyan 1997).

(6) *Electroconvulsive shock induced amnesia*: This test was used to assess the memory function (Itoh *et al.* 1990). Electroconvulsive shock (150 mA, 0.2 s) was administered immediately after the learning trial on day 1.

To assess the protective effect of IHp on the above-mentioned amnesic models, retention of the previously learned task was scored on day 2 and day 9, using all three methods detailed above.

Anti-inflammatory activity

(1) *Carrageenan-induced pedal edema in rats*: Male albino rats were injected with 0.1 ml of a 1% carrageenan solution in saline into the sub-plantar region of the left hind paw (Winter *et al.* 1962). The paw was marked with ink at the level of the lateral malleolus and immersed in mercury up to this mark. The paw volume was measured before and 1, 2, 3, 4 and 6 h after the

injection of carrageenan by the mercury displacement method plethysmographically. The edema volume was determined and expressed as percentage swellings, compared with the initial hind paw volume of each rat.

(2) *Cotton pellet induced granuloma in rats*: Subacute inflammation was produced by cotton pellet induced granuloma in rats (Winter and Porter 1957). Sterile cotton (50 ± 1 mg) soaked in 0.2 ml of distilled water containing penicillin (0.1 mg) and streptomycin (0.13 mg) was implanted subcutaneously bilaterally in axilla under the ether anaesthesia. The animals were sacrificed on the seventh day. The granulation tissue with cotton pellet was dried at 60°C overnight and then dry weight was taken. The weight of the cotton pellet before implantation was subtracted from the weight of the dried, dissected pellets.

Analgesic activity

(1) *Tail flick latent period*: The technique used was described by Davies and co-workers (1946), using a Techno analgesiometer. The rat was placed in a rat holder, with its tail coming out through a slot in the lid. The tail was kept on the bridge of the analgesiometer (called jacket) with an electrically heated nichrome wire underneath. The tail received radiant heat from the wire, heated by passing current of 6 mA. Through the water jacket, cold water was continuously passed, so that the bridge did not get heated and the tail could be conveniently placed over the bridge. The time taken for the withdrawal of the tail after switching on the current was taken as the latent period, in s, of 'tail flicking' response. This latent period was considered as the index of nociception. The cutoff time for determination of the latent period was taken as 30 s to avoid injury to the skin (Bhattacharya *et al.* 1971). Three tail flick latencies were measured per rat at each time interval and the means of the tail flick latencies were used for statistical analysis. Pentazocine (10 mg/kg, i.p.) was used as a reference standard.

(2) *Hot plate reaction time in mice*: Mice were screened by placing them on a hot plate maintained at 55 ± 1°C and recording the reaction time in seconds for forepaw licking or jumping (Turner 1965). Only mice which reacted within 15 s and which did not show large variation when tested on four separate occasions, each 15 min apart, were taken for the test. Pentazocine (10 mg/kg, i.p.) was used as a reference standard. The time for forepaw licking or jumping on the heated plate of the analgesiometer maintained at 55°C was taken as the reaction time.

(3) *Acetic acid induced writhing response in mice*: Acetic acid solution (15 mg/ml) at the dose of 300 mg/kg body weight was injected and the number of writhings in the following 30 min period was observed (Turner 1965). A significant reduction in the number of writhings by any treatment as compared to vehicle treated animals was considered as a positive analgesic response. The percentage inhibition of writhing was calculated. Aspirin (25 mg/kg, i.p.) was used as a reference standard.

Anti-stress activity

Induction of chronic stress: The method of Armando *et al.* (1993) was used. The rats were randomly assigned to the unstressed control, stress and drug treated stress groups. Those assigned to the vehicle or drug treated groups were subjected daily (including Sundays) to 1 h of footshock through a grid floor in a standard conditioning chamber with the escape route closed. The duration of each shock (2 mA) and the intervals between the shocks were randomly programmed between 3 and 5 s and 10 and 110 s, respectively in order to make them unpredictable. Animals were sacrificed on day 14, 1 h after the last shock procedure on completion of the test procedure involved.

Techniques used for assessment of stress intensity

The following parameters were used to assess the intensity of stress-induced effects:

(1) *Gastric ulceration*: The stomach was removed and split open along the greater curvature. The number of discrete ulcers were noted by the help of a magnifying glass. The severity of the ulcers was scored after histological confirmations, 0 = no ulcers, 1 = changes limited to superficial layers of the mucosa with no congestion, 2 = half the mucosal thickness showing necrotic changes and congestion, 3 = more than two-thirds of mucosal thickness showing necrotic changes and congestion, and 4 = complete destruction of the mucosa with marked haemmarhage. Thereafter, the period ulcer severity score was calculated (Bhargava and Singh 1981).

(2) *Adrenal gland and spleen weights*: The adrenal gland and spleen were removed and weighed (Bhattacharya *et al.* 2000).

Methods used to assess stress-induced perturbations

Stress-induced 'behavioural depression': The following methods were used to assess behavioural depression:

(3) *Stress-induced 'behavioural despair' test*: Rats were forced to swim individually in a polypropylene vessel (45 × 40 × 30 cm) with a water level of 20 cm, which ensured that the rat's feet did not touch the floor of the vessel and that it could not climb out of it. The rat was allowed to swim for 10 min. Thereafter, during the next 5 min, the total period of immobility, characterised by complete cessation of swimming with the head floating above water level, was noted. This immobility period, after initial frenzied attempts to escape, is postulated to represent 'behavioural despair' as an experimental model of endogenous depression (Porsolt *et al.* 1977).

(4) *Learned helplessness test*: On day 12 of the investigation, rats were subjected to footshock (60 scrambled shocks, 15 s duration, 0.8 mA, every min) in a two compartment jumping box (Techno) with the escape door to the adjoining unelectrified compartment closed. The exercise continued for 1 h. On day 14, 48 h later, the rats were subjected to avoidance training, using the same apparatus but keeping the escape route to the unelectrified chamber open. During this avoidance training the rats were placed in the electrified chamber and allowed to acclimatise for 5 min before being subjected to 30 avoidance trials, with an inter-trial interval of 30 s. During the first 3 s of the trial, a buzzer stimulus (conditioned stimulus, CS) was present followed by electroshock (unconditioned stimulus, UCS) (0.8 mA) delivered via the grid floor for the next 3 s. The avoidance response was characterised by escape to the adjoining 'safe' chamber during CS. Failure to escape during UCS within 15 s assessed as 'escape' failure which is postulated to indicate despair or depression (Thiebot *et al.* 1992).

(5) *Stress-induced suppresion of sexual behaviour in male rats*: A male rat was placed in a cage in a dimly lit room for 10 min with 2 oestrinised (sequentially treated with oestradiol valerate 5 μg/rat, followed 48 h later by hydroxyprogesterone 1.5 mg/rat, s.c.) female rats. The total numbers of mounts were counted (Morishita *et al.* 1993a).

Stress-induced cognitive dysfunction: The following parameters were used to assess the effect of stress on retention of a learned task as memory:

(6) *Active avoidance test*: Rats were trained for an active avoidance task before subjecting them to stress. During training, the rat was placed in the right electrified compartment of a shuttle box (Techno) and allowed to acclimatise for 5 min. Thereafter, the animal was subjected to 15 s of a buzzer stimulus (conditioned stimulus) which was followed by electric shock (1 mA, 50 Hz) given through the grid floor (unconditioned stimulus). The rats were given at least 10 trials, with an inter-trial interval of 60 min, until they reached the criterion of 100% avoidance response of jumping to the unelectrified left chamber of the shuttle box during conditioned

stimulus. The test was repeated on day 14 in order to assess the retention of the active avoidance learning (Jaiswal *et al.* 1989).

(7) *Passive avoidance test*: The test apparatus was a rectangular box (45 × 30 × 40) with an elec-trified grid floor. An 8 cm high platform (17 × 12 cm) was fixed to the centre of the floor. A rat was placed on the platform and allowed to step down. 24 h later, on day 1 of the experiment, the rat was again placed on the platform and on stepping down, received footshock (0.75 mA, 2 s) through the grid floor. The rat was given three more trials until the latency of step down had stabilised. The test was repeated on day 14 and retention of learning as memory, for each rat was recorded (Sen and Bhattacharya 1991).

Neurochemical study

(1) *Assay of brain monoamines and their metabolites*: The rats were sacrificed by decapitation and the heads were immersed in dry ice ethanol solution. The brains were then removed and hippocam-pus, hypothalamus, ponsmedulla, striatum and frontal cortex were dissected out. The tissues were quickly frozen in dry ice/ethanol solution, weighed and stored at $-80°C$ until assay. Frozen tissues were put into 1.5 ml Eppendorf microtubes and homogenised with an ultra sonicater, in 150 µl of ice chilled 0.1 M perchloric acid (PCA), containing 0.1 mM EDTA. After centrifugation at 12,000 × g for 15 min at 4°C, the clear supernatants were filtered through a 0.45 µm filter and 30–50 µl of filtrate were injected into HPLC system. HPLC were carried out according to the method described by Murai and co-workers (1988). Briefly, the mobile phase was 0.02 M sodium acetate/0.0125 M citric acid buffer, pH 3.2 containing 16% (v/v) methanol, 0.033% heptanesulphonic acid and 0.1 mM EDTA. The column temperature was 30°C, flow rate 2.5 ml/min and the back pressure was 205 kg/cm^2. The working standard solu-tion were prepared in 0.1 M PCA containing 0.1 mM EDTA and stored at $-80°C$. The amounts of standards per injection volume of 20 µl were each 40 pmol of NE, DA, 5-HT, MHPG, DOPAC and 5-HIAA. The samples and standards were detected by electrochemical detector with glossy carbon at a voltage setting of 10.83 V versus an Ag–AgCl reference electrode.

(2) *Determination of MAO inhibitory activity*: MAO inhibitory activity was determined by a radiochemical technique (Bhattacharya *et al.* 1991) in the whole brain. Brain samples were acidified (100 µl 2 M HCl) prior to extraction into two volumes of ethyl acetate. The ethyl acetate was then reduced to dryness under a stream of nitrogen and the residue was reconstituted in 1 ml of 100 mM phosphate buffer, pH 7.4 MAO-A and MAO-B inhibitory activities were determined separately by incubation (30 min at 37°C) of aliquots (80 µl) of extract with 20 µl 0.5 (w/v) rat brain homogenate and either 20 µl ^{14}C-5HT (final concentration 83 µM, specific activity 55 mcl/mmol) or ^{14}C-phenylethylamine (PEA, final concentration 5 µM, specific activity 56 mCl/mmol) as substrates, respectively.

Radioligand receptor binding study

High affinity binding assay: The brains of the rats were removed after 45 min of the final admin-istration of the drugs and the crude synaptic membranes were prepared as described earlier (Seth *et al.* 1981). Briefly, brain regions were homogenised in 19 volumes of 0.32 M sucrose and cen-trifuged at 50,000 × g for 10 min. The resulting pellet was homogenised in distilled water and recentrifuged at the same speed. The final pellet was suspended in 40 mM Tris–HCl buffer, pH 7.4, at a concentration representing 50 mg of the original tissue/ml.

Binding incubations were carried out in triplicate in a final volume of 1 ml containing 40 mM Tris–HCl buffer, pH 7.4, and the appropriate labelled and unlabelled pharmacological agents. The amount of tissue used per tube corresponded to 5–15 mg of the original wet weight and contained 300–400 µg of membrane protein as determined by the method of Lowry *et al.* (1951). At the end of a 15 min incubation at 37°C, samples were filtered on glass fiber discs (25-mm diameter, 0.3-µm-pore size, Gelman Inc., Ann Arbor, MI) and rapidly washed twice with 5 ml of Tris buffer. The filter discs were then dried and counted in 5 ml of scintillation mixture using a Tricarb 2660 scintillation counter (Packard Instrument Co., Downers Grove, IL) at an efficiency of 38–43%, to determine membrane bound radioactivity. Control incubations, containing unlabelled competing ligand, were carried out simultaneously with the experimental series to determine the extent of non-specific binding. The final concentration of unlabelled competing compounds in control incubations was 1×10^{-6} M. Specific binding was taken to be that binding which was displaced in the presence of this large excess of the competing compound. The assay for dopamine receptors was performed by using 1×10^{-9} M [1-phenyl-4-^{3}H] spiroperidol (20 Ci/mmol) as the binding ligand and haloperidol as the competing compound in control tubes. In a parallel manner, 1×10^{-9} M DL-[benzillic-4,4'-^{3}H] quinuclidinyl benzilate (QNB) (45 Ci/mmol) was used to measure muscarinic sites with atropine sulphate as a competitor. 0.7×10^{-9} M [methyl-^{3}H] flunitrazepam (94 Ci/m mol) was used for benzodiazepine sites with unlabelled diazepam as competitor and 3×10^{-9} M ^{3}[H] ketanserin (32.5 Ci/mmol) was used for serotonin sites with unlabelled cenanserin as competitor. The method used was essentially similar to other filtration binding methods (Yamamura *et al.* 1978) and satisfies the requirements for saturability, specificity, reversibility and regional distribution (Bondy 1981, Seth *et al.* 1982). The values presented are representative of three separate runs, each in triplicate, performed on the pooled samples of five animals in each series. All individual values of control and experimental groups were found to be within the 95% confidence limits.

Statistical analysis: The data were expressed as means ±SD for each treatment group. The data obtained from each response measure were subjected to Kruskal-Wallis one way analysis of variance (ANOVA) and inter group comparisons were made by Mann–Whitney-*U*-test (two-tailed) for only those responses which yielded significant treatment effects in the ANOVA test.

Observations and results

General neuropharmacological screening

Potentiation of pentobarbital-induced sleeping time: IHp at 100 and 200 mg/kg, p.o. dose did not significantly potentiate pentobarbitone (40 mg/kg, i.p.)-induced onset and duration of sleep while the standard anxiolytic agent diazepam (2 mg/kg, i.p.) exhibited a significant potentiation of pentobarbitone response and produced early onset and prolonged duration of sleep. The results are summarised in Table 13.1.

Table 13.1 Effect of IHp on pentobarbitone sodium hypnosis in rats

Treatment (Dose)	Onset of action (min)	Duration of action (min)	Potentiation (%)
Vehicle	3.44 ± 0.18	117.87 ± 6.10	—
IHp (100 mg/kg)	4.15 ± 0.14	129.14 ± 10.16	9.56
IHp (200 mg/kg)	4.90 ± 0.17	134.66 ± 9.49	14.24

Note
None of the treatments show any statistically significant results.

Locomotor activity: IHp at 200 mg/kg reduced locomotor activity but the same was unaffected by 100 mg/kg dose. Diazepam (2 mg/kg, i.p.) produced a significant decrease in locomotor activity. The results are summarised in Table 13.2.

Effect on muscle grip performance of mice: The IHp extract seemed to be devoid of any motor incoordination effect in the rota-rod test. IHp (100 and 200 mg/kg, p.o.) failed to produce muscle relaxant effect, while diazepam (2 mg/kg, i.p.) produced a significant ataxia. The results are summarised in Table 13.3.

Maximal electroshock (MES) seizures in rats: IHp (100 and 200 mg/kg, p.o.) has got no anticonvulsant activity of its own. However, higher dose, that is, 200 mg/kg, has potentiated the anticonvulsant action of phenobarbitone (2.5 mg/kg, i.p.). The results are summarised in Table 13.4.

Pentylenetetrazole (PTZ) induced convulsions in mice: IHp (100 and 200 mg/kg, p.o.) did not offer any protection against PTZ-induced clonic convulsions. Diazepam (10 mg/kg, i.p.), produced significant protection against convulsions produced by pentylenetetrazole (100 mg/kg, i.p.). However, the higher dose of IHp has potentiated the anticonvulsant activity of diazepam (10 mg/kg, i.p.), which is statistically significant. The results are summarised in Table 13.5.

Table 13.2 Effect of IHp on locomotor activity in mice

Treatment	Locomotor activity counts in 10 min	
	Counts	Change in activity (%)
Vehicle(12)	169.72 ± 12.25	—
IHp(8) (100 mg/kg, p.o)	162.73 ± 13.04	−4.12
IHp(8) (200 mg/kg, p.o)	147.68 ± 13.48[a]	−12.98
Lorazepam(8) (0.5 mg/kg, i.p.)	132.47 ± 14.20[a]	−21.94

Notes
Values in parentheses indicate number of animals.
a $P < 0.01$ in comparison to vehicle.

Table 13.3 Effect of IHp on muscle-grip performance of mice

Treatment	Fall off time (min)	
	Before drug	After drug
Vehicle(12)	4.23 ± 0.18	4.35 ± 0.18
IHp(8) (100 mg/kg, p.o)	4.14 ± 0.20	4.28 ± 0.18
IHp(8) (200 mg/kg, p.o)	3.94 ± 0.24	4.19 ± 0.18
Lorazepam(8) (0.5 mg/kg, i.p.)	4.10 ± 0.22	0.06 ± 0.02[a]

Notes
Values in parentheses indicate number of animals.
a $P < 0.01$ in comparison to vehicle.

Table 13.4 Effect of IHp on MES and anticonvulsant action of phenobarbitone in rats

Treatment	Dose (mg/kg)	Convulsions (%)
IHp(12)	100	100
IHp(8)	200	82.5
Phenobarbitone(8)	2.5	41.25[a]
IHp + Phenobarbitone(8)	100 + 2.5	41.25[a]
IHp + Phenobarbitone(8)	200 + 2.5	29.65[a]

Notes
Values in parentheses indicate number of animals.
a $P < 0.01$ in comparison to vehicle.

Table 13.5 Effect of IHp on pentylenetetrazole induced convulsions in mice

Treatment	Dose (mg/kg)	Incidence of convulsions	Convulsions (%)
Vehicle(12)	—	8	100
IHp(8)	100	6	75
IHp(8)	200	6	75
Diazepam(8)	10	4	50[a]
IHp + Diazepam(8)	100 + 10	4	50[a]
IHp + Diazepam(8)	200 + 10	2	25[a]

Notes
Values in parentheses indicate number of animals.
a $P < 0.01$ in comparison to vehicle.

Table 13.6 Effect of IHp on behavioural despair test in rats

Treatment	n	Duration of immobility (s)
Vehicle	12	112.34 ± 4.65
IHp (100 mg/kg, p.o.)	8	78.53 ± 4.16[a]
IHp (200 mg/kg, p.o.)	8	67.86 ± 2.11[a]
Imipramine (15 mg/kg, i.p.)	8	66.74 ± 3.56[a]

Note
a $P < 0.01$ in comparison to vehicle.

Antidepressant activity

Behaviour despair test: In the initial experiments acute administration of even high doses of IHp extract did not reveal any antidepressant-like effects in this test. Repeated oral administration of IHp extract for three consecutive days did, however, dose dependently reduce the immobility time in rats. Imipramine also showed similar activity and the effects were comparable to that of higher doses of IHp extract. The results are summarised in Table 13.6.

Learned helplessness test: Control rats with prior experiences of inescapable shocks exhibited marked increase in escape failures as compared to those with no such prior experiences. The escape failures significantly and dose dependently decreased in rats treated with both the dose of IHp. As in the behaviour despair test, oral route of IHp (200 mg/kg) was almost equi-effective of 15 mg/kg, i.p. dose of imipramine. The results are summarised in Table 13.7.

Tail suspension test with mice: IHp extract caused a significant and dose dependent decrease in immobility time in the tail suspension test. This effect is regarded as indicative for antidepressant activity. Imipramine also showed significant antidepressant activity and the effects were comparable to that of IHp. The results are summarised in Table 13.8.

Reserpine induced hypothermia: The results demonstrate that IHp extract completely antagonised reserpine induced hypothermia. Imipramine also showed complete antagonism of reserpine induced hypothermia and its effects were comparable to that of IHp. The results are summarised in Table 13.9.

5-Hydroxytryptophan (5-HTP) induced head twitches in mice: IHp significantly inhibited 5-HTP induced head twitches response. However, the lower dose was statistically more significant in inhibiting the response than the higher dose. The data have been summarised in Table 13.10.

L-dopa induced hyperactivity and aggressive behaviour in mice: Effect of IHp (100 and 200 mg/kg, p.o.) was seen on the L-dopa induced excitatory behaviour in mice. Both the doses significantly augmented the hyperactivity scores as is shown in Table 13.11.

Table 13.7 Effect of IHp on learned helplessness test in rats

Treatment	n	Escape failures (N)		
		Day 1	Day 2	Day 3
Vehicle	12	20.66 ± 2.06	11.66 ± 1.77	9.83 ± 0.93
IHp (100 mg/kg, p.o.)	8	17.33 ± 1.86^a	9.66 ± 1.21^a	8.50 ± 1.04^a
IHp (200 mg/kg, p.o.)	8	14.83 ± 1.16^{aa}	8.16 ± 0.75^{aa}	7.16 ± 0.98^{aa}
Imipramine (15 mg/kg, i.p.)	8	12.50 ± 1.87^{aa}	8.66 ± 1.03^{aa}	7.33 ± 0.81^{aa}

Note
a and aa $P < 0.05$ and 0.01 respectively in comparison to vehicle.

Table 13.8 Effect of IHp on tail suspension test in mice

Treatment	n	Duration of immobility (s)
Vehicle	12	118.92 ± 7.16
IHp (100 mg/kg, p.o.)	8	103.18 ± 6.18^a
IHp (200 mg/kg, p.o.)	8	92.46 ± 8.28^{aa}
Imipramine (15 mg/kg, i.p.)	8	91.04 ± 3.84^{aa}

Note
a and aa $P < 0.05$ and 0.01 respectively in comparison to vehicle.

Table 13.9 Effect of IHp on reserpine induced hypothermia in rats

Treatment	Change in rectal temperature (°C)						
	18h	19h	20h	21h	22h	23h	24h
Vehicle(12)	0.01	−0.01	0.03	−0.03	−0.01	−0.05	−0.01
Vehicle + Reserpine(VR)(6)	−1.6	−1.1	−0.4	−0.1	0.01	−0.01	0.03
IHp-100 + Reserpine	−0.53[aa]	−0.25[aa]	−0.1[aa]	−0.08	−0.01	−0.01	0.01
IHp-200 + Reserpine	0[aa]	0.01[aa]	0.03[aa]	0.01[a]	0.01	−0.01	−0.01
IMP + Reserpine	−0.16[aa]	−0.1[aa]	−0.03[aa]	−0.01[a]	0.01	−0.03	0.01

Notes

Values in parentheses indicate number of animals; IMP = Imipramine

a and aa $P < 0.05$ and 0.01, respectively in comparison to VR.

Table 13.10 Effect of IHp on 5-HTP induced head twitches in mice

Treatment	n	Number of head twitches
Vehicle	12	15.78 ± 0.89
IHp (100 mg/kg, p.o.)	8	1.28 ± 0.15[a,b]
IHp (200 mg/kg, p.o.)	8	7.64 ± 0.50[a]
Imipramine (15 mg/kg, i.p.)	8	7.50 ± 0.48[a]

Notes

a $P < 0.01$ in comparison to vehicle.

b $P < 0.01$ in comparison to IHp-200.

Table 13.11 Effect of IHp on L-dopa induced hyperactivity and aggressive behaviour in mice

Treatment	n	Degree of excitation
Vehicle	12	2.95 ± 0.16
IHp (100 mg/kg, p.o.)	8	0.94 ± 0.2[a,b]
IHp (200 mg/kg, p.o.)	8	1.70 ± 0.25[a]

Notes

a $P < 0.01$ in comparison to vehicle.

b $P < 0.01$ in comparison to IHp-200.

Anxiolytic activity

Open-field exploratory behaviour: Rats treated with both the doses of IHp extract showed dose dependent significant increase in open field ambulation, rearings, self grooming and activity in centre in comparison to vehicle treated rats, evincing significant anxiolytic activity of IHp. However, the open-field faecal droppings remain unchanged. Lorazepam (LR) also induced significant anxiolytic activity and the effects were found to be more than that of IHp extract. The data have been summarised in Table 13.12.

Elevated plus maze behaviour: IHp treated rats exhibited dose-dependent significant increase in time spent in open arms, entries made in open arms and a significant decrease in time spent in enclosed arms and entries in enclosed arms in comparison to control rats. The result obtained by open/closed time and entries ratios also indicated significant anxiolysis in rats by IHp extract. LR caused more anxiolysis in comparison to IHp extract. The results have been summarised in Table 13.13.

Elevated zero maze behaviour: The rats treated with IHp extract showed anxiolysis in terms of significant increase in time spent in open arms, entries in open arms and number of head dips on elevated zero maze. However the response stretched attend postures remain unchanged. LR also caused significant anxiolytic activity and the effects were comparable to that of IHp extract. The results have been summarised in Table 13.14.

Social interaction: The rats treated with IHp extract spent significantly more time in social interaction in comparison to control rats and effects of IHp extract was found to be dose dependent. LR also caused significant increase in social interaction in rats and its effects was comparable to that of higher dose (200 mg/kg) of IHp. The results have been shown in Table 13.15.

Table 13.12 Effect of IHp on open-field exploratory behaviour in rats

Treatment (Dose)	n	Ambulation (N)	Rearings (N)	Self groomings (N)	Activity in centre (N)	Faecal droppings (N)
Vehicle	12	46.08 ± 5.40	9.66 ± 2.01	7.58 ± 1.67	1.25 ± 1.21	3.58 ± 1.37
IHp (100 mg/kg)	6	$74.00 \pm 14.87^{aa,cc}$	10.00 ± 2.60^{cc}	8.16 ± 1.16^{cc}	2.66 ± 1.50^{cc}	2.83 ± 0.75
IHp (200 mg/kg)	6	$83.83 \pm 6.11^{aa,b}$	11.33 ± 2.80^{cc}	7.66 ± 1.50^{cc}	4.66 ± 2.16	3.16 ± 0.75
Lorazepam (0.5 mg/kg)	6	85.66 ± 3.55^{aa}	16.66 ± 2.50^{aa}	12.16 ± 1.83^{aa}	$6.00^{aa} \pm 1.41$	2.33 ± 0.81

Notes

a,b,c indicate statistical significance respectively in comparison to vehicle, IHp (100 mg/kg) and lorazepam treatments. b and aa,cc denote $P < 0.05$ and 0.01, respectively.

Table 13.13 Effect IHp on the elevated plus maze behaviour in rats

Treatment (Dose)	n	Time spent on (s)		Entries on		Ratio of open/ enclosed arms	
		Enclosed arms	Open arms	Enclosed arms (N)	Open arms (N)	Entries (N)	Time (s)
Vehicle	12	214.94 ± 5.43	28.72 ± 1.79	7.83 ± 2.79	2.75 ± 0.75	0.37 ± 0.11	0.13 ± 0.01
IHp (100 mg/kg)	6	165.89^{aa} ± 5.47	$64.96^{aa,c}$ ± 7.37	11.66^{a} ± 1.36	$4.66^{aa,cc}$ ± 1.21	0.40^{cc} ± 0.12	$0.38^{aa,c}$ ± 0.03
IHp (200 mg/kg)	6	$158.46^{aa,b,c}$ ± 5.27	$69.74^{aa,b}$ ± 2.18	$14.33^{aa,b,cc}$ ± 2.42	$6.66^{aa,b}$ ± 1.50	0.46^{cc} ± 0.06	$0.44^{aa,bb,c}$ ± 0.01
Lorazepam (0.5 mg/kg)	6	164.14^{aa} ± 4.65	70.16^{aa} ± 3.57	9.83 ± 1.94	7.66^{aa} ± 1.63	0.78^{aa} ± 0.09	0.42^{aa} ± 0.01

Notes

a,b,c indicate statistical significance respectively in comparison to vehicle, IHp (100 mg/kg) and lorazepam treatments. a,b,c and aa,bb,cc denote $P < 0.05$ and 0.01, respectively.

Novelty induced suppressed feeding latency: IHp caused dose dependent significant attenuation of novelty induced feeding latency in rats in comparison to vehicle treatment. LR also induced similar effects, however, its effect was observed to be more than that of IHp extract. The results have been shown in Table 13.16.

Table 13.14 Effect of IHp on the elevated zero maze behaviour in rats

Treatment (Dose)	n	Time spent on open arms (s)	Head dips (N)	Stretched attend postures (N)	Entries in open arms (N)
Vehicle	12	51.28 ± 3.55	8.75 ± 1.71	3.83 ± 1.02	4.75 ± 1.35
IHp (100 mg/kg)	6	70.14 ± 3.62[aa]	12.00 ± 3.03[a]	3.00 ± 0.89	8.00 ± 1.67 [aa,cc]
IHp (200 mg/kg)	6	72.94 ± 2.59[aa]	12.66 ± 4.71[a]	2.50 ± 1.37	9.66 ± 1.36 [aa,b,c]
Lorazepam (0.5 mg/kg)	6	73.14 ± 1.89[aa]	14.16 ± 1.60[aa]	3.66 ± 0.81	11.33 ± 1.21[aa]

Notes
a,b,c indicate statistical significance respectively in comparison to vehicle, IHp (100 mg/kg) and lorazepam treatments. a,b,c and aa,cc denote $P < 0.05$ and 0.01 respectively.

Table 13.15 Effect of IHp on social interaction test in rats

Treatment	n (pair)	Social interaction time (s)
Vehicle	12	157.75 ± 8.73
IHp (100 mg/kg, p.o)	6	213.92 ± 9.62[a,c]
IHp (200 mg/kg, p.o.)	6	226.07 ± 17.37[a]
Lorazepam (0.5 mg/kg, i.p.)	6	235.87 ± 6.37[a]

Notes
a,c indicate statistical significance respectively in comparison to vehicle and IHp-100 treatments. a and c denote $P < 0.01$.

Table 13.16 Effect of IHp on novelty induced suppressed feeding latency in rats

Treatment	n	Latency to feed (s)
Vehicle (home cage)	12	78.35 ± 4.23
Vehicle (novel cage)	12	146.70 ± 6.06[a]
IHp (100 mg/kg, p.o.)	6	134.54 ± 1.69[b,d]
IHp (200 mg/kg, p.o.)	6	113.92 ± 2.17[b,c,d]
Lorazepam (0.5 mg/kg, i.p.)	6	90.14 ± 2.48[b]

Notes
a,b,c,d indicate statistical significance respectively in comparison to vehicle (home cage), vehicle (novel cage), IHp-100 and lorazepam treatments. a,b,c and d denote $P < 0.01$.

Nootropic activity

Transfer latency on elevated plus-maze: TL was the time elapsed between the time the animal was placed in the open arm end and the time when it fully entered (with all four paws) into the enclosed arm. Scopolamine, sodium nitrite and electroconvulsive shock (ECS) treatment significantly increased the TL on second and ninth day. Both doses of IHp and piracetam when given alone, shortened the TL on the first, second as well as the ninth day and also antagonised the effect of scopolamine, sodium nitrite and ECS on TL significantly. The results are summarised in Tables 13.17, 13.18 and 13.19.

Passive avoidance behaviour: The results indicate that both doses of IHp had no significant *per se* effect on the retention of the PA in rats. Only the higher dose (IHp 200 mg/kg) produced a significant reversal of scopolamine impaired PA retention. However, IHp did not attenuate the sodium nitrite impaired PA retention. IHp at both doses produced significant reversal of PA retention deficits induced by ECS. Piracetam not only significantly facilitated the retention but

Table 13.17 Effect of IHp extract on scopolamine induced amnesia on elevated plus maze in rats

Treatment	Dose (mg/kg)	Transfer latencies (s)		
		Day 1	Day 2	Day 9
Vehicle(12)	—	35.38 ± 3.93	26.85 ± 2.82^{cc}	28.80 ± 5.11^{cc}
IHp(6)	100	28.42 ± 4.25	$20.25 \pm 2.96^{aa,cc}$	$19.75 \pm 3.72^{aa,cc}$
IHp(6)	200	22.35 ± 3.46	18.20 ± 3.34^{aa}	$17.50 \pm 2.78^{aa,c}$
Piracetam(6)	500	28.50 ± 2.83	$20.86 \pm 2.98^{aa,cc}$	$20.04 \pm 3.79^{aa,cc}$
Scopolamine(6)	1	40.75 ± 3.65	$46.94 \pm 4.41^{aa,c}$	37.14 ± 4.46^{aa}
IHp + Scopl(6)	100 + 1	26.91 ± 2.40	$20.15 \pm 3.35^{bb,cc}$	$19.00 \pm 2.53^{bb,cc}$
IHp + Scopl(6)	200 + 1	23.85 ± 3.51	$19.25 \pm 2.29^{bb,cc}$	$18.40 \pm 1.99^{bb,cc}$
Pira + Scopl(6)	500 + 1	28.25 ± 2.80	$19.80 \pm 4.19^{bb,cc}$	$18.95 \pm 3.47^{bb,cc}$

Notes
Values in parentheses indicate number of animals; Scopl = scopolamine; Pira = piracetam.
aa and bb $P < 0.01$ in comparison to vehicle and scopolamine respectively.
c and cc $P < 0.05$ and 0.01 in comparison to day 1.

Table 13.18 Effect of IHp extract on sodium nitrite induced amnesia on elevated plus maze in rats

Treatment	Dose (mg/kg)	Transfer latencies (s)		
		Day 1	Day 2	Day 9
Vehicle(12)	—	33.95 ± 4.60	25.42 ± 4.34^{cc}	27.41 ± 3.44^{cc}
IHp(6)	100	27.15 ± 4.15	$21.08 \pm 4.22^{a,c}$	$20.75 \pm 3.64^{aa,c}$
IHp(6)	200	21.55 ± 4.13	18.00 ± 4.76^{aa}	17.64 ± 4.46^{aa}
Piracetam(6)	500	27.65 ± 5.52	$20.05 \pm 3.74^{a,c}$	$19.16 \pm 4.42^{aa,c}$
Sodium nitrite (6)	25	38.40 ± 3.60	$43.95 \pm 3.17^{aa,c}$	35.75 ± 3.42^{aa}
IHp + so nit(6)	100 + 25	27.80 ± 2.66	$21.72 \pm 3.25^{bb,cc}$	$20.16 \pm 4.06^{bb,cc}$
IHp + so nit(6)	200 + 25	20.25 ± 3.15	17.65 ± 3.21^{bb}	16.70 ± 2.82^{bb}
Pira + so nit(6)	500 + 25	30.46 ± 2.90	$21.70 \pm 5.68^{bb,cc}$	$19.55 \pm 2.92^{bb,cc}$

Notes
Values in parentheses indicate number of animals; so nit = sodium nitrite; Pira = piracetam.
a $P < 0.05$ and aa $P < 0.01$ in comparison to vehicle.
bb $P < 0.01$ in comparison to sodium nitrite.
c and cc $P < 0.05$ and 0.01 in comparison to day 1.

Table 13.19 Effect of IHp extract on elecrtroconvulsive shock (ECS) induced amnesia on elevated plus maze in rats

Treatment	Dose (mg/kg)	Transfer latencies (s)		
		Day 1	Day 2	Day 9
Vehicle(12)	—	32.70 ± 4.23	23.47 ± 2.45^{cc}	24.75 ± 2.60^{cc}
IHp(6)	100	30.87 ± 3.79	22.53 ± 2.80^{cc}	22.30 ± 2.85^{cc}
IHp(6)	200	29.15 ± 2.90	22.00 ± 3.70^{cc}	21.65 ± 3.00^{cc}
Piracetam(6)	500	28.95 ± 2.82	$18.68 \pm 2.85^{a,cc}$	$17.14 \pm 2.70^{a,cc}$
ECS(6)	—	31.55 ± 4.51	$41.87 \pm 4.70^{aa,cc}$	32.87 ± 4.15^{a}
IHp + ECS(6)	100	31.64 ± 3.72	$23.70 \pm 2.80^{bb,cc}$	$22.42 \pm 2.60^{bb,cc}$
IHp + ECS(6)	200	31.20 ± 3.65	$24.16 \pm 3.05^{bb,cc}$	$23.50 \pm 3.15^{bb,cc}$
Pira + ECS(6)	500	29.14 ± 3.01	$21.45 \pm 2.58^{bb,cc}$	$20.27 \pm 2.90^{bb,cc}$

Notes

Values in parentheses indicate number of animals; Pira = piracetam.

a and aa indicate statistical significance at $P < 0.05$ and 0.01 in comparison to vehicle respectively.

b and bb indicate $P < 0.05$ and 0.01 in comparison to ECS respectively.

c and cc indicate $P < 0.05$ and 0.01 in comparison to day 1, respectively.

Table 13.20 Effect of IHp extract on scopolamine induced passive avoidance retention deficits in rats

Treatment	Dose (mg/kg)	Step through latencies (s)		
		Day 1	Day 2	Day 9
Vehicle(12)	—	11.40 ± 2.70	12.85 ± 2.63	11.58 ± 2.83
IHp(6)	100	10.80 ± 3.11	11.65 ± 4.45	11.33 ± 3.25
IHp(6)	200	10.16 ± 2.56	13.75 ± 2.68	13.00 ± 2.76
Piracetam(6)	500	10.45 ± 2.21	$18.70 \pm 2.71^{aa,cc}$	$19.05 \pm 2.68^{aa,cc}$
Scopolamine(6)	1	12.19 ± 4.08	09.89 ± 3.12^{a}	09.37 ± 2.90
IHp + Scopl(6)	100 + 1	09.05 ± 2.80	10.20 ± 2.50	10.03 ± 2.38
IHp + Scopl(6)	200 + 1	11.75 ± 2.40	14.10 ± 2.14^{b}	12.95 ± 2.30^{b}
Pira + Scopl(6)	500 + 1	10.82 ± 2.57	$22.65 \pm 3.87^{bb,cc}$	$21.37 \pm 3.10^{bb,cc}$

Notes

Values in parentheses indicate number of animals; Scopl = scopolamine; Pira = piracetam.

a and aa $P < 0.05$ and 0.01 respectively in comparison to vehicle.

b and bb $P < 0.05$ and 0.01, respectively in comparison to scopolamine.

cc $P < 0.01$, in comparison to day 1.

also reversed the scopolamine, sodium nitrite and ECS impaired PA retention in rats. The results are summarised in Tables 13.20, 13.21 and 13.22.

Active avoidance behaviour: The results indicate that rats treated with IHp in both doses required significantly less trials and time to learn the conditioned avoidance response (CAR) task in comparison to vehicle treated rats. IHp also had a dose-related facilitatory effect on the 24 h and 1 week retentions of the previously learned active avoidance. Rats pretreated with IHp and piracetam required significantly less trials, shocked trials and total time to re-learn the task as compared to vehicle treated rats. Furthermore, both the doses of IHp and piracetam significantly attenuated the scopolamine, sodium nitrite and ECS induced CAR retention deficits as indicated by significantly less trials, shocked trials and time taken to reach the criterion of learning and relearning of the CAR task by rats in comparison to rats treated with scopolamine and sodium nitrite. The results are summarised in Tables 13.23, 13.24 and 13.25.

Table 13.21 Effect of IHp extract on sodium nitrite induced passive avoidance retention deficits in rats

Treatment	Dose (mg/kg)	Step through latencies (s)		
		Day 1	Day 2	Day 9
Vehicle(12)	—	10.05 ± 1.98	11.46 ± 2.39	10.85 ± 2.12
IHp(6)	100	10.75 ± 3.45	11.58 ± 4.33	11.16 ± 4.34
IHp(6)	200	11.00 ± 2.88	14.25 ± 2.95^a	14.09 ± 2.69^a
Piracetam(6)	500	10.90 ± 2.57	19.25 ± 3.84^{aa,cc}	19.20 ± 3.71^{aa,cc}
Sodium nitrite(6)	25	11.55 ± 2.03	09.22 ± 1.52^c	08.31 ± 1.42^{a,cc}
IHp + so nit(6)	100 + 25	08.60 ± 2.11	09.80 ± 1.80	09.10 ± 2.25
IHp + so nit(6)	200 + 25	09.47 ± 1.92	13.43 ± 4.04^c	12.19 ± 3.83
Pira + so nit(6)	500 + 25	10.65 ± 3.93	21.08 ± 7.75^{bb,c}	20.88 ± 7.55^{bb,c}

Notes

Values in parentheses indicate number of animals; so nit = sodium nitrite; Pira = piracetam.

a and aa $P < 0.05$ and 0.01 respectively in comparison to vehicle.

bb $P < 0.01$ in comparison to sodium nitrite.

c and cc $P < 0.05$ and 0.01 in comparison to day 1.

Table 13.22 Effect of IHp extract on electroconvulsive shock (ECS) induced passive avoidance retention deficits in rats

Treatment	Dose (mg/kg)	Step through latencies (s)		
		Day 1	Day 2	Day 9
Vehicle(12)	—	12.05 ± 2.80	13.45 ± 3.00	12.32 ± 2.50
IHp(6)	100	11.46 ± 2.45	11.85 ± 4.10	11.95 ± 2.65
IHp(6)	200	10.64 ± 2.54	12.06 ± 2.80	11.40 ± 2.40
Piracetam(6)	500	10.80 ± 2.75	19.87 ± 2.00^{aa,cc}	20.25 ± 1.85^{aa,cc}
ECS(6)	—	11.95 ± 2.80	06.75 ± 1.15^{aa,cc}	06.14 ± 1.10^{aa,cc}
IHp + ECS(6)	100	11.77 ± 2.40	12.14 ± 2.00^{bb}	12.00 ± 1.54^{bb}
IHp + ECS(6)	200	11.24 ± 2.65	13.95 ± 3.15^{bb}	12.78 ± 2.00^{bb}
Pira + ECS(6)	500	11.85 ± 2.54	18.64 ± 2.25^{bb,cc}	17.87 ± 2.10^{bb,cc}

Notes

Values in parentheses indicate number of animals; Pira = piracetam;

aa, bb and cc indicate statistical significance at $P < 0.01$ in comparison to vehicle, ECS and day 1, respectively.

Table 13.23 Effect of IHp extract on scopolamine induced conditioned avoidance response (CAR) acquisition and retention deficits in rats

Treatment	Dose (mg/kg)	Total trials	Shocked trials	CR1	CR2	Total time (s)
Acquisition						
Vehicle(12)	—	5.16 ± 0.75	2.83 ± 0.75	2.16 ± 0.75	0.16 ± 0.40	134.74 ± 5.19
IHp(6)	100	4.16 ± 0.81^a	2.33 ± 0.51	2.33 ± 0.51	0.00 ± 0.00	123.54 ± 6.13^{aa}
IHp(6)	200	4.33 ± 0.51^a	2.16 ± 0.40	2.00 ± 0.63	0.16 ± 0.40	119.85 ± 7.26^{aa}
Piracetam(6)	500	4.00 ± 0.63^a	1.66 ± 0.51^a	2.00 ± 0.00	0.33 ± 0.51	110.45 ± 4.77^{aa}
Scopol(6)	1	5.33 ± 0.51	3.00 ± 0.63	2.16 ± 0.75	0.16 ± 0.40	140.55 ± 5.37^{aa}
IHp + Scopl(6)	100 + 1	5.00 ± 0.63	2.50 ± 0.54	2.50 ± 0.54	0.00 ± 0.00	127.86 ± 6.39^{bb}
IHp + Scopl(6)	200 + 1	4.66 ± 0.51^b	2.33 ± 0.59	2.16 ± 0.75	0.16 ± 0.40	125.54 ± 5.39^{bb}
Pira + Scopl(6)	500 + 1	4.33 ± 0.51^{bb}	2.00 ± 0.63	2.16 ± 0.75	0.16 ± 0.40	120.10 ± 4.58^{bb}

(*Continued*)

Table 13.23 (Continued)

Treatment	Dose (mg/kg)	Total trials	Shocked trials	CR1	CR2	Total time (s)
Retention after 24 h						
Vehicle(12)	—	3.83 ± 0.75^{cc}	1.66 ± 0.51^{cc}	1.83 ± 0.41	0.33 ± 0.51	109.35 ± 7.60^{cc}
IHp(6)	100	3.16 ± 0.41^{c}	1.00 ± 0.63^{cc}	1.66 ± 0.51^{c}	0.50 ± 0.54^{c}	$85.16 \pm 5.36^{aa,cc}$
IHp(6)	200	$3.00 \pm 0.63^{a,cc}$	$0.83 \pm 0.40^{aa,cc}$	1.50 ± 0.83	0.66 ± 0.51^{c}	$80.64 \pm 4.54^{aa,cc}$
Piracetam(6)	500	$2.66 \pm 0.51^{aa,cc}$	$0.66 \pm 0.51^{aa,cc}$	1.50 ± 0.83	0.50 ± 0.54	$75.80 \pm 3.76^{aa\ cc}$
Scopol(6)	1	4.86 ± 0.51^{a}	2.66 ± 0.51^{aa}	1.83 ± 0.40	0.16 ± 0.40	$128.40 \pm 5.42^{aa\ cc}$
IHp + Scopl(6)	100 + 1	$3.66 \pm 0.51^{bb,cc}$	$1.00 \pm 0.63^{bb,cc}$	2.50 ± 0.54^{bb}	0.16 ± 0.40	$93.15 \pm 5.67^{bb,cc}$
IHp + Scopl(6)	200 + 1	$3.50 \pm 0.54^{bb,cc}$	$0.83 \pm 0.40^{bb,cc}$	2.16 ± 0.98	0.50 ± 0.54	$90.85 \pm 4.66^{bb,cc}$
Pira + Scopl(6)	500 + 1	$2.83 \pm 0.40^{bb,cc}$	$0.83 \pm 0.40^{bb,cc}$	1.66 ± 0.51	0.33 ± 0.51	$80.75 \pm 4.06^{bb,cc}$
Retention after 1 week						
Vehicle(12)	—	4.00 ± 0.63^{cc}	1.83 ± 0.40^{cc}	1.83 ± 0.41	0.33 ± 0.51	115.47 ± 4.75^{cc}
IHp(6)	100	$2.83 \pm 0.75^{a,c}$	$0.66 \pm 0.51^{aa,cc}$	1.50 ± 0.54^{cc}	0.66 ± 0.51^{cc}	$83.87 \pm 4.48^{aa,cc}$
IHp(6)	200	$2.66 \pm 0.51^{aa,cc}$	$0.50 \pm 0.54^{aa,cc}$	1.33 ± 0.51	0.83 ± 0.37^{cc}	$80.05 \pm 5.58^{aa,cc}$
Piracetam(6)	500	$2.33 \pm 0.51^{aa,cc}$	$0.50 \pm 0.54^{aa,cc}$	1.16 ± 0.75^{c}	0.66 ± 0.51	$74.06 \pm 3.54^{aa,cc}$
Scopol(6)	1	4.83 ± 0.75^{a}	2.83 ± 0.41^{aa}	1.66 ± 0.81	0.33 ± 0.51	133.80 ± 6.26^{aa}
IHp + Scopl(6)	100 + 1	$3.33 \pm 0.51^{bb,cc}$	$0.83 \pm 0.40^{bb,cc}$	2.00 ± 0.63	0.50 ± 0.54^{c}	$90.60 \pm 4.93^{bb,cc}$
IHp + Scopl(6)	200 + 1	$3.16 \pm 0.41^{bb,cc}$	$0.83 \pm 0.40^{bb,cc}$	1.83 ± 0.75	0.50 ± 0.54	$87.55 \pm 4.55^{bb,cc}$
Pira + Scopl(6)	500 + 1	$2.66 \pm 0.51^{bb,cc}$	$0.66 \pm 0.51^{bb,cc}$	1.66 ± 0.51	0.33 ± 0.51	$80.00 \pm 3.79^{bb,cc}$

Notes

Values in parentheses indicate number of animals; Scopl = scopolamine; Pira = piracetam.

a and aa $P < 0.05$ and 0.01 respectively in comparison to vehicle.

bb $P < 0.01$ in comparison to scopolamine.

c and cc $P < 0.05$ and 0.01 in comparison to acquisition.

Table 13.24 Effect of IHp extract on sodium nitrite induced conditioned avoidance response (CAR) acquisition and retention deficits in rats

Treatment	Dose (mg/kg)	Total trials	Shocked trials	CR1	CR2	Total time (s)
Acquisition						
Vehicle(12)	—	5.33 ± 0.52	3.00 ± 0.63	2.33 ± 0.51	0.00 ± 0.00	137.80 ± 3.78
IHp(6)	100	4.83 ± 0.75	2.50 ± 0.54	2.16 ± 0.41	0.16 ± 0.40	125.16 ± 4.10^{aa}
IHp(6)	200	4.50 ± 0.55^{aa}	2.33 ± 0.51^{a}	2.16 ± 0.41	0.00 ± 0.00	121.55 ± 3.84^{aa}
Piracetam(6)	500	4.16 ± 0.41^{aa}	1.83 ± 0.41^{aa}	2.00 ± 0.63	0.33 ± 0.51	114.40 ± 4.40^{aa}
so nit (6)	25	5.50 ± 0.55	3.16 ± 0.75	2.16 ± 0.41	0.16 ± 0.40	142.50 ± 5.35
IHp + so nit(6)	100 + 25	5.16 ± 0.75	2.66 ± 0.51	2.83 ± 0.75	0.00 ± 0.00	129.95 ± 4.93^{bb}
IHp + so nit(6)	200 + 25	4.83 ± 0.75	2.16 ± 0.40^{b}	2.50 ± 0.83	0.16 ± 0.40	128.20 ± 4.82^{bb}
Pira + so nit(6)	500 + 25	4.50 ± 0.54	2.16 ± 0.41^{b}	2.13 ± 0.75	0.16 ± 0.40	123.25 ± 4.34^{bb}
Retention after 24 h						
Vehicle(12)	—	4.16 ± 0.75^{cc}	2.00 ± 0.63^{cc}	2.00 ± 0.63	0.16 ± 0.40	111.70 ± 4.20^{cc}
IHp(6)	100	$3.33 \pm 0.52^{a,cc}$	$1.16 \pm 0.40^{aa,cc}$	1.83 ± 0.41	0.33 ± 0.51	$86.55 \pm 4.12^{aa,cc}$
IHp(6)	200	$3.16 \pm 0.41^{aa,cc}$	$1.00 \pm 0.00^{aa,cc}$	1.83 ± 0.41	0.33 ± 0.51	$83.50 \pm 3.63^{aa,cc}$
Piracetam(6)	500	$2.83 \pm 0.41^{aa,cc}$	$0.83 \pm 0.37^{aa,cc}$	1.66 ± 0.51	0.33 ± 0.51	$77.94 \pm 5.01^{aa,cc}$
so nit(6)	25	5.00 ± 0.63^{a}	2.66 ± 0.81	2.33 ± 0.81	0.00 ± 0.00	$131.40 \pm 4.50^{aa,cc}$
IHp + so nit(6)	100 + 25	$4.00 \pm 0.63^{b,cc}$	$1.00 \pm 0.63^{bb,cc}$	2.66 ± 0.51	0.33 ± 0.51	$95.01 \pm 5.49^{bb,cc}$
IHp + so nit(6)	200 + 25	$3.66 \pm 0.51^{bb,cc}$	$0.83 \pm 0.40^{bb,cc}$	2.50 ± 0.83	0.33 ± 0.51	$92.35 \pm 5.88^{bb,cc}$
Pira + so nit(6)	500 + 25	$3.16 \pm 0.40^{bb,cc}$	$1.00 \pm 0.00^{bb,cc}$	1.83 ± 0.40	0.33 ± 0.51	$86.54 \pm 4.92^{bb,cc}$

(*Continued*)

Table 13.24 (Continued)

Treatment	Dose (mg/kg)	Total trials	Shocked trials	CR1	CR2	Total time (s)
Retention after 1 week						
Vehicle(12)	—	4.16 ± 0.41^{cc}	2.00 ± 0.00^{cc}	2.00 ± 0.00	0.16 ± 0.40	118.20 ± 3.84^{cc}
IHp(6)	100	$3.00 \pm 0.63^{aa,cc}$	$0.83 \pm 0.40^{aa,cc}$	1.66 ± 0.51	0.50 ± 0.54	$85.00 \pm 3.70^{aa,cc}$
IHp(6)	200	$2.83 \pm 0.41^{aa,c}$	$0.66 \pm 0.51^{aa,cc}$	$1.50 \pm 0.54^{a,c}$	$0.66 \pm 0.51^{aa,c}$	$82.94 \pm 4.65^{aa,cc}$
Piracetam(6)	500	$2.50 \pm 0.55^{aa,cc}$	$0.66 \pm 0.51^{aa,cc}$	1.33 ± 0.51^{aa}	0.50 ± 0.54	$76.05 \pm 3.23^{aa,cc}$
so nit(6)	25	5.16 ± 0.75^{aa}	2.83 ± 0.75^{aa}	2.16 ± 0.75	0.16 ± 0.40	$134.70 \pm 4.93^{aa,cc}$
IHp + so nit(6)	100 + 25	$3.83 \pm 0.75^{b,cc}$	$0.83 \pm 0.40^{bb,cc}$	2.66 ± 1.03	0.33 ± 0.51	$93.10 \pm 4.07^{bb,cc}$
IHp + so nit(6)	200 + 25	$3.33 \pm 0.51^{bb,cc}$	$0.83 \pm 0.40^{bb,cc}$	2.00 ± 1.09	0.50 ± 0.54	$91.55 \pm 3.88^{bb,cc}$
Pira + so nit(6)	500 + 25	$2.83 \pm 0.40^{bb,cc}$	$0.83 \pm 0.40^{bb,cc}$	1.66 ± 0.51	0.33 ± 0.51	$84.95 \pm 3.74^{bb,cc}$

Notes

Values in parentheses indicate number of animals; so nit = sodium nitrite; Pira = piracetam.
a and aa $P < 0.05$ and 0.01, respectively in comparison to vehicle.
b and bb $P < 0.05$ and 0.01, respectively in comparison to sodium nitrite.
c and cc $P < 0.05$ and 0.01 in comparison to acquisition.

Table 13.25 Effect of IHp extract on electroconvulsive shock (ECS) induced CAR acquisition and retention deficits in rats

Treatment	Dose (mg/kg)	Total trials	Shocked trials	CR1	CR2	Total time (s)
Acquisition						
Vehicle(12)	—	5.00 ± 0.71	2.66 ± 0.72	2.33 ± 0.51	0.00 ± 0.00	131.54 ± 7.10
IHp(6)	100	4.83 ± 0.65	2.50 ± 0.51	2.16 ± 0.63	0.16 ± 0.40	128.29 ± 8.54
IHp(6)	200	4.66 ± 0.51	2.33 ± 0.40	2.16 ± 0.75	0.16 ± 0.40	127.68 ± 7.65
Piracetam(6)	500	4.00 ± 0.63^{a}	2.16 ± 0.51	1.50 ± 0.54	0.33 ± 0.51	121.77 ± 6.90^{aa}
ECS(6)	—	5.66 ± 0.51	2.50 ± 0.63	2.83 ± 0.54	0.33 ± 0.51	143.80 ± 7.50^{aa}
IHp + ECS(6)	100	5.00 ± 0.63	2.66 ± 0.54	2.16 ± 0.75	0.16 ± 0.40	125.60 ± 6.65^{bb}
IHp + ECS(6)	200	4.83 ± 0.51^{b}	2.83 ± 0.63	2.00 ± 0.51	0.00 ± 0.00	123.59 ± 6.70^{bb}
Pira + ECS(6)	500	4.16 ± 0.51^{bb}	2.33 ± 0.59	1.66 ± 0.54^{bb}	0.16 ± 0.40	117.48 ± 5.60^{bb}
Retention after 24 h						
Vehicle(12)	—	3.83 ± 0.63^{c}	2.00 ± 0.63	1.66 ± 0.51	0.16 ± 0.40	110.85 ± 8.25^{cc}
IHp(6)	100	3.66 ± 0.51^{c}	1.83 ± 0.40	1.50 ± 0.83	0.33 ± 0.51	107.42 ± 7.10^{cc}
IHp(6)	200	3.16 ± 0.41^{c}	1.66 ± 0.51	1.00 ± 0.63	0.50 ± 0.54	103.78 ± 8.00^{cc}
Piracetam(6)	500	$2.50 \pm 0.54^{a,cc}$	1.00 ± 0.63	1.16 ± 0.41	0.33 ± 0.51	$86.16 \pm 6.10^{aa,cc}$
ECS(6)	—	4.83 ± 0.51	2.83 ± 0.40	1.83 ± 0.40	0.16 ± 0.40	136.55 ± 7.10^{aa}
IHp + ECS(6)	100	$3.83 \pm 0.51^{b,c}$	1.16 ± 0.41	2.16 ± 0.98	0.50 ± 0.54	$92.40 \pm 6.25^{bb,cc}$
IHp + ECS(6)	200	$3.66 \pm 0.51^{b,c}$	1.00 ± 0.63	2.00 ± 0.54	0.66 ± 0.51	$89.74 \pm 6.65^{bb,cc}$
Pira + ECS(6)	500	$3.16 \pm 0.41^{bb,c}$	0.83 ± 0.40	1.66 ± 0.51	0.66 ± 0.51	$80.16 \pm 5.50^{bb,cc}$
Retention after 1 week						
Vehicle(12)	—	4.16 ± 0.63^{c}	2.16 ± 0.40	1.66 ± 0.75	0.33 ± 0.51	117.22 ± 6.90^{cc}
IHp(6)	100	3.83 ± 0.51^{c}	2.00 ± 0.41	1.00 ± 0.41	0.83 ± 0.37	114.35 ± 7.10^{cc}
IHp(6)	200	3.83 ± 0.5^{c}	1.83 ± 0.40	1.33 ± 0.51	0.66 ± 0.51	112.95 ± 6.40^{cc}
Piracetam(6)	500	$2.33 \pm 0.51^{aa,cc}$	0.83 ± 0.41	1.16 ± 0.75	0.33 ± 0.51	$76.44 \pm 5.10^{aa,cc}$
ECS(6)	—	5.00 ± 0.75	3.16 ± 0.54	1.51 ± 0.54	0.33 ± 0.51	137.60 ± 6.54^{aa}
IHp + ECS(6)	100	$3.66 \pm 0.51^{b,c}$	1.00 ± 0.41	2.16 ± 0.75	0.50 ± 0.54	$92.50 \pm 6.70^{bb,cc}$
IHp + ECS(6)	200	$3.16 \pm 0.41^{bb,c}$	0.83 ± 0.40	1.67 ± 0.81	0.66 ± 0.51	$89.67 \pm 7.90^{bb,cc}$
Pira + ECS(6)	500	$2.83 \pm 0.75^{bb,cc}$	0.66 ± 0.51	1.51 ± 0.54	0.66 ± 0.51	$84.75 \pm 7.67^{bb,cc}$

Notes

Values in parentheses indicate number of animals; Pira = piracetam.
a and aa indicate statistical significance at $P < 0.05$ and 0.01 in comparison to vehicle respectively.
b and bb indicate $P < 0.05$ and 0.01 in comparison to ECS respectively.
c and cc indicate $P < 0.05$ and 0.01 in comparison to acquisition respectively.

Table 13.26 Effect of IHp on carrageenan-induced acute paw edema in rats

Treatment	Dose (mg/kg)	Mean % edema at 3 h	Inhibition (%)
Vehicle(10)	—	175.64 ± 6.70	—
IHp(6)	100	71.90 ± 2.80^a	59.06
IHp(6)	200	62.48 ± 3.10^a	64.43
Indomethacin(6)	20	60.87 ± 2.45^a	65.34
IHp + Indomethacin(6)	100 + 20	55.35 ± 1.90^a	68.49
IHp + Indomethacin(6)	200 + 20	51.24 ± 1.84^a	70.83

Notes
Values in parentheses indicate number of animals.
a $P < 0.01$ in comparison to vehicle.

Table 13.27 Effect of IHp on cotton pellet induced granuloma in rats

Treatment	Dose (mg/kg)	Weight of dry cotton pellet granuloma (mg)	Inhibition (%)
Vehicle(10)	—	75.00 ± 6.10	—
IHp(6)	100	34.00 ± 3.75^a	56.40
IHp(6)	200	30.00 ± 2.10^a	61.75
Indomethacin(6)	20	28.00 ± 2.25^a	64.26
IHp + Indomethacin(6)	100 + 20	26.00 ± 3.15^a	67.48
IHp + Indomethacin(6)	200 + 20	23.00 ± 2.05^a	72.15

Notes
Values in parentheses indicate number of animals.
a $P < 0.01$ in comparison to vehicle.

Anti-inflammatory and analgesic activity

Carrageenan-induced pedal edema in rats: The results are given in Table 13.26. IHp exerted a significant anti-inflammatory effect against carrageenan-induced inflammation in doses of 100 and 200 mg/kg but was less than that of indomethacin (20 mg/kg, i.p.). IHp at both doses potentiated the anti-inflammatory activity of indomethacin.

Cotton pellet induced granuloma in rats: The weight of the granulation tissue was significantly reduced by both doses of IHp extract and indomethacin. IHp at both doses potentiated the anti-inflammatory activity of indomethacin as shown by further reduction of granulation tissue (Table 13.27).

Tail flick latent period: IHp at both the dose levels had significant analgesic activity and potentiated the analgesic activity of pentazocine. The results are summarised in Table 13.28.

Hot plate reaction time in mice: IHp extract showed significant analgesic activity at both dose levels. Additionally, IHp potentiated the analgesic activity of pentazocine (10 mg/kg, i.p.). The results are summarised in Table 13.29.

Acetic acid induced writhing response in mice: Both doses of IHp and aspirin (25 mg/kg, i.p.) showed a significant decrease in writhing response induced by acetic acid. In addition, IHp at both doses potentiated the analgesic activity of aspirin as shown by a further decrease in writhing response, when given in combination. The results are summarised in Table 13.30.

Table 13.28 Effect of IHp on tail flick latent period in rats

Treatment	Dose (mg/kg)	Mean latent period of tail flick response (s)	
		Initial	After 30 min
Vehicle(10)	—	09.07 ± 1.42	09.84 ± 1.15
IHp(6)	100	10.45 ± 0.67	14.92 ± 0.80[a]
IHp(6)	200	10.54 ± 1.10	15.85 ± 0.96[a]
Pentazocine(6)	10	10.15 ± 0.62	16.08 ± 1.26[a]
IHp + Pentazocine(6)	100 + 10	09.75 ± 0.83	23.40 ± 1.63[a]
IHp + Pentazocine(6)	200 + 10	09.67 ± 0.87	25.64 ± 1.33[a]

Notes
Values in parentheses indicate number of animals.
a $P < 0.01$ in comparison to vehicle.

Table 13.29 Effect of IHp on hot plate reaction time in mice

Treatment	Dose (mg/kg)	Mean latent time (s)	
		Initial	After 30 min
Vehicle(10)	—	11.45 ± 0.95	10.78 ± 1.10
IHp(6)	100	12.04 ± 1.05	45.28 ± 2.70[a]
IHp(6)	200	11.80 ± 1.38	48.90 ± 2.45[a]
Pentazocine(6)	10	11.40 ± 1.90	51.65 ± 3.15[a]
IHp + Pentazocine(6)	100 + 10	10.85 ± 0.91	53.20 ± 3.68[a]
IHp + Pentazocine(6)	200 + 10	11.28 ± 1.17	55.84 ± 4.15[a]

Notes
Values in parentheses indicate number of animals.
a $P < 0.01$ in comparison to vehicle.

Table 13.30 Effect of IHp on acetic acid induced writhing in mice

Treatment	Dose (mg/kg)	Mean number of writhes in 30 min	Inhibition (%)
Vehicle(10)	—	72.41 ± 3.40	—
IHp(6)	100	46.50 ± 3.27[a]	35.78
IHp(6)	200	31.00 ± 1.41[a]	57.19
Aspirin(6)	25	26.83 ± 2.40[a]	62.95
IHp + Aspirin(6)	100 + 25	22.16 ± 1.94[a]	69.40
IHp + Aspirin(6)	200 + 25	19.83 ± 1.72[a]	72.61

Notes
Values in parentheses indicate number of animals.
a $P < 0.01$ in comparison to vehicle.

Anti-stress activity

Gastric ulceration: Chronic stress markedly increased the incidence, number and severity of gastric ulcers. IHp (100 and 200 mg/kg, p.o.) and PG (100 mg/kg, p.o.) significantly reduced these stress-induced gastric indices. The results are summarised in Table 13.31.

Table 13.31 Effect of IHp on chronic stress-induced gastric ulcerations in rats

Treatment	Dose (mg/kg)	Ulcer incidence (%)	Number of ulcers	Severity of ulcers
Vehicle + Stress(VS)(10)	—	100	8.7 ± 1.7	16.8 ± 3.3
IHp + VS(6)	100	67^a	4.5 ± 2.3^a	8.7 ± 4.5^a
IHp + VS(6)	200	33^a	2.3 ± 1.7^{aa}	4.8 ± 3.5^a
PG + VS(6)	100	67^a	2.3 ± 1.0^a	4.8 ± 2.0^a

Notes
Values in parentheses indicate number of animals.
a $P < 0.01$ in comparison to VS.

Table 13.32 Effect of IHp and PG on chronic stress-induced changes in adrenal gland and spleen weights in rats

Treatment (mg/kg, p.o.)	n	Adrenal gland weight (mg/100 g)	Spleen weight (mg/100 g)
Vehicle	12	24.15 ± 2.10	195.70 ± 11.00
Vehicle + Stress(VS)	8	41.10 ± 3.75^a	128.65 ± 8.54^a
IHp-100 + VS	8	30.74 ± 2.90^b	157.75 ± 7.05^b
IHp-200 + VS	8	27.05 ± 2.54^b	178.48 ± 6.10^b
PG-100 + VS	8	28.45 ± 2.70^b	170.10 ± 9.15^b

Notes
a indicates difference with vehicle treated group.
b indicates difference with VS at $P < 0.01$.

Table 13.33 Effect of IHp on and PG on chronic stress-induced increase in swim stress immobility in rats

Treatment (mg/kg, p.o.)	n	Duration of immobility (s)
Vehicle	12	117.05 ± 10.15
Vehicle + Stress(VS)	12	260.92 ± 7.35^a
IHp-100 + VS	8	181.47 ± 8.65^b
IHp-200 + VS	8	139.28 ± 6.10^b
PG-100 + VS	8	142.45 ± 6.58^b

Notes
a indicates difference with vehicle treated group.
b indicates difference with VS at $P < 0.01$.

Adrenal gland and spleen weights: Chronic stress significantly increased adrenal gland weight and reduced that of spleen. These stress-induced changes were attenuated by IHp (100 and 200 mg/kg, p.o.) and PG (100 mg/kg, p.o.). The results are shown in Table 13.32.

Stress-induced 'behavioural despair' and 'learned helplessness' test: Chronic stress increased the duration of immobility test, while increasing escape failures with a concomitant decrease in avoidance response in the learned helplessness test-features indicative of depression. IHp (100 and 200 mg/kg, p.o.) and PG (100 mg/kg, p.o.) tended to reverse the stress-induced behavioural changes. The results are shown in Table 13.33 and Table 13.34 respectively.

Table 13.34 Effect of IHp on chronic stress-induced changes in learned helplessness test in rats

Treatment	Dose (mg/kg)	Escape failures (N)	Avoidance response (N)
Vehicle(10)	—	14.75 ± 1.80	3.70 ± 0.71
Vehicle + Stress(VS)(10)	—	26.08 ± 1.95^a	0.83 ± 0.18^a
IHp + VS(6)	100	18.35 ± 1.55^b	1.45 ± 0.55^b
IHp + VS(6)	200	15.70 ± 1.24^b	2.70 ± 0.64^b
PG + VS(6)	100	15.95 ± 1.20^b	2.65 ± 0.58^b

Notes

Values in parentheses indicate number of animals.

a indicates difference with vehicle treated group.

b indicates difference with VS at $P < 0.01$.

Table 13.35 Effect of IHp and PG on chronic stress-induced suppression of sexual behaviour in male rats

Treatment (mg/kg, p.o.)	n	Number of mounting (N)
Vehicle	12	6.15 ± 0.80
Vehicle + Stress(VS)	12	1.70 ± 0.74^a
IHp-100 + VS	8	3.25 ± 0.78^b
IHp-200 + VS	8	4.00 ± 0.95^{bb}
PG-100 + VS	8	4.15 ± 0.86^{bb}

Notes

a indicates difference with vehicle treated group at $P < 0.01$.

b and bb indicate difference with VS at $P < 0.05$ and 0.01 respectively.

Table 13.36 Effect of IHp and PG on chronic stress-induced memory deficits in active avoidance response in rats

Treatment (mg/kg, p.o.)	n	Active avoidance response on day 14 (%)
Vehicle	12	82
Vehicle + Stress(VS)	12	20^a
IHp-100 + VS	8	55
IHp-200 + VS	8	70^b
PG-100 + VS	8	70^b

Notes

a indicates difference with vehicle treated group.

b indicates difference with VS at $P < 0.01$.

Stress-induced suppresion of sexual behaviour in male rats: Chronic stress significantly decreased the sexual behaviour of male rats, as indicated by a decrease in the number of mountings. This stress effect was reversed by IHp (100 and 200 mg/kg, p.o.) and PG (100 mg/kg, p.o.). The results are shown in Table 13.35.

Active and passive avoidance tests: Chronic stress produced a significant decrease in the retention of acquired active and passive learning. These stress-induced memory deficits were reduced by IHp (100 and 200 mg/kg, p.o.) and PG (100 mg/kg, p.o.). The results are shown in Tables 13.36 and 13.37 respectively.

Neurochemical study

Assay of brain monoamines and their metabolites: IHp extract, 200 mg/kg, p.o., significantly modulated the levels of the neurotransmitters in various regions of the brain under the study. IHp treatment significantly reduced the levels of 5-HT in the hypothalamus, hippocampus, striatum, frontal cortex and pons medulla. Similarly, levels of 5-HIAA, a metabolite of 5-HT, were also found to be decreased in the hypothalamus, hippocampus, striatum and frontal cortex. The ratio of 5-HIAA/5-HT (turnover) was significantly attenuated only in the frontal cortex. IHp treatment significantly augmented the levels of NE in the hypothalamus, hippocampus, striatum, frontal cortex and pons medulla. Likewise, the levels of MHPG, a metabolite of NE, were also increased in all the brain regions assayed, namely hypothalamus, hippocampus, striatum, frontal cortex and pons medulla. The turnover was also augmented significantly in the striatum and frontal cortex. IHp treatment also elicited a significant increase in the levels of DA in the hypothalamus, striatum and frontal cortex. The levels of DOPAC, a metabolite of DA, were also significantly augmented in the hippocampus and frontal cortex. Similarly, the ratio of DOPAC/DA (turnover) was also found to be increased in the hypothalamus. The results are summarised in Tables 13.38, 13.39 and 13.40.

Table 13.37 Effect of IHp and PG on chronic stress-induced memory deficits in passive avoidance response in rats

Treatment (mg/kg, p.o.)	n	Step-down latency inflexion ratio on day 14
Vehicle	12	8.9 ± 1.2
Vehicle + Stress(VS)	12	2.6 ± 0.9^a
IHp-100 + VS	8	4.1 ± 0.8
IHp-200 + VS	8	6.2 ± 0.6^b
PG-100 + VS	8	5.0 ± 0.8^b

Notes
a indicates difference with vehicle treated group.
b indicates difference with VS at $P < 0.01$.

Table 13.38 Effect of IHp (200 mg/kg, p.o.) on 5-HT and 5-HIAA levels in rat brain

Brain region	5-HT (ng/g)	5-HIAA (ng/g)	Turnover (%)
Hypothalamus			
Control(8)	1.096 ± 0.084	0.604 ± 0.048	56.18 ± 5.08
IHp(6)	0.694 ± 0.023^a	0.404 ± 0.023^a	58.91 ± 5.15
Hippocampus			
Control(8)	0.396 ± 0.032	0.249 ± 0.022	63.25 ± 5.90
IHp(6)	0.226 ± 0.029^a	0.149 ± 0.018^a	66.40 ± 6.20
Striatum			
Control(8)	0.362 ± 0.041	0.598 ± 0.036	165.74 ± 5.80
IHp(6)	0.219 ± 0.028^a	0.374 ± 0.029^a	170.95 ± 11.34
Pons medulla			
Control(8)	0.839 ± 0.052	0.529 ± 0.041	64.10 ± 5.71
IHp(6)	0.704 ± 0.032^a	0.403 ± 0.034^a	57.84 ± 5.10
Frontal cortex			
Control(8)	0.402 ± 0.029	0.298 ± 0.018	75.38 ± 6.35
IHp(6)	0.321 ± 0.033^a	0.168 ± 0.022^a	53.62 ± 4.40^a

Notes
Values in parentheses indicate number of animals.
a denotes statistical significance in comparison to control at $P < 0.01$.

Table 13.39 Effect of IHp (200 mg/kg, p.o.) on NE and MHPG levels in rat brains

Brain region	NE (ng/g)	MHPG (ng/g)	Turnover (%)
Hypothalamus			
Control(8)	2.78 ± 0.162	1.66 ± 0.193	60.73 ± 5.70
IHp(6)	3.23 ± 0.177^a	2.12 ± 0.169^a	65.94 ± 6.15
Hippocampus			
Control(8)	0.49 ± 0.031	0.29 ± 0.084	59.69 ± 4.98
IHp(6)	0.66 ± 0.098^a	0.44 ± 0.062^a	67.48 ± 6.24
Striatum			
Control(8)	0.36 ± 0.062	0.22 ± 0.026	61.89 ± 5.26
IHp(6)	0.52 ± 0.033^a	0.39 ± 0.016^a	75.60 ± 6.38^a
Pons medulla			
Control(8)	0.98 ± 0.096	0.58 ± 0.052	60.97 ± 5.72
IHp(6)	1.23 ± 0.072^a	0.72 ± 0.084^a	58.64 ± 4.94
Frontal cortex			
Control(8)	0.29 ± 0.029	0.21 ± 0.019	73.05 ± 6.28
IHp(6)	0.039 ± 0.028^a	0.36 ± 0.013^a	92.86 ± 8.10^a

Notes

Values in parentheses indicate number of animals.

a denotes statistical significance in comparison to control at $P < 0.01$.

Table 13.40 Effect of IHp (200 mg/kg, p.o.) on DA and DOPAC levels in rat brains

Brain region	DA (ng/g)	DOPAC (ng/g)	Turnover (%)
Hypothalamus			
Control(8)	0.56 ± 0.033	0.24 ± 0.016	43.78 ± 3.90
IHp(6)	0.69 ± 0.082^a	0.42 ± 0.083^a	61.10 ± 5.78^a
Hippocampus	Not Done		
Striatum			
Control(8)	9.86 ± 0.512	4.09 ± 0.199	42.83 ± 3.92
IHp(6)	10.53 ± 0.642^a	4.43 ± 0.444^a	43.25 ± 4.04
Pons medulla	Not Done		
Frontal cortex			
Control(8)	0.32 ± 0.029	0.16 ± 0.011	51.16 ± 4.75
IHp(6)	0.39 ± 0.056^a	0.21 ± 0.034^a	54.45 ± 5.10

Notes

Values in parentheses indicate number of animals.

a denotes statistical significance in comparison to control at $P < 0.01$.

Determination of MAO inhibitory activity: IHp treatment had no significant MAO-A or MAO-B inhibitory activity. The results are summarised in Table 13.41.

Radioligand receptor binding study

High affinity binding assay: IHp extract significantly reduced the [^{3}H] spiroperidol binding to corpus striatal membranes whereas significant increases in [^{3}H] ketanserin and [^{3}H] flunitrazepam bindings in frontal cortex membranes were observed. However, no significant alteration in the binding level of [^{3}H] QNB was observed in the hippocampus. The results are summarised in Table 13.42.

Table 13.41 Effect of IHp on MAO-A and MAO-B inhibitory
activity in rat brain

Treatment	MAO-A inhibition (%)	MAO-B inhibition (%)
Vehicle(8)	16.9 ± 0.34	19.6 ± 0.42
IHp(6)	18.4 ± 0.86	20.5 ± 0.57

Notes
Values in parentheses indicate number of animals. None of the
treatments show any statistically significant results.

Table 13.42 Effect of IHp on high affinity neurotransmitter binding sites in brain regions of rats

Brain region	Receptor (radioligand)	Treatment (mg/kg)	p moles bound/g protein	Change (%)
Corpus striatum(6)	DA-D$_2$ (^{3}[H] Spiroperidol)	Vehicle	463.62 ± 61.61	—
		IHp-100	345.77 ± 22.15	-25.42
		IHp-200	300.32 ± 34.40^a	-35.22
Frontal cortex(6)	5-HT$_{2A}$ (^{3}[H] Ketanserin)	Vehicle	127.95 ± 13.29	—
		IHp-100	139.92 ± 13.95	$+09.35$
		IHp-200	146.70 ± 13.95	$+14.65$
Frontal cortex(6)	BDZ (^{3}[H]Flunitrazepam)	Vehicle	41.46 ± 02.27	—
		IHp-100	61.25 ± 03.92^{aa}	$+47.33$
		IHp-200	70.87 ± 03.74^{aa}	$+70.93$
Hippocampus(6)	M$_2$ (^{3}[H] QNB)	Vehicle	202.12 ± 06.14	—
		IHp-100	224.76 ± 07.14^a	$+10.07$
		IHp-200	180.87 ± 13.95	-10.46

Notes
Values in parentheses indicate number of animals.
a and aa indicate statistical significance at $P < 0.05$ and 0.01, respectively in comparison to vehicle.

Discussion

HP, also known as St John's wort, has been used therapeutically for more than two centuries in Europe. HP has been mentioned in *Ayurveda* and is known as *Bassant*, though its clinical use does not appear to include nervous disorders (Satyavati *et al.* 1987). The therapeutic uses of *Hypericum* are multiple, ranging from an antidepressant to use in cancer. Several herbal products widely used and recommended in Germany for the treatment of nervous disorders and sleeplessness, contain hydroalcoholic extracts of the medicinal plant HP (Lohse and Muller 1995). Such therapeutic uses of the extracts have long been known and recent well-designed clinical trials have demonstrated their antidepressant activities (Linde *et al.* 1996, Volz 1997). A meta-analysis of the existing double-blind studies (Linde *et al.* 1996) demonstrates that these extracts are more effective than placebo and similarly effective as some standard antidepressant drugs like amitriptyline, imipramine and maprotiline for the treatment of mild to moderately severe depressive disorders (Linde *et al.* 1996, Volz 1997, Wheatley 1997). Furthermore, it has also been shown that treatments with such extracts seem to be devoid of major side effects typical for the tricyclic antidepressants (TCA) or for the specific serotonin reuptake inhibitors (SSRI) and less expensive (Woelk *et al.* 1994, De Smet and Nolen 1996, Linde *et al.* 1996, Vorbach *et al.* 1997, Wheatley 1997). Available reports clearly suggest that exploitation of this medicinal plant could not only eventually lead to more rational use of the *Hypericum* extracts for the treatment of depression, but may also be helpful in the search of antidepressant molecules with potentially novel mechanism(s) or mode(s) of action. The extracts used for such purposes are, in general, standardised on their contents of hypericins. These naphthodianthrones are characteristic components of HP and they are also considered to be active antidepressive components of the extracts (Holzl *et al.* 1994, Butterweck *et al.* 1997). However, recent detailed studies indicate that several chemical classes of *Hypericum* extract constituents could be involved in their clinical efficacies and that the effects of hypericins themselves are modulated by other extract component(s). Thus a systematic activity-guided fractionation study demonstrated that antidepressant effects of hypericins in animal models are observed or enhanced after co-administering them with another extract fraction devoid of them (Butterweck *et al.* 1997). Although the structure(s) of the potentiating component(s) remains unknown, it seems to be a flavonic molecule. Another interesting finding in this study is that dopamine antagonists can inhibit the central effects of a *Hypericum* extract and their components.

In addition, independent efforts in our laboratories (Chatterjee *et al.* 1998a) have shown that a *Hypericum* extract devoid of hypericins and flavonics but enriched in the acylphloroglucinol derivative hyperforin also possesses antidepressant activities in two behavioural models. In addition, antidepressant activities of isolated pure hyperforin have also been demonstrated and other studies conducted with this constituent indicate that serotoninergic 5-HT$_3$/5-HT$_4$ receptor mediated processes could be involved in its mode of action (Chatterjee *et al.* 1996, 1998b). Results of some *in vitro* studies also revealed that like many therapeutically used antidepressants, this major *Hypericum* extract component is an inhibitor of synaptosomal uptakes of biogenic amines (Muller *et al.* 1997, Chatterjee *et al.* 1998a). Taken together, these observations confirm that several constituents and mechanisms are responsible for the clinically observed antidepressant activities of *Hypericum* extracts and that complex interactions between diverse pharmacologically active principles could be involved in their observed central effects. It is becoming clear, therefore, that standardising the therapeutically used extracts on one single class of constituents is not sufficient and that efforts should be made to identify and evaluate the relative therapeutic importance of various extract components and to study possible pharmacological interactions between them. IHp contains more amount of hyperforin than German *Hypericum* (S. Ghosal unpublished data). Therefore, to rationalise hyperforin as a possible antidepressant component of *H.* extracts, we have used IHp extract standardised for 4.5–5% hyperforin. However, IHp has not been investigated for its neuropsychopharmacological properties earlier. The present investigation was undertaken to investigate the neuropsychopharmacological profile of this widely used plant drug in rodents. If IHp has similar efficacy as its European counterpart, it could prove to be a major export item, in view of the US$ 3–4 billion market of *Hypericum* in the West.

General neuropharmacological screening

The effect of pentobarbitone sodium on righting reflex (hypnosis) is used to elucidate CNS-active properties of drugs (Vogel and Vogel 1997). The loss of righting reflex is measured as criterion for the duration of pentobarbitone-induced sleeping time. IHp produced a dose related potentiation of pentobarbitone hypnosis indicating that the IHp has a sedative action in the doses used.

Interruption of light beams as lateral movements of rats or mice in a cage has been used by many authors (Dews 1953, Salens *et al.* 1968, Nakatsu and Owen 1980). Most of the central nervous system acting drugs influence the locomotor activities in man and animals. The locomotor activity can be an index of wakefulness of mental activity (Kulkarni 1993). IHp attenuated motor activity possibly by its central sedative action, and it is quite possible that the antianxiety effect of the drug (discussed later) is responsible.

The rota-rod test is used to evaluate the activity of drugs interfering with motor coordination. In 1957, Dunham and Miya suggested that the skeletal muscle relaxation induced by a test compound could be evaluated by testing the ability of mice or rats to remain on a revolving rod. Many central depressive drugs are active in this test. The test does not really differentiate between anxiolytics and neuroleptics but can evaluate the muscle relaxant potency in a series of compounds such as the benzodiazepines. Moreover, the test has been used in the toxicology for testing neurotoxicity. The IHp extract seemed to be devoid of any motor incoordination effect in the rota-rod test. IHp (100 and 200 mg/kg, p.o.) failed to produce a muscle relaxant effect, while the benzodiazepine (BDZ) derivative diazepam (2 mg/kg, i.p.) produced a significant ataxia due to its significant sedation. Thus IHp may have an advantage over BDZs.

The maximal electroshock (MES) assay in animals is used primarily as an indication for compounds, which are effective in grandmal epilepsy. Tonic hind extensions are evoked by electric stimuli, which are suppressed by anti-epileptics but also by other centrally active drugs (Vogel and Vogel 1997). IHp did not alter the flexor, extensor and clonic phases of the MES seizures, and did not offer any protection against pentylenetetrazole-induced convulsions. Pentylenetetrazole (PTZ) induced convulsions have been used primarily to evaluate antiepileptic drugs likely to be useful in petitmal epilepsy. However, it has been shown that most anxiolytic agents are also able to prevent or antagonise PTZ-induced convulsions. These methods are widely accepted as screening procedure (Vogel and Vogel 1997). IHp at both dose levels potentiated anticonvulsant effect of phenobarbitone (2.5 mg/kg, i.p.) and diazepam (10 mg/kg, i.p.) probably by its sedative/anxiolytic action (Kumar *et al.* 2000a).

Antidepressant activity

Amongst a wide variety of proposed and critically assessed *in vivo* models of depression (Willner 1984, 1991, Bhattacharya *et al.* 1999) the two most commonly used paradigms are learned helplessness (Seligman and Maier 1967, Seligman and Beagley 1975) and behaviour despair forced swim tests (Porsolt *et al.* 1977). A few reports have also demonstrated the efficacies of standardised *Hypericum* extracts in these animal models (Winterhoff *et al.* 1995, Ozturk *et al.* 1996, Butterweck *et al.* 1997, Bhattacharya *et al.* 1998, Chaatterjee *et al.* 1998a, Kumar *et al.* 1999). In learned helplessness tests, rodents are exposed to inescapable and unavoidable electric shocks in one situation and later fail to escape shock in a different situation when escape is possible (Overmier and Seligman 1967, Maier and Seligman 1976, Bhattacharya *et al.* 1999). This phenomenon was evaluated as a potential animal model of depression (Sherman *et al.* 1979). A drug is considered to be effective, if the learned helplessness is reduced and the number of failures to escape is decreased (Vogel and Vogel 1997). Behavioural despair was proposed as a model to test for antidepressant activity by Porsolt *et al.* (1977, 1978). It was suggested that mice or rats forced to swim in a restricted space from which they cannot escape, exhibit a characteristic immobility. This behaviour reflects a state of despair, which can be reduced by several agents, which are therapeutically effective in human depression (Vogel and Vogel 1997). Apart from these two paradigms, the observed results in tail suspension and reserpine induced hypothermia tests provide additional measures for assessing antidepressant activity. In tail suspension tests, the immobility displayed by rodents when subjected to an unavoidable and inescapable stress has been hypothesised to reflect depressive disorders in humans. Clinically effective antidepressants reduce the immobility that mice display after active and unsuccessful attempts to escape when suspended by their tails (Vogel and Vogel 1997). The reserpine induced hypothermia test has been proven as a simple and reliable method to detect antidepressant activity. However, the reversal of hypothermia is not specific for antidepressants. Amphetamines and some antipsychotic agents (chlorpromazine) can also antagonise the fall in body temperature. The different time course of antidepressants (slow onset of action, long lasting effect) and amphetamine-like drugs (quick onset of action, short lasting effect) allows differentiation between two groups of drugs (Vogel and Vogel 1997). As mentioned earlier, IHp inhibited 5-HTP induced head twitch response, although the inhibitory effect of the higher dose used was less than that of the lower dose. The 5-HTP induced head twitch response in mice is indicative of central serotonergic activity (Ernst 1972). Serotonin is known to be involved in the sleep stages of the sleep wake cycle and IHp induced inhibition of a central 5-HTP mediated action can explain its effect on arousal and increase in physical performance. However, its effect on pentobarbitone sleeping time is apparently contradictory to this conjecture. Antidepressants like imipramine are known to potentiate barbiturate hypnosis (Bhattacharya *et al.* 1999).

IHp induced a dose related increase in L-dopa induced hyperactivity. This observation tends to support the observed anti 5-HT activity but is contrary to the observed hypnotic potentiating effect of the drug. Controversy exists on the relative importance on the central noradrenergic and dopaminergic mediation in L-dopa induced excitatory behaviour. This controversy has not been resolved despite the extensive use of a number of pharmacological drugs which selectively affect either neurotransmitter system. However, the general consensus appears to be that the central excitatory response of L-dopa is mediated by noradrenaline rather than dopamine (Mukhopadhyay *et al.* 1987). It is thus possible that IHp has a facilitatory effect on the central noradrenergic neurotransmitter system, which may explain its beneficial effect on its physical endurance.

Recently, *Hypericum* extracts containing hyperforin have been reported to exhibit anxiolysis in rats on various paradigms of anxiety (Bhattacharya *et al.* 1998, Kumar *et al.* 2000a). A standardised extract of HP has been reported to possess psychotropic activities like antidepressants in a water wheel and isolation induced aggression tests in mice (Okpanyi and Weischer 1987). Other researchers have also reported similar antidepressant activity in HP extract using tail suspension and forced swim tests (Butterweck *et al.* 1996, 1997). If HP shares a similar mechanism with currently used antidepressants, this is not apparent so far. The available reports indicate that HP appears to affect multiple neurotransmitters without fitting easily into known antidepressant categories (Cott 1997). Although HP demonstrates monoamine oxidase (MAO) inhibition *in vitro*, this effect has not been demonstrated *in vivo*, nor have there been any reported cases of MAO inhibitor-associated hypertensive crisis in humans using HP (Cott 1997). While previous studies showed that hypericin inhibits MAO in concentrations of 50 μg/ml (Suzuki *et al.* 1984) others have failed to confirm this effect (Demisch *et al.* 1989, Bladt and Wagner 1994). Bladt and Wagner (1994) reported that *Hypericum* fractions with the greatest MAO inhibition contain the highest concentration of flavonoids. Computer modelling of *Hypericum* constituents also suggests flavonoids to be the most likely MAO inhibitor fraction, due to structural similarity to taloxotone and brofaromine, two known inhibitors of MAO-A (Holtje and Walper 1993). In another study, the xanthone fraction was found to be a particularly strong inhibitor of MAO-A *in vitro* (Sparenberg *et al.* 1993).

Earlier studies suggested that hypericin, a MAO inhibitor, may be and active constituent of HP. However, many recent studies indicate that the major active constituent of HP may be the acylphloroglucinol, hyperforin. Extracts rich in hyperforin and devoid of hypericin have shown significant antidepressant activity in animal models of clinical depression. However, although hyperforin may be the major antidepressant component of HP extract, there may be other unidentified antidepressant components (Chatterjee *et al.* 1996, Chatterjee *et al.* 1998a). Hyperforin-containing HP extracts have been shown to inhibit synaptosomal uptake of serotonin, norepinephrine and dopamine with similar affinities and to lead to a significant down-regulation of cortical β-adrenoceptors and 5-HT$_2$-receptors after subchronic treatment of rats. These reuptake inhibiting properties are not shown by hypericin (Muller *et al*. 1998). Preliminary investigations of IHp indicate that it contains significant amounts of hyperforin and is devoid of hypericin. A comparative chemical evaluation of the European and Indian varieties of HP is in progress.

Depressive disorders are now regarded as a major health problem (NIH 1992, Eisenberg 1992). Despite considerable progress made during the last five decades, successful treatment of clinical depression with currently available therapeutic agents can be achieved only in 65–75% of patients, of which only 40–50% achieve complete recovery (Keith and Mathews 1993). Such a situation necessitates the development of more effective antidepressants (Broekkamp *et al*. 1995, Moller and Volz 1996). The first generation of antidepressants, the TCA, discovered only after fortuitous clinical findings with imipramine (Kumar *et al*. 1999), are still widely used because of their familiarity and low cost. The introduction of second-generation antidepressants may have reduced the risks of adverse effects of the first generation tricyclic antidepressants, but made little impact on improving the effectiveness of treatment (Polter *et al*. 1992, Broekkamp *et al*. 1995, Moller and Volz 1996). The search for new molecules as targets for antidepressant drug discovery, therefore, remains a continuing challenge for modern psychiatry. It has been pointed out that, as in various other therapeutic areas (Kinghorn and Balandrin 1993, Pezzuto 1997), investigations of traditional herbal products may provide a good chance for novel treatments for affective and other CNS disorders (Duke 1995, Cott 1996).

It has been pointed out that although the majority of the world's health care services use herbal medicines (Jonas 1997), the wide acceptance and rational uses of such botanical medicine is possible only when the active constituents and their modes of action are known. The case of Indian *Hypericum* extract does not also seem to be much different. Our experimental findings, thus, not only confirm the previous studies but also provide a rationale for the use of IHp as an antidepressant.

Anxiolytic activity

Most of the animal models of anxiety now in use were developed for benzodiazepines (BDZ) and, since these compounds also exhibit significant muscle relaxant and anticonvulsant effects, evaluation of anxiolytic activity, even with non-BDZ compounds, invariably now includes tests for these neuropharmacological actions (Wada *et al*. 1989). The sedative, amnesic and ataxic effects of BDZ and non-BDZ anxiolytics are definite drawbacks when these drugs are used for the treatment of anxiety. However, since the question of reliability and validity is foremost in establishing animal tests, recourse has to be taken to compare the pharmacological profile of activity of a putative anxiolytic agent with that elicited by a BDZ (File 1985). As such, despite the additional effects that lorazepam is known to have, it was used to validate the anxiolytic activity of IHp.

In the open field test, when animals are taken from their home cage and placed in a novel environment, they express their anxiety and fear by a decrease in ambulation and exploration, freezing, rearing and grooming behaviours, and an increase in defaecation due to heightened autonomic activity. These behavioural changes are attenuated by classical anxiolytics and augmented by anxiogenic agents (Bhattacharya and Satyan 1997). Likewise, the elevated plus maze and elevated zero maze tests are based on the principle that exposure to an elevated and open maze arm leads an approach-conflict that is considerably stronger than that of evoked by exposure to an enclosed arm of the maze. Thus, open/enclosed arms entries and time ratios provide a measure of fear-induced inhibition of exploratory activity. These responses are increased by anxiolytics and reduced by anxiogenic agents (Pellow *et al*. 1985). Furthermore, anxiolytics increase the social interaction and decrease the feeding latency respectively in the social interaction and novelty induced suppressed feeding latency tests in a novel environment (Bhattacharya and Satyan 1997).

Overall, the results of the present study indicate that IHp treatment caused significant dose related anxiolysis in rats tested on all the behavioural paradigms, namely, open field exploratory behaviour, elevated plus maze behaviour, elevated zero maze behaviour, social interaction and novelty induced suppressed feeding latency tests. However, the anxiolytic activity of IHp was found to be less marked than that of the common BDZ anxiolytic agent lorazepam.

Recently, *Hypericum* extracts containing hyperforin have been reported to exhibit anxiolysis in rats on elevated plus maze test (Bhattacharya *et al*. 1998). *Hypericum* has also been reported to be a sedative (Nahrstedt and Butterweck 1997) and useful in the treatment of chronic tension headaches (Heinz and Gobel 1996). The most potent effect thus so far reported is for the GABA receptors, with effects shown at IC_{50} approximately 75 ng/ml for $GABA_A$ and 6 ng/ml for $GABA_B$ (Cott 1997). The crude extract of *Hypericum* had significant receptor affinity for adenosine, $GABA_A$, $GABA_B$, serotonin, benzodiazepine, inositol triphosphate (IP_3) and monoamine oxidase (MAO-A, B) (Cott and Misra 1997). These data are consistent with recent pharmacological evidence suggesting that several constituents of this plant may be important for the reported psychotherapeutic activities. It is conceivable that the very high affinity of *Hypericum* extract for GABA receptors may be important for its anxiolytic activity. The significance of this GABA binding is unknown, but there is considerable literature on the role of GABA in affective disorders. $GABA_B$ stimulation has been found to enhance receptor down regulation during imipramine treatment (Enna *et al*. 1986). Fengabine, a GABAergic agent, has also been reported to be an effective antidepressant (Nielsen *et al*. 1990). GABA plasma levels have been reported to be low in both bipolar and unipolar depression (Petty *et al*. 1992, 1993) and benzodiazepines, which enhance $GABA_A$ activity, may be effective antidepressants as well as anxiolytics (Petty *et al*. 1995). GABA has also been found to be one of the major constituents of HP (Nahrstedt and Butterweck 1997) and crude extract of *Hypericum* has been observed to have significant affinity for BDZ receptors in *in vitro* studies (Cott 1997). Therefore, the observed anxiolytic effect of IHp in the present study may be attributed to its high affinity for GABA and BDZ receptors. However, further studies should be planned to characterise the *Hypericum* effects. Studies with GABA agonists and antagonists with *Hypericum* should also be carried out to specify its intrinsic activity.

Nootropic activity

As science learns more about how the brain works and fails to work, the possibility for developing a 'cognition enhancer' becomes more plausible. The discovery and preparation of drugs with psychological effects has occupied the interest and energy of humans since the beginning of recorded history (Singh and Dhawan 1992). The demand for drugs that can help us think faster, remember more and focus more keenly has already been demonstrated by the market success of drugs like Ritalin, which tames the attention span and Prozac, which ups the competitive edge (Whitehouse *et al*. 1997). Nootropics represent a new class of psychotropic agents with selective facilitatory effect on integrative functions on the central nervous system, particularly on intellectual performance, learning capacity and memory (Giurgea 1973). A number of drugs, including piracetam, have now been introduced in therapy to ameliorate cognitive deficits (Jaiswal and Bhattacharya 1992).

The findings of this investigation indicate that IHp can be regarded as a nootropic agent in view of its facilitatory effect on retention of acquired learning and retention though it had minimal effect on learning acquisition on all paradigms used in the present study. In the plus-maze

test, rats show natural aversion to open and high spaces and therefore, spend more time in enclosed arms. Itoh *et al.* (1990) suggested that transfer latency (TL) might be shortened if the animal had previous experience of entering the open arm and the shortened TL could be related to memory. TLs on the first and second day are taken as acquisition and retrieval respectively. Both 100 and 200 mg/kg doses of IHp and piracetam (500 mg/kg) reduced TL on the first, second as well as ninth day and significantly reversed scopolamine induced amnesia, suggesting an underlying cholinergic mechanism. The drug-induced increase in cortical muscarinic acetylcholine receptor capacity (Kulkarni and Joseph 1998) might partly explain the cognition-enhancing and memory-improving effects of IHp observed in animals. The increase in muscarinic receptors with cholinergic antagonists like scopolamine, represent receptor upregulation (Jaiswal and Bhattacharya 1992) as a part of the physiological response to overcome decreased cholinergic activity. IHp also produced a similar effect (in the absence of an antagonist), which may indicate drug induced increase in cholinergic function. The shortening of TL by IHp as well as piracetam indicated improvement in memory. Both IHp and piracetam meet major criteria for nootropic activity, namely, improvement of memory in absence of cognitive deficit (Poschel 1988).

Passive avoidance test uses normal behaviour of rats. These animals avoid bright light and prefer dim light illumination. When placed in a brightly illuminated space connected to a dark enclosure, they rapidly enter the dark compartment and remain there. It is widely used in testing drug effects on memory (Hock 1994). The time to step-through during the learning phase was measured. In this test a prolongation of the step-through latencies is specific to the experimental situation. An increase of the step-through latency is defined as learning (Vogel and Vogel 1997). Although both doses of IHp had no significant *per se* effect, IHp (200 mg/kg, p.o.) and piracetam significantly reversed scopolamine, and electroconvulsive shock (ECS) induced impaired retention though no significant reversal was observed with sodium nitrite amnesia.

Active avoidance learning is a fundamental behaviour phenomenon (D'Amato 1970). As in other instrumental conditioning paradigms the animal learns to control the administration of the unconditioned stimulus by appropriate reaction to the conditioned stimulus preceding the noxious stimulus. The first stage of avoidance learning is usually escape, whereby a reaction terminates the unconditioned stimulus. The results indicate that IHp in both doses, and piracetam, resulted in less trials and time to reach the criterion of conditioned avoidance learning. When tested after 24 h and 1 week, IHp and piracetam treated rats required significantly fewer trials, shocked trials and total time to re-learn the task as compared to vehicle treated rats. Both the doses of IHp and piracetam significantly attenuated the scopolamine, sodium nitrite and ECS induced retarded learning.

The administration of antimuscarinic agent scopolamine to young human volunteers produces transient memory deficits (Drachman and Leavitt 1974). Analogously, scopolamine has been shown to impair memory retention when given to mice shortly before training in a dark avoidance task (Schindler *et al.* 1984). In spite of the fact that the pathogenesis of primary degenerative dementia (Alzheimer's disease) in man has been only partially elucidated, the scopolamine amnesia test is widely used as primary screening test for anti-Alzheimer drugs (Vogel and Vogel 1997). Sodium nitrite is known to induce cerebral anoxia and memory deficits (Satyan 1997). ECS has been reported to alter central nervous system functions. A decrease in acetylcholine (ACh) levels (Spignoli and Pepeu 1986b) and increase in acetylcholinesterase activity (Appleyard *et al.* 1987), inhibition of GABA synthesis and increase in noradrenaline and serotonin release (Green *et al.* 1987a, 1987b) have been reported.

The neurochemical basis of learning and memory remains controversial, despite extensive experimental and clinical studies (Jaiswal and Bhattacharya 1992). Although the role of the central cholinergic system is fairly well established, its deficiency being implicated in memory deficits, the role of other neurotransmitter systems cannot be ignored (Hollander *et al.* 1986). Several studies have indicated that increase in serotonergic neurotransmission can interfere with learning acquisition and memory consolidation (Jaffard *et al.* 1989). The role of 5-HT in anxiety is now well established and it has been conclusively shown that an increase in central serotonergic activity invariably leads to anxiety, whereas a decrease in brain 5-HT activity results in anxiolysis (Kahn *et al.* 1988). Thus the neurochemical effects indicated by IHp can explain its nootropic and anxiolytic actions, particularly the induced decrease in 5-HT turnover, as indicated by a decrease in 5-HT and 5-HIAA levels in rat brain (Kumar *et al.* 2001a). Conversely, the increase in DA turnover, as evidenced by the induced increase in the levels of DA and its metabolites, HVA and DOPAC, can contribute to the observed nootropic activity. Piracetam, the classical nootropic agent, has been reported to augment rat brain dopaminergic activity (Nyback *et al.* 1979). One of our studies has also shown that there is augmentation of rat brain dopaminergic activity after IHp treatment (Kumar *et al.* 2001a).

Some newer anxiolytics like buspirone (Molodavkin and Voronina 1997) and tianeptine (File and Mabbutt 1991), have been reported to exhibit both anxiolytic and nootropic activity, probably by inducing a reduction in central serotonergic functions. However, there is no reason to believe that improvement in memory is secondary to anxiolysis, since benzodiazepines are known to have an adverse effect on learning and memory (File *et al.* 1990). In our earlier studies, IHp has shown antidepressant (Kumar *et al.* 1999) and anxiolytic activities (Kumar *et al.* 2000a). The two major components likely to be responsible for the antidepressant activity of HP appear to have different neurochemical activities. Thus hypericin appears to have MAO inhibitory action (Suzuki *et al.* 1984) whereas hyperforin is a potent uptake inhibitor of 5-HT and catecholamines (Bhattacharya *et al.* 1998).

The present investigation indicates that IHp may have nootropic activity. However, further investigations using more experimental paradigms, are required before the nootropic action of IHp can be affirmed.

Anti-inflammatory and analgesic activity

This study demonstrated that IHp extract was effective in animal models of acute inflammation. Among the many methods used for screening of anti-inflammatory drugs, one of the most commonly employed techniques is based upon the ability of such agents to inhibit the edema produced in the hind paw of the rat after injection of phlogistic agents. The time course of edema development in carrageenan-induced paw edema model in rats is generally represented by a biphasic curve (Winter *et al.* 1962). The first phase occurs within an hour of injection and is partly due to the trauma of injection and also to the serotonin component (Crunkhorn and Meacock 1971). Prostaglandins (PGs) play a major role in the development of the second phase of reaction, which is measured around 3 h (Di Rosa 1972). The presence of PGE_2 in the inflammatory exudates from the injected foot can be demonstrated at 3 h and periods thereafter (Vinegar *et al.* 1969). Carrageenan induced paw edema model in rats is known to be sensitive to cyclooxygenase inhibitors and has been used to evaluate the effect of non-steroidal anti-inflammatory agents which primarily inhibit the enzyme cyclooxygenase in prostaglandin synthesis (Phadke and Anderson 1988). Based on these reports it can be inferred that the inhibitory effect of IHp extract on carrageenan-induced inflammation in rats could be due to inhibition of the enzyme cyclooxygenase leading to inhibition of prostaglandin synthesis. However, this likely action of IHp needs to be elucidated.

In the cotton pellet induced granuloma model of sub-acute inflammation, IHp extract significantly reduced the weight of granulation tissue and potentiated the anti-inflammatory activity of indomethacin. This method was described first by Meier *et al.* (1950) who showed that foreign body granulomas were provoked in rats by subcutaneous implantation of pellets of compressed cotton. This method has been useful for evaluation of steroidal and non-steroidal anti-inflammatory drugs (Vogel and Vogel 1997). It has been considered that cotton pellet induced granuloma is closely related to the formation of antibodies. Thus, from the findings of the test it may be inferred that the IHp is effective in the delayed immunological response.

Hypericin has been reported to inhibit the growth of glioma cell lines *in vitro* and to be a patent inducer of glioma cell death due to inhibition of protein kinase C (PKC) as measured by [^{3}H] thymidine uptake. In this regard the glioma inhibitory activity is reported to be equal to or greater than tamoxifen and additionally was reported to be increased by approximately 13% upon exposure to visible light (Couldwell *et al.* 1994). Other researchers report PKC-inhibitory activity with both hypericin and pseudohypericin (Takahashi *et al.* 1989). Receptor tyrosine kinase activity of epidermal growth factor has been reported to inhibit by hypericin (de Witte *et al.* 1993). The PKC inhibition may contribute to the anti-inflammatory effects associated with *Hypericum* as hypericin has been found to inhibit the release of arachidonic acid and leukotriene B$_4$ (Panossian *et al.* 1996).

IHp extract also exhibited analgesic activity in rodents and synergies with the analgesic activity of pentazocine. The extract was found to significantly increase the tail flick reaction time in rats. Originally, the tail flick method was developed by Schumacher *et al.* (1940) and Wolff *et al.* (1940) for quantitative measurement of pain threshold in man against radiation and for evaluation of analgesic opiates. Later on, the procedure has been used by many authors to evaluate analgesic activity in animal experiments by measuring drug-induced changes in the sensitivity of mice or rats to heat stress applied to their tails. This test is very useful for discriminating between centrally acting morphine-like analgesics and non-opiate analgesics.

The hot plate method was originally described by Woolfe and MacDonald (1944). This test has been found to be suitable for evaluation of centrally but not of peripherally acting analgesics. The validity of this test has been shown even in the presence of substantial impairment of motor performance (Plummer *et al.* 1991). Mixed opiate agonists–antagonists can be evaluated if the temperature of the hot plate is lowered to 49.5°C (O'Callaghan and Holtzman 1975, Zimer *et al.* 1986). It is known that centrally acting analgesic drugs elevate the pain threshold of rodents towards heat. The above findings indicate that IHp may be centrally acting.

In order to distinguish between the central and peripheral analgesic action of IHp, acetic acid-induced writing response in mice was used. This method is not only simple and reliable but also affords rapid evaluation of the peripheral type of analgesic action. In this test, the animals react with a characteristic stretching behaviour, which is called writhing. It was found that IHp significantly inhibited the acetic acid-induced writing response and potentiated the anti-inflammatory activity of aspirin as well. The abdominal constriction is related to the sensitisation of nociceptive receptors to prostaglandins. It is therefore possible that IHp exerts an analgesic effect probably by inhibiting synthesis or action of prostaglandins and leukotrienes (Panossian *et al.* 1996).

Based on the results of the present study it can be concluded that IHp has potential anti-inflammatory activity against both exudative-proliferative and sub-chronic phases of inflammation. The extract also has analgesic activity, which is both centrally and peripherally mediated.

Anti-stress activity

Stress research in laboratory animals has assumed an important role in understanding the biological and behavioural consequences of external or internal stressors, which threaten to perturb homeostasis, and may induce a number of clinical diseases when the body fails to counter the stress situation (McCarty 1989). One difficulty with the concept of stress, is that it has become so broad as to include virtually every type of environmental changes. It has now been proposed that stressors be classified based upon dimensions of intensity, frequency of exposure and the duration of stress exposure. A variety of stressful situations have been employed and the lack of consistency of the stress protocols is astounding (McCarty 1989). Likewise, there is wide variation in the physiological consequences of the stressors utilised in animal research (Vogel 1985). However, it is now widely accepted that chronic intermittent stress, particularly of an unpredictable pattern, is more likely to induce neural, endocrine and biochemical perturbations than either acute or chronic stress of a predictable nature (McCarty 1989). In addition, the factor of coping and control over the aversive stimulation, plays an important part in stress research since stress responses are minimal in such situations (McCarty 1989). The validity of the method used in present study is demonstrated by the biological effects induced by it, which include gastric ulcerations, increase in adrenal gland weight and decrease in the weight of the spleen. All these parameters have been conclusively shown to be stress-induced effects (Natelson 1981).

The prevention and management of stress disorders remains a major clinical problem. Benzodiazepines (BDZs) appear to be effective against acute stress but fail to prevent the consequences of chronic stress (McCarty 1989). In addition, the problems of tolerance and physical dependence exhibited by BDZs, on prolonged use, limit their utility (McCarty 1989). An answer to this vexing problem was first provided when Brekhman and Dardymov (1969) reported that some plant-derived agents could induce a state of non-specific increase of resistance to affect internal homeostasis. These agents, named adaptogens, appeared to be effective only when the physiological perturbations were discernible following prolonged illness, old age and exposure to chronic stress (Brekhman and Dardymov 1969). A number of such plants, the most important one being PG, were extensively used in the erstwhile USSR and the Far East, for promoting physical and mental health, while helping the body to resist internal and external stressors. These adaptogens were shown to be effective in attenuating stress induced adverse effects in astronauts, soldiers and athletes in the USSR (Brekhman and Dardymov 1969). PG, the first clinically used adaptogen, has been extensively investigated experimentally and clinically for its stress-attenuating activity (Shibata *et al.* 1985).

Both IHp and PG prevented chronic stress-induced gastric ulcerations, in the term of the incidence and severity of the ulcers. Involution of the spleen and increase in adrenal gland weight, are also consequences of chronic stress (McCarty 1989), both responses being reversed by IHp and PG.

There is considerable experimental and clinical evidence to suggest that chronic stress induces endogenous depression (Bhattacharya *et al.* 2000). A number of animal models of depression are based on the use of uncontrollable stress and the biochemical correlates of such tests are consonant with those seen in chronic stress, including monoamine deficiency and increased activity of the corticotrophin-releasing factor (Bhattacharya *et al.* 2000). Both IHp and PG were able to reverse chronic stress-induced indices validated as animal models of depression. Chronic stress is known to affect other endocrine responses as well, which can induce sexual debility in males (Saito *et al.* 1984) and perturb glucose metabolism (Shoji *et al.* 1992). Maturity-onset diabetes mellitus may represent a state of stress-induced disturbance in glucose homeostasis (Shoji *et al.* 1992). IHp and PG reversed chronic stress-induced inhibition of male sexual behaviour.

Stress is known to interfere with cognitive functions, tending to retard the memory engram rather than the acquisition of learning (Bhattacharya 1993). The mechanisms involved in the memory-attenuating effect of stress remain conjectural but a similar neurochemical basis

operating in the induction of stress-induced depression may be responsible (Bhattacharya, 1993). IHp and PG attenuated the stress-induced deficit of retention of learned tasks, both in the active and passive avoidance parameters, thus facilitating memory and its recall.

The findings indicate that, like the standard adaptogen PG, IHp can attenuate chronic stress-induced biochemical, behavioural and physiological perturbations in rats. PG has earlier been reported to reverse chronic stress induced effects in humans (Brekhman and Dardymov 1969). Japanese traditional medicinal plant formulations, like *Gosya-jinki-gan, Kyushin* and *Reiousan* and essentially based on *Ginkgo biloba*, have been reported to reduce the adverse effects of chronic hanging stress on sexual and learning behaviours in mice (Morishita 1993b). In a recent study, the model of chronic stress used in this investigation (chronic, unpredictable and inescapable footshock stress) has been shown to induce marked gastric ulceration, significant increase in adrenal gland weight and plasma corticosterone levels, with concomitant decrease in spleen weight, and concentration of adrenal gland ascorbic acid and corticosterone in rats. These effects were significantly attenuated by PG and by an Ayurvedic formulation (Siotone) of medicinal plants (Bhattacharya *et al.* 2000).

Increased generation of oxidative free radicals (OFR), or impaired antioxidant defence mechanisms, have been implicated in chronic stress induced perturbed homeostasis including immunosupression, inflammation, diabetes mellitus, peptic ulceration and other stress-related diseases (Maxwell 1995). IHp has been shown to exert significant antioxidant activity induced by augmented activity of OFR scavenging enzymes, superoxide dismutase, catalase and glutathione peroxide (Tripathi *et al.* 1999, Tripathi and Pandey 2000). Thus, the observed adaptogenic antistress effect of IHp may be at least partly due to its antioxidant activity.

The present investigation indicates that IHp has significant adaptogenic activity as shown by its mitigating effects on several chronic stress induced physiological and behaviour perturbations, comparable to that induced by the well accepted adaptogenic agent, PG.

Neurochemical study

IHp treatment differentially modulated the monoamine neurotransmitters (5-HT, NE and DA) and their metabolites (5-HIAA, MHPG and DOPAC) in the various regions of rodent brain. Earlier, we observed the anxiolytic (Kumar *et al.* 2000a) and antidepressant (Kumar *et al.* 1999) activity of the IHp extract in rodents. A significant decrease in the 5-HT and 5-HIAA levels in all the brain regions assayed may explain the anxiolytic activity of the IHp extract. There is now considerable evidence to implicate the serotonergic system with anxiety. Increased brain 5-HT activity has been linked to anxiety (Iverson 1984, Soderpalm and Engel 1990, Graeff 1993). On the contrary, depletion of 5-HT by synthesis inhibitors or selective neurotoxins, blockade of 5-HT$_2$ and 5-HT$_3$ receptors, and inhibition of neuronal release of 5-HT by 5-HT$_{1A}$ receptor agonists, induced anxiolytic effects (Marsden 1989, Soderpalm and Engel 1990, Costall *et al.* 1992, Handley and McBlane 1993, Coplan *et al.* 1995). The clinically active anxiolytic drugs (benzodiazepines and buspirone) elicit their action by selectively decreasing the 5-HT release in dorsal raphe nuclei (DRN) from 5-HT neurons (Crespi *et al.* 1990, Deakin 1992). Reports also indicate that raised 5-HT function is involved in attenuated cognitive performance, partially due to decrease in acetylcholine release (Costall *et al.* 1992) and in anxiety (Iverson 1984). It is also well documented that serotonin is inhibitory to the firing rate of dopamine neurons projecting into the striatum (Soubrie *et al.* 1984). IHp, presumably, acts on the raphe nuclei, which are enriched with serotonergic innervation (Taylor 1990) leading to decreased serotonergic activity and augmented catecholamine function. The catecholamine deficiency postulate of depression is well accepted (Stahl 1998).

Literature reports also suggest attenuated levels of NE and DA in stress-induced depression (Anisman *et al.* 1991). Clinically used tricyclic antidepressants are known to increase DA and NE functions in the mesolimbic system (Willner 1991). Augmented levels of NE and DA in all the brain regions studied could explain the putative antidepressant activity of the IHp extract. There is a major controversy regarding the active chemical entity responsible for the antidepressant activity. Hypericin has MAO inhibiting action, whereas hyperforin, the other major constituent, is devoid of MAO inhibiting activity and now appears to be the major antidepressant principle of HP (Bhattacharya *et al.* 1998, Chatterjee *et al.* 1998a, Muller 1998). The present study supports this contention.

Radioligand receptor binding study

This study shows that IHp is effective in neurotransmitter receptor mechanisms relevant for antidepressant and anxiolytic activities in man. Additionally, it shows possible neurotransmitter receptor mechanisms for nootropic activity of IHp extract. IHp significantly decreased the binding of [^{3}H] spiroperone in corpus striatum indicating the down regulation of DA-D$_2$ receptors. The role of the dopaminergic system is well established in rewarded behaviour and in anhedonia (Fibiger 1995), and upregulation of dopaminergic receptors was reported in depression (Dhaenen and Bossuyt 1994). Down regulation may then account for the antidepressant activity. The binding of [^{3}H] ketanserin to 5-HT$_{2A}$ receptors was increased with IHp treatment. However, most of the clinically used antidepressants are reported to down regulate the 5-HT$_{2A}$ receptors (Fraser *et al.* 1988). Reports also suggest that down regulation of 5-HT$_{2A}$ receptors was not essential for the activity as seen in the case of ECS treatment, which increases the responsiveness of 5-HT$_{2A}$ receptors in rats (Green and Heal 1985). 5-HT$_{2A}$ receptors also appear to play a role in appetite control, thermoregulation, sleep and sexual behaviour (Glennon and Dukat 1995). In an earlier study, HP extract was also reported to up regulate the 5-HT$_{2A}$ receptors in the cortical regions (Muller *et al.* 1997). The observed effect of IHp on 5-HT$_{2A}$ receptors may thus, at least partly account for the antidepressant activity (Muller *et al.* 1997).

IHp treatment significantly increased the binding of [^{3}H] flunitrazepam to the cortical regions indicating the upregulation of the benzodiazepine receptors (BDZ-R). A decrease in the binding of [^{3}H] diazepam in the cortex of rats exposed to standard conflict behaviour (Lippa *et al.* 1978) and inbred mice strain characterised by high 'emotionality or anxiety' (Robertson 1979), has been documented. Reports also indicated that administration of single dose of diazepam (a benzodiazepine anxiolytic agent) increased the [^{3}H] flunitrazepam binding sites by above 140% in the total brain homogenate (Speth *et al.* 1979). As such, increased binding of [^{3}H] flunitrazepam with IHp treatment, observed in this study, supports the anxiolytic activity reported in our earlier study with IHp (Kumar *et al.* 2000a, 2002).

Conclusion

The investigations indicate that IHp has significant antidepressant, anxiolytic, nootropic, anti-stress, anti-inflammatory and analgesic actions. Some of these actions, including antidepressant, anxiolytic and nootropic, can be rationalised on the basis of the neurochemical data emanating from this study. The present study indicates that IHp can be clinically useful not only in depression but also in anxiety disorders and cognitive dysfunction. Clinical studies are required to confirm these actions and several laboratories have shown such an interest following publications of our data. If IHp has a similar efficacy to its European counterparts, it could prove to be a major export item in view of the 3–4 billion US$ market of Hp in the West.

Acknowledgement

Vikas Kumar is grateful to the Indian Council of Medical Research, New Delhi for the award of Senior Research Fellowship. Thanks are due to Prof. S. Ghosal, Consultant, Drug Research and Development Centre, Kolkata and to Indian Herbs Ltd, Saharanpur, for their help.

References

Andrew, C. (1996) *The Encyclopedia of Medicinal Plants*, Dorling Kindersley Ltd, London, p. 104.

Anisman, H., Zalcman, S., Shanki, N. and Za-Charko, R. (1991) Multisystem regulation of performance deficits induced by stressors. In A. Boulton, G. Baker and M.M. Iverson (eds), *Neuromethods: animals models in psychiatry*, vol 19, pp. 1–15.

Appleyard, M.E., Green, A.R. and Greenfield, S.A. (1987) Acetylcholinesterase activity rises in rat cerebrospinal fluid post-ictally: effect of a substantial nigra lesion on this rise and on seizure threshold. *Br. J. Pharmacol.* 91, 149–54.

Armando, I., Lemoine, A.P., Segura, E.T. and Barontini, M.B. (1993) The stress-induced reduction in monoamine oxidase (MAO) an activity is reversed by benzodiazepines: role of peripheral benzodiazepine receptors. *Cellular Mol. Neurobiol.* 13, 593–600.

Askew, B.M. (1963) A simple screening procedure for imipramine-like antidepressant agents. *Life Sci.* 10, 725–30.

Benigni, R., Capra, C. and Cattorini, P.E. (1971) *Hypericum, Plante Medicinali, Chimica, Farmacologiae Terapia*, Milano: Inverni and Della Beffa.

Bhargava, K.P. and Singh, N. (1981) Anti-stress activity of *Ocimum sanctum* Linn. *Indian J. Med. Res.* 10, 443–51.

Bhattacharya, S.K., Raina, M.K., Banerjee, D. and Neogy, N.C. (1971) Potentiation of morphine and pethidine analgesia by some monoamine oxidase inhibitor. *Indian J. Exp. Biol.* 9, 257–9.

Bhattacharya, S.K., Clow, A., Przyborowska, A., Halket, J., Glover, V. and Sandler, M. (1991) Effect of aromatic amino acids, pentylenetetrazole and yohimbine on isatin and tribulin activity in rat brain. *Neurosci. Lett.* 132, 44–6.

Bhattacharya, S.K. (1993) Evaluation of adaptogenic activity of Trasina, an ayurvedic herbal formulation. In B. Mukherjee (ed.), *Traditional medicine*. Oxford and IBH Publishers, New Delhi, pp. 320–6.

Bhattacharya, S.K. and Satyan, K.S. (1997) Experimental methods for evaluation of psychotropic agents in rodents: I-Anti-anxiety agents. *Indian J. Exp Biol.* 35, 565–75.

Bhattacharya, S.K., Chakrabarti, A. and Chatterjee, S.S. (1998) Activity profiles of two hyperforin-containing *Hypericum* extracts in behavioural models. *Pharmacopsychiatry* 31, 22–9.

Bhattacharya, S.K., Satyan, K.S. and Ramanathan, M. (1999) Experimental methods for evaluation of psychotropic agents in rodents: II-Antidepressants. *Indian J. Exp. Biol.* 37, 117–23.

Bhattacharya, S.K., Bhattacharya, A. and Chakrabarti, A. (2000) Adaptogenic activity of Siotone, a polyherbal formulation of Ayurvedic rasayanas. *Indian J. Exp. Biol.* 38, 119–28.

Bladt, S. and Wagner, H. (1994) Inhibition of MAO by fractions and constituents of *Hypericum* extract. *J. Geriatr. Psychiatry Neurol.* 7, S57–9.

Bodnoff, S.R., Suranyi-Cadotte, B., Aitken, D.H., Quirion, R. and Mesney, M.J. (1988) The effects of chronic antidepressant treatment on an animal model of anxiety. *Psychopharmacology* 95, 298–302.

Bombardelli, E. and Morazzoni, P. (1995) *Hypericum perforatum*. *Fitoterapia* 66, 43–68.

Bondy, S.C. (1981) Neurotransmitter binding interaction as a screen for neurotoxicity. In A. Vernadakis and K.N. Prasad (eds), *Mechanisms of actions of neurotoxic substances*, Raven Press, New York, pp. 21–50.

Brekhman, I.I. and Dardymov, I.V. (1969) New substances of plant origin which increase nonspecific resistance. *Ann. Rev. Pharmacol. Toxicol.* 9, 419–30.

Broekkamp, C.L., Leysen, D., Peeters, B.W. and Pinder, R.M. (1995) Prospects for improved antidepressants. *J. Med. Chem.* 38, 4615–33.

Bronstein, P.M. (1972) Open field behaviour of the rat a function of age: cross sectional and longitudinal investigations. *J. Comparative Physiol. Psychol.* 80, 335–41.

Butterweck, V., Winterhoff, H. and Nahrstedt, A. (1996) An extract of *Hypericum perforatum* L. decreases immobility time in rats in the forced swimming test. *2nd International Congress on Phytomedicine*, Munich.

Butterweck, V., Wall, A., Lieflandes-Wolf, U., Winterhoff, H. and Nahrstedt, A. (1997) Effects of total extract and fractions of *Hypericum perforatum* in animal assays for antidepressant activity. *Pharmacopsychiatry* 30, 117–24.

Challem, J. (2001) St. John's wort vs Drugs. *Nutrition Science News* 6, 212–18.

Chatterjee, S.S., Koch, E., Nolder, M., Biber, A. and Erdelmeier, C. (1996) Hyperforin and *Hypericum* extract: interactions with some neurotransmitter systems. *Phytomedicine* 3, 106.

Chatterjee, S.S., Bhattacharya, S.K., Wonnemann, M., Singer, A. and Muller, W.E. (1998a) Hyperforin as a possible antidepressant component of *Hypericum* extracts. *Life Sci.* 63, 499–510.

Chatterjee, S.S., Nolder, M., Koch, E. and Erdelmeier, C. (1998b) Antidepressant activity of *Hypericum perforatum* and hyperforin: the neglected possibility. *Pharmacopsychiatry* 31, 7–15.

Chermat, R., Thierry, B., Mico, J.A., Steru, L. and Simon, P. (1986) Adaptation of the tail suspension test to the rat. *J. Pharmacol.* 17, 348–50.

Coplan, J.D., Wolk, S.I. and Klein, D.F. (1995) Anxiety and serotonin$_{1A}$ receptor. In F.E. Bloom and D.J. Kupfer (ed.), *Psychopharmacology: the fourth generation of progress*, Raven press, New York, pp. 1301–10.

Costall, B., Domency, A.M., Kelly, M.E. and Naylor, R.J. (1992) Influence of 5-HT on cognitive performance. *Adv. Biosci.* 85, 147–64.

Cott, J. (1996) Natural product formulations available in Europe for psychotropic indications. *Psychopharmacol. Bull.* 31, 745–51.

Cott, J. and Misra, R. (1997) Medicinal plants: a potential source for new psychotherapeutic drugs. In S. Kanba and E. Richelson (eds), *New drug development from herbal medicines in neuropsychopharmacology*, Brunner/Mazel Inc., New York.

Cott, J.M. (1997) *In vitro* receptor binding and enzyme inhibition by *Hypericum perforatum* extract. *Pharmacopsychiatry* 30, 108–12.

Couldwell, W.T., Gopalakirshna, R. and Hinton, D.R. (1994) Hypericin: a potential antiglioma therapy. *Neurosurgery* 35, 705–10.

Crespi, F., Garratt, J.C. and Sleigut, A.J. (1990) *In vivo* evidence that 5-HT neuronal firing and release are not necessarily correlated with 5-HT metabolism. *Neuroscience* 35, 139–44.

Crunkhorn, P. and Meacock, S.C.R. (1971) Mediators of the inflammation induced in the rat paw by carrageenan. *Br. J. Pharmacol.* 42, 392–402.

D'Amato, M.R. (1970) *Experimental Psychology: Methodology, Psychophysics and Learning*, McGraw-Hill, New York, pp. 381–416.

David, C.P. (2000) Dietary supplements: scientific evidence of efficacy. *American Association of Pharmaceutical Scientists Newsmagazine* 3, 22–43.

Davies, O.L., Raventos, J. and Walpole, A.L. (1946) A method for evaluation of analgesic activity using rats. *Br. J. Pharmacol.* 1, 255–64.

Deakin, W.J.F. (1992) Amine mechanisms in anxiety and depression. In J.M. Elliot, D.J. Heal and C.A. Marsden (eds), *Experimental approaches to anxiety and depression*, John Wiley and sons Ltd, New York, p. 124.

Demisch, L., Holzl, J., Golnik, B. and Kacynarczyk. (1989) Identification of MAO-type-A inhibitors in *Hypericum perforatum* L. (hyperforat). *Pharmacopsychiatry* 22, 194.

De Smet, P.A. and Nolen, W.A. (1996) St. John's wort as an antidepressant (editorial comment), *Br. J. Pharmacol.* 313, 241–7.

de Witte, P., Agostinis, P., Van Lint, J., Merlevede, W. and Vandenheede, J.R. (1993) Inhibition of epidermal growth factor receptor tyrosine kinase activity by hypericin. *Biochem. Pharmacol.* 46, 1929–36.

Dews, P.B. (1953) The measurement of the influence of drugs on voluntary activity in mice. *Br. J. Pharmacol.* 8, 46–8.

Dhaenen, H. and Bossuyt, A. (1994) Dopamine D_2 receptors in the brain measured with SPECT. *Biol. Psychiatry* 35, 128–32.

Di Rosa, M. (1972) Biological properties of carageenan. *J. Pharmacy Pharmacol.* 24, 89–102.

Drachman, D.A. and Leavitt, J. (1974) Human memory and the cholinergic system. *Arch. Neurol.* **30**, 113–21.

Duke, J.A. (1995) Commentary-novel psychotherapeutic drugs: a role for ethnobotany. *Psychopharmacol. Bull.* **31**, 177–84.

Eisenberg, L. (1992) Treating depression and anxiety in primary care. Closing the gap between knowledge and practice. *N. E. J. Med.* **326**, 1080–4.

Enna, S.J., Karbon, E.W. and Duman, R.S. (1986) $GABA_B$ agonist and imipramine-induced modifications in rat brain beta-adrenergic receptor binding and function. In G. Bartholini, K.G. Lloyd and P.L. Morsell (eds), *GABA and mood disorders: experimental and clinical research*, Raven Press, New York, pp. 23–49.

Ernst, A.M. (1972) Relationship of the central effect of dopamine gnawing compulsion syndrome in rats and the release of serotonin. *Archives Internationales de Pharmacodynamie et de Therapie*, **199**, 219–25.

Ernst, E. (1995) St John's wort, an antidepressant? A systematic, criteria based study. *Phytomedicine* **2**, 67–71.

Fibiger, H.C. (1995) Neurobiology of depression: focus on dopamine. *Adv. Biochem. Psychopharmacol.* **49**, 1–17.

File, S.E. and Hyde, J.R. (1978) Can social interaction be used to measure anxiety? *Br. J. Pharmacol.* **62**, 19–24.

File, S.E. (1985) Animal models for predicting clinical efficacy of anxiolytic drugs: social behaviour. *Neuropsychobiology* **13**, 55–62.

File, S.E., Mabbutt, P.S. and Toth, E. (1990) A comparison of the effects of diazepam and scopolamine in two positively reinforced learning tasks. *Pharmacol. Biochem. Behav.* **37**, 587–92.

File, S.E. and Mabbutt, P.S. (1991) Effects of tianeptine in animal models of anxiety and on learning and memory. *Drug Dev. Res.* **23**, 47–56.

Fraser, A., Offord, S.I. and Lucki, I. (1988) Regulation of serotonin receptors and responsiveness in brain. In Sanders-Bush (ed.), *The serotonin receptors*, Humana press, New Jersey, pp. 319–62.

Frichsen Brown, C. (1989) *Medicinal and Other Uses of North America Plants: A Historical Survey with Reference to the Eastern Indian Tribes*, Dover Publications, New York, p. 382.

Gaster, B. and Holroyd, J. (2000) St. John's wort for depression: a systemic review. *Arch. Intern. Med.* **160**, 152–6.

Giurgea, C. (1973) The nootropic approach to the pharmacology of the integrative action of the brain. *Conditional Reflex* **8**, 108–15.

Glennon, R.A. and Dukat, M. (1995) Serotonin receptor subtypes. In F.E. Bloom and D.J. Kupfer (eds), *Psychopharmacology: the fourth generation of progress*, Raven Press, New York, pp. 415–29.

Graeff, F.G. (1993) Role of 5-HT in defensive behaviour and anxiety. *Rev. Neurosci.* **4**, 181–211.

Green, A.R. and Heal, D.J. (1985) The effects of drugs on serotonin mediated behaviour models. In A.R. Green (ed.), *Neuropharmacology of Serotonin*, Oxford University Press, Oxford, pp. 326–68.

Green, A.R., Metz, A., Minchin, M.C.W. and Vincent, N.D. (1987a) Inhibition of the rate of GABA synthesis in regions of rat brain following a convulsion. *Br. J. Pharmacol.* **92**, 5–11.

Green, A.R., Heal, D.J. and Vincent, N.D. (1987b) The effects of single and repeated electroconvulsive shock administration on the release of 5-hydroxytryptamine and noradrenaline from cortical slice of rat brain. *Br. J. Pharmacol.* **92**, 25–30.

Gruenwald, J. (2001) St. John's wort – The international debate. *Neutraceuticals World* **6**, 26–8.

Hahn, G. (1992) *Hypericum perforatum* (St John's wort–a medicinal herb used in antiquity and still of interest today). *J. Naturopathic Med.* **3**, 94–6.

Handley, S.L. and Mc Blane, J.W. (1993) 5-HT drugs in animal models of anxiety. *Psychopharmacology* **112**, 13–20.

Heinze, A. and Gobel, H. (1996) *Hypericum* perforans in the treatment of chronic tension-type headache, *2nd International Congress on Phytomedicine*, Munich.

Hock, F.J. (1994) Involvement of nitric oxide-formation in the action of Iosartan (DUP 753): effects in an animal inhibitory avoidance model. *Behav. Brain Res.* **61**, 163–7.

Hoelzl, J. (1993) Inhaltsstoffe und Weitschrift des johanniskratutes. *Zeitschrift fur Phytotherapie* **14**, 255–64.

Hoelzl, J., Sattler, S. and Schutt, H. (1994) Johanniskraut: eine alternative zu synthetischen antidepressiva? *Pharmazeutische Zeit. Wiss.* **46**, 3959–77.

Hollander, E., Mohs, R.S. and Davis, K.L. (1986) Cholinergic approaches to the treatment of Alzheimer's disease. *Br. Med. Bull.* 42, 97–100.

Holtje, H.D. and Walper, A. (1993) Molecular modeling of the antidepressive mechanism of Hypericum ingredients. *Nervenheilkunde* 12, 339–40.

Hubner, W.D., Lande, S. and Podzuweit, H. (1994) *Hypericum* treatment of mild depression with somatic symptoms. *J. Geriatric, Psychiatry Neurol.* 7, S12–14.

Hutchins, R. (1991) *Indian Herbology of North America.* Shambhala, Boston, pp. 257–60.

Itoh, J., Nabeshima, T. and Kameyama, T. (1990) Utility of an elevated plus maze for the evaluation of memory in mice: effects of nootropics, scopolamine and electroconvulsive shock. *Psychopharmacology* 101, 27–33.

Iverson, S.D. (1984) 5-HT and anxiety. *Neuropharmacology* 23, 1553–60.

Jaffard, R., Mocaer, E., Alaoui, F., Beracochead, D., Marighetto, A. and Meunier, M. (1989) Effects de iatianeptine sur l'apprentissage et la memoire chez la souris. *J. Psychiatry, Biol. Therapeut.*, edition speciale, 37–9.

Jaiswal, A.K., Upadhyay, S.N. and Bhattacharya, S.K. (1989) Effect of piracetam, a nootropic agent, on discrimination learning deficits induced by parental undernutrition and environmental impoverishment in young rats. *Indian J. Exp. Biol.* 279, 269–73.

Jaiswal, A.K. and Bhattacharya, S.K. (1992) Effects of Shilajit on memory, anxiety and brain monoamines in rats. *Indian J. Pharmacol.* 24, 12–17.

Jonas, W.B. (1997) Researching alternative medicine. *Nature Med.* 3, 824–7.

Kahn, R.S., Van Praag, H.M., Wizler, S., Asnis, G.M. and Barr, G. (1988) Serotonin and anxiety revisited. *Biol. Psychiatry* 23, 189–208.

Keith, S.J. and Mathews, S.M. (1993) The value of psychiatric treatment: its efficacy in severe mental disorders. *Psychopharmacol. Bull.* 29, 427–30.

King, R.A. and Glasser, R.L. (1970) Duration of electroconvulsive shock-induced retrograde amnesia in rats. *Physiol. Behav.* 5, 335–9.

Kinghorn, A.D. and Balandrin, M.F. (1993) *ACS Symposium* Series No. 534. American Chemical Society Books, Washington, p. 2.

Kulkarni, S.K. (1993) *Handbook of Experimental Pharmacology*, Vallabh Prakashan, Delhi, p. 43.

Kulkarni, S.K. and Joseph, P. (1998) Psychopharmacological profile of siotone granules, a herbal preparation. *Indian Drugs* 35, 536–44.

Kumar, V., Singh, P.N., Jaiswal, A.K. and Bhattacharya, S.K. (1999) Antidepressant activity of Indian *Hypericum perforatum* Linn in rodents, *Indian J. Exp. Biol.* 37, 1171–6.

Kumar, V., Jaiswal, A.K., Singh, P.N. and Bhattacharya, S.K. (2000a) Anxiolytic activity of Indian *Hypericum perforatum* Linn: an experimental study, *Indian J. Exp. Biol.* 38, 36–41.

Kumar, V., Singh, P.N., Muruganandam, A.V. and Bhattacharya, S.K. (2000b) Effect of Indian *Hypericum perforatum* Linn on animal models of cognitive dysfunction. *J. Ethnopharmacol.* 72, 119–28.

Kumar, V., Singh, P.N., Muruganandam, A.V. and Bhattacharya, S.K. (2000c) *Hypericum perforatum*: nature's mood stabilizer. *Indian J. Exp. Biol.* 38, 1077–85.

Kumar, V., Singh, P.N. and Bhattacharya, S.K. (2001a) Neurochemical studies on Indian *Hypericum perforatum* Linn. *Indian J. Exp. Biol.* 39, 334–8.

Kumar, V., Singh, P.N. and Bhattacharya, S.K. (2001b) Anti-inflammatory and analgesic activity of Indian *Hypericum perforatum* L. *Indian J. Exp. Biol.* 39, 339–43.

Kumar, V., Singh, P.N. and Bhattacharya, S.K. (2001c) Anti-stress activity of Indian *Hypericum perforatum* Linn. *Indian J. Exp. Biol.* 39, 344–9.

Kumar, V., Khanna, V.K., Seth, P.K., Singh, P.N. and Bhattacharya, S.K. (2002) Brain neurotransmitter receptor binding and nootropic studies on Indian *Hypericum perforatum* Linn. *Phytotherapy Res.* 16, 210–16.

Linde, K., Ramirez, G., Mulrow, C.D., Pauls, A., Weidenhammer, W. and Melchart, D. (1996) St John's wort for depression–an overview and meta-analysis of randomized clinical trials. *Br. Med. J.* 313, 253–8.

Lippa, A.S., Klepner, C.A., Yunger, L., Sano, M.C., Smith, W.V. and Beer, B. (1978) Relationship between benzodiazepine receptors and experimental anxiety in rats. *Pharmacol. Biochem. Behav.* 9, 853–6.

Lohse, M.J. and Muller, B. (1995) *Arzneiverordnungs Report '95. Aktuelle Daten, Kosten, Trends und Kommentare*. U. Schwabe and D. Paffrath (eds), Gustav Fischer, Stuttgart, pp. 360–77.

Lowry, O.H., Rosebrough, N.J., Farr, A.L. and Randall, R.J. (1951) Protein measurement with the Folin phenol reagent. *J. Biol. Chem.* 193, 265–75.

Maier, S.F. and Seligman, M.E.P. (1976) Learned helplessness: theory and evidence. *J. Exp. Psychol.* 105, 3–46.

Marsden, C.A. (1989) 5-Hydroxy tryptamine receptor subtypes and new anxiolytic agents: an appraisal. In P. Tyner (ed.), *Psychopharmacology of anxiety*, Oxford University Press, Oxford, pp. 3–27.

Maxwell, S.R.J. (1995) Prospects for the use of antioxidants therapies. *Drugs* 49, 345–61.

McCarty, R. (1989) In G.R. Van Loon and R.R. Kvetnansky (eds), *Stress neurochemical and humoral mechanisms*, Gordan and Breach Science Publishers, New York, p. 3.

Meier, R., Schuler, W. and Desaulles, P. (1950) Zur Frage des Mechananismus der Hemmung des Bindegewebswachstums durch Cortisone. *Experientia* 6, 469–71.

Mitra, S.K. (1990) Neuropharmacological investigations on *Panax ginseng*. PhD thesis, Banaras Hindu University, Varanasi, India.

Moller, H.J. and Volz, H.P. (1996) Drug treatment of depression in the 1990s: An overview of achievements and future possibilities. *Drugs* 52, 625–38.

Molodavkin, G.M. and Voronina, T.A. (1997) Buspirone–a wide spectrum preparation. *Eksperimentalnaia-l-Klinicheskaia-Farmakologiia* 60, 3–6.

Morishita, S., Shoji, M., Oguni, Y., Hirai, Y., Sugimoto, C. and Ito, C. (1993a) Effects of crude drugs derived from animal sources on sexual and learning behaviour in chemically stressed mice. *Phytotherapy Res.* 7, 57–63.

Morishita, S., Shoji, M., Oguni, Y., Hirai, Y., Sugimoto, C. and Ito, C. (1993b) Effects of traditional medicines, *Gosya-jinki-gan, Kyushin* and *Reiousan* on sexual and learning behaviour in chemically stressed mice. *Phytotherapy Res.* 7, 179–84.

Mukhopadhayay, M., Upadhayay, S.N. and Bhattacharya, S.K. (1987) Neuropharmacological studies on selective monoamine oxidase A and B inhibitors. *Indian J. Exp. Biol.* 25, 761–70.

Muldner, H. and Zoller, M. (1984) Antidepressive Wirkung eines auf den Wirkstoffkomplex Hypericin standardisierten *Hypericum*-Extraktes. *Arznemittel Forchung* 34, 918–20.

Muller, W.E. and Rossol, R. (1994) Effects of *Hypericum* extracts on the suppression of serotonin receptors. *J. Geriatric, Psychiatry Neurol.* 7, S63–4.

Muller, W.E., Rolli, M., Schafer, C. and Hafner, U. (1997) Effects of *Hypericum* extract (L1 160) in biochemical models of antidepressant activity. *Pharmacopsychiatry* 30, 102–7.

Muller, W.E. (1998) St John's wort (*Hypericum*), the story goes on. *Pharmacopsychiatry* 31, 1.

Muller, W.E., Singer, A., Wonnemann, M., Hafner, U., Rolli, M. and Schafer, C. (1998) Hyperforin represents the neurotransmitter reuptake inhibiting constituent of *Hypericum* extract. *Pharmacopsychiatry* 31, 16–21.

Murai, S., Saito, H.M., Masuda, Y. and Itoh, T.J. (1988) Rapid determination of norepinephrine, dopamine, serotonin, their precursor amino acids and related metabolites in discrete brain areas of mice within ten minutes by HPLC with electrochemical detection. *Neurochemistry* 50, 473–9.

Nahrstedt, A. and Butterweck, V. (1997) Biologically active and other chemical constituents of the herb *Hypericum perforatum* L. *Pharmacopsychiatry* 30, 129–34.

Nakatsu, K. and Owen, J.A. (1980) A microprocessor-based animal monitoring system. *J. Pharmacol. Meth.* 3, 71–82.

Natelson, B.H., Tapp, W.N., Adamus, J.E., Mittler, J.C. and Levin, B.E. (1981) Humoral indices of stress in rats. *Physiol. Behav.* 26 (6), 1049–54.

Nielsen, N.P., Cesana, B., Zizolfi, S., Ascalone, V., Priore, P. and Morselli, P.L. (1990) Therapeutic effects of fengabine, a new GABAergic agent, in depressed outpatients: a double-blind study versus clomipramine. *Acta Psychiatrica Scand.* 82, 366–71.

NIH consensus conference: diagnosis and treatment of depression in late life. (1992) *J. Am. Med. Assoc.* 268, 1018–24.

Nyback, F., Wiesel, A. and Skett, P. (1979) Effects of piracetam on brain monoamine metabolism and serum prolactin levels in rat. *Psychopharmacol.* 61, 235–8.

O'Callaghan, J.P. and Holtzman, S.G. (1975) Quantification of the analgesic activity of the narcotic antagonists by a modified hot plate procedure. *J. Pharmacol. Exp. Therapeutics* 192, 497–505.

Ojima, K., Matsumoto, K., Tohda, M and Watanabe, H. (1995) Hyperactivity of central noradrenergic and CRF systems is involved in social isolation-induced decrease in pentobarbital sleep. *Brain Res.* 684, 87–94.

Okpanyi, S.N. and Weischer, M.L. (1987) Animal experiments on the psychotropic action of *Hypericum* extract. *Arznemittel Forchung* 37, 10–13.

Overmier, J.B. and Seligman, M.E. (1967) Effects of inescapable shock upon subsequent escape and avoidence responding. *J. Comparative Physiol. Psychol.* 63, 28–33.

Ozturk, Y., Audin, S., Beis, R., Baser, K.H.C. and Bergeroglu, H. (1996) Effects of *Hypericum perforatum* L. and *Hypericum calycinum* L. extracts on the central nervous system in mice. *Phytomedicine* 3, 139–46.

Panossian, A.G., Gabrielian, E., Manvelian, V., Jurcic, K. and Hagner, H. (1996) Immunosuppressive effects of hypericinon stimulated human leucocytes: inhibition of arachidonic acid release, leukotriene B4 and interleukin–1 production and activation of nitric oxide formation. *Phytomedicine* 3, 19–28.

Pellow, S., Chopin, P., File, S.E. and Briley, M. (1985) Validation of open: closed arm entries in an elevated plus-maze as a measure of anxiety in the rat. *J. Neurosci. Meth.* 35, 565–75.

Pellow, S. and File, S.E. (1986) Anxioytic and anxogenic drug effects on exploratory activity in an elevated plus-maze: a novel test of anxiety in the rat. *Pharmacol. Biochem. Behav.* 24, 525–9.

Petty, F., Kramer, G.L., Gullion, C.M. and Rush, A.J. (1992) Low plasma gama-aminobutyric acid levels in male patients with depression. *Biol. Psychiatry* 32, 354–63.

Petty, F., Krammer, G.L., Fulton, M., Moeller, F.G. and Rush, A.J. (1993) Low plasma GABA is a trait-like marker for bipolar illness. *Neuropsychopharmacol.* 9, 125–32.

Petty, F., Trivedi, M.H., Fulton, M. and Rush, A.J. (1995) Benzodiazepines as antidepressants: does GABA play a role in depression? *Biol. Pyschiatry* 38, 578–91.

Pezzuto, J.M. (1997) Plant-derived anticancer agents. *Biochem. Pharmacol.* 53, 121–33.

Phadke, J.D. and Anderson, L.A. (1988) Ethnopharmacology and western medicine. *J. Ethnopharmacol.*, 25, 61–72.

Plummer, J.L., Cmielewski, P.L., Gourlay, G.K., Owen, H. and Cousins, M.J. (1991) Assessment of antinocipetive drug effects in the presence of impaired motor performance. *J. Pharmacol. Meth.* 26, 79–87.

Polter, W.Z., Manji, H.K., Osman, O.T. and Rudorfer, M.V. (1992) New prospects in psychiatry: the bioclinical interface, P.J. Macher and Crocq (eds), Elsevier, New York, p. 113.

Porsolt, R.D., Bertin, A. and Jalfre, M. (1977) Behavioral despair in mice: a primary screening test for antidepressants. *Archives Internationales de Pharmacodynamie et de Therapie* 229, 327–36.

Porsolt, R.D., Anton, G., Blavet, N. and Jalfre, M. (1978) Behavioural despair in rats: a new model sensitive to antidepressant treatments. *Eur. J. Pharmacol.* 47, 379–91.

Poschel, B.P.H. (1988) *Handbook of Psychopharmacology*, Vol 20, Plenum Press, New York, p. 437.

Ramanathan, M. (1997) Behavioural effects during experimentally induced diabetes mellitus. PhD thesis, Banaras Hindu University, Varanasi, India.

Ramanathan, M., Khanna, V.K., Seth, P.K., Jaiswal, A.K. and Bhattacharya, S.K. (1999) Central neurotransmitter receptor binding and behaviour during streptozotocin-induced diabetes mellitus in rats. *Biogenic Amines* 15, 355–65.

Robertson, H.A. (1979) Benzodiazepine receptors in emotional and non-emotional mice: comparison of four strains. *Eur. J. Pharmacol.* 56, 163–6.

Rudzik, A.D., Hester, J.B., Tang, A.H., Stray, R.N. and Friss, W. (1973) The benzodiazepines, Raven Press, New York, pp. 285–97.

Saito, H., Nishiyama, N., Fujimori, H., Hinata, K., Kamegaya, T., Kato, Y. and Bao, T. (1984) In F. Usdin, R. Kvetnansky and I.J. Kopin (eds), *Stress: the role of catecholamines and neurotransmitters*, Gordan and Breach Science Publishers, New York, p. 467.

Salens, J.K., Kovacsic, G.B. and Allen, M.P. (1968) The influence of the adrenergic system on the 24-hours locomotor activity pattern in mice. *Archives Internationales de Pharmacodynamie et de Therapie* 173, 411–16.

Satyan, K.S. (1997) Chemical and neuropharmacological investigations on Indian *Gingko biloba* Linn. PhD thesis, Banaras Hindu University, Varanasi, India.

Satyavati, G.V., Gupta, A.K. and Tandon, N. (1987) *Medicinal plants of India*, Vol II, Indian Council of Medical Research, New Delhi, pp. 56–9.

Schindler, U., Rush, D.K. and Fielding, S. (1984) Nootropic drugs: animal models for studying effects on cognition. *Drug Dev. Res.* 4, 567–76.

Schumacher, G.A., Goodell, H., Hardy, J.D. and Wolff, H.G. (1940) Uniformity of the pain threshold in man. *Science* 92, 110–12.

Seligman, M.E. and Maier, S.F. (1967) Failure to escape traumatic shock. *J. Exp. Psychol.* 74, 1–9.

Seligman, M.E. and Beagley, G. (1975) Learned helplessness in the rat. *J. Compar. Physiol. Psychol.* 88, 534–41.

Sen, A.P. and Bhattacharya, S.K (1991) Effect of selective muscarinic receptor agonists and antagonists on active-avoidance learning acquisition in rats. *Indian J. Exp. Biol.* 29, 136–9.

Seth, P.K., Agrawal, A.K. and Bondy, S.C. (1981) Biochemical changes in the brain consequent to dietary exposure of developing and mature rats to chlordecone (Kepone). *Toxicol. Appl. Pharmacol.* 59, 262–7.

Seth, P.K., Alleva, F.R. and Balaz, T. (1982) Alterations of high-affinity binding sites of neurotransmitter receptors in rats after neonatal exposure to streptomycin. *Neurotoxicol.* 3, 13–20.

Shepherd, J.K., Grewal, S.S., Fletcher, A., Bill, D.J. and Dourish, C.T. (1994) Behavioural and pharmacological characterization of the elevate "zero maze" as an animal model of anxiety. *Psychopharmacology* 116, 56–64.

Sherman, A.D., Allers, G.L., Petty, F. and Henn, F.A. (1979) A neuropharmacologically-relevant animal model of depression. *Neuropharmacology* 18, 891–3.

Shibata, S., Tanaka, O., Shoji, J. and Saito, H. (1985) In B. Wagner, H. Hikino and N.R. Fransworth (eds), *Economic and medicinal plant research*, Academic Press, London, p. 217.

Shoji, M., Sato, H., Hirai, Y., Oguni, Y., Sugimoto, C., Morishita, S. and Ito, C. (1992) Pharmacological effects of Gosha-jinki-gan-ryo extract: effects on experimental diabetes. *Nippon Yakurigaku Zasshi* 99, 143–52.

Singh, H.K. and Dhawan, B.N. (1992) Drugs affecting learning and memory. In P.N. Tandon, V. Bijlani and S. Wadhwa (eds), *Lectures in neurobiology*, Wiley Eastern, New Delhi, pp. 189–202.

Soderpalm, B. and Engel, J.A. (1990) Serotonergic involvement in conflict behaviour. *Eur. J. Neuropsychopharmacol.* 1, 7–13.

Sommer, H. and Harrer, G. (1994) Placebo-controlled double-blind study examining the effectiveness of a *Hypericum* preparation in 105 mildly depressed patients. *J. Geriatric Psychiatry Neurol.* 7, S9–11.

Soubrie, P., Reisine, T.D. and Glowinski, J. (1984) Functional aspects of serotonin transmission in the basal ganglia, a review and an *in vivo* approach using the pushpull cannula technique. *Neurosci.* 131, 605–24.

Sparenberg, B., Demisch, L. and Holzl, J. (1993) Investigations of the antidepressive effects of St John's wort. *Pharmazeutische Zeitung Wissenschaften* 6, 50–4.

Speth, R.C., Bresolin, N. and Yamamura, H.I. (1979) Acute diazepam administration produces rapid increase in the brain benzodiazepine receptor density. *Eur. J. Pharmacol.* 59, 159–60.

Spignoli, G., Pedata, F., Giovannelli, I., Banfi, S., Moroni, F. and Pepeu, G.C. (1986a) Effect of oxiracetam and piracetam on central cholinergic mechanism and active avoidance acquisition. *Clin. Neuropharmacol.* 9, 539–57.

Spignoli, G. and Pepeu, G.C. (1986b) Oxiracetam prevents electroshock-induced decrease in brain acetylcholine and amnesia. *Eur. J. Pharmacol.* 126, 253–7.

Stahl, S.M. (1998) *Essential Psychopharmacology*. First South Asia edition, Cambridge University Press, UK, p. 326.

Suzuki, O., Katsumata Y., Oya, M., Bladt, S. and Wagner, H. (1984) Inhibition of monoamine oxidase by hypericin. *Planta Med.* 50, 272–4.

Takahashi, I., Nakanishi, S., Kobyashi, E., Nakano, H., Suzuki, K. and Tamaoki, T. (1989) Hypericin and pseudohypericin specifically inhibit protein kinase C: possible relation to their antiretroviral activity. *Biochem. Biophys. Res. Commun.* 165, 1207–12.

Taylor, D.P. (1990) Serotonin agents in anxiety. *Ann. N. Y. Acad. Sci.* 600, 545–57.

Thiebot, M.H., Martin, P. and Puech, A.J. (1992) Animal behavioural studies in the evaluation of antide-pressant drugs. *Br. J. Psychiatry* 160, 44–50.

Tripathi, Y.B., Pandey, E. and Dubey, G.P. (1999) Antioxidant property of *Hypericum perforatum* (L.) of Indian origin and its comparison with established Medhya rasayanas of Ayurvedic medicine. *Curr. Sci.* 76, 27–9.

Tripathi, Y.B. and Pandey, E. (2000) Role of alcoholic extract of shoot of *Hypericum perforatum* Linn on lipid perioxidation and various species of free radicals in rats. *Indian J. Exp. Biol.* 37, 567–71.

Turner, R.A. (1965) *Screening Methods in Pharmacology*, Academic Press, New York, p. 100.

Upton, R. (1997) American Pharmacopoeia and Therapeutics Compendium. *St John's wort monograph.* American Herbal Pharmacopoeia, Santa Cruz, CA, USA, p. 4, 27.

Vinegar, R., Schrieber, W. and Hugo, R. (1969) Biphasic development of carrageenan edema in rats. *J. Pharmacol. Exp. Therapeutics* 166, 96–103.

Vogel, W.H. (1985) Coping stress, stressors and health consequences. *Neuropsychobiology* 13, 129–35.

Vogel, G.H. and Vogel, W.H. (1997) *Drug Discovery and Evaluation: Pharmacological Assays*, Springer Verlag, Berlin, pp. 292–316, 413.

Volz, H.P. (1997) Controlled clinical trials of *Hypericum* extracts in depressed patients–an overview. *Pharmacopsychiatry* 30, 72–6.

Vorbach, E.U., Arnoldt, K.H. and Hubner, W.D. (1997) Efficacy and tolerability of St John's wort extract LI 160 versus imipramine in patients with severe depressive episodes according to ICD-10. *Pharmacopsychiatry* 30, 81–5.

Vorbach, E.U., Hubner, W.D. and Arnoldt, K.H. (1994) Effectiveness and tolerance of the *Hypericum* extract LI 160 in comparision with imipramine: randomized double-blind study with 135 outpatients. *J. Geriatric Psychiatry Neurol.* 7, S19–23.

Wada, T., Nakajima, R., Kurihara, E., Narumi, S., Masuoka, Y., Goto, G., Saji, Y. and Fukuda, N. (1989) Pharmacological characterization of a novel non-benzodiazepine selective anxiolytic. *J. J. Pharmacol.* 49, 337–49.

Wheatley, D. (1997) LI 160, an extract of St John's wort, versus amitriptyline in mildly to moderately depressed outpatients-controlled 6-week clinical trial. *Phamacopsychiatry* 30, 77–80.

Whitehouse, P.J., Sciulli, C.G. and Mason, R.M. (1997) Dementia drug development: use of information systems to harmonize global drug development. *Psychopharmacol. Bull.* 33, 129–33.

Willner, P. (1984) The validity of animal models of depression. *Psychopharmacology* (Berlin) 83, 1.

Willner, P. (1991) Anxiety, depression and mania animal models. In P. Soubrie (ed.), *Psychiatric disorders*, Karger, Basel, p. 71.

Winter, C.A. and Porter, C.C. (1957) Effect of alteration in side chain upon anti-inflammatory and liver glycogen activities of hydrocortisone ester. *J. Am. Pharm. Assoc. Scient.*, 46, 515–19.

Winter, C.A., Risley, E.A. and Nuss, G.W. (1962) Carrageenan-induced edema in hind paw of the rats as an assay for anti-inflammatory drugs. *Proc. Soc. Exp. Biol. Med.* 11, 544–7.

Winterhoff, H., Butterweck, V., Nahrstedt, A., Gumbinger, H.G., Schulz, S., Erping, S., Bosshammer, F. and Wieligmam, A. (1995) *Phytopharmaka in Forschung und klinischer anwendung*, Steinkopff-Verlag, Darmstadt, p. 39.

Woelk, H., Burkard, G. and Grunweeld, J. (1994) Benefits and risks of the *Hypericum* extract LI 160: drug-monitoring study with 3250 patients. *J. Geriatric, Psychiatry Neurol.* 7, S34–8.

Wolff, H.G., Hardy, J.D. and Goodell, H. (1940) Studies on pain. Measurement of the effect of morphine, codeine and other opiates on the pain threshold and an analysis of their relation to the pain experience. *J. Clin. Inv.* 19, 659–80.

Woolfe, G. and MacDonald, A.D. (1944) The evaluation of the analgesic action of pethidine hydrochloride (DEMEROL). *J. Pharmacol. Exp. Therapeut.* 80, 300–7.

Yamamura, H.I., Enna, S.J. and Kuhar, M.J. (eds) (1978) *Neurotransmitter Receptor Binding*. Raven Press, New York.

Zimer, P.O., Wynn, R.L., Ford, R.D. and Rudo, F.G. (1986) Effect of hot plate temperature on the antinociceptive activity of mixed opioid agonist antagonist compounds. *Drug Dev. Res.*, 7, 277–80.

14 *Hypericum* in the treatment of depression

C. Stevinson and E. Ernst

Following its long history of use as a folk remedy for depression, the herb *Hypericum perforatum* (St John's wort) has received a great deal of attention from scientists investigating its antidepressant effects. Over the past two decades, a growing number of clinical trials have been conducted in order to determine the efficacy of *Hypericum* in treating patients with depressive disorders.

Evidence from systematic reviews

In one of the first attempts to evaluate this evidence in a systematic manner, Ernst located 18 randomised trials of which 11 met criteria for adequate methodological quality (Ernst 1995). Based on these studies, the author concluded that *Hypericum* appeared to represent an effective antidepressant therapy and warranted further research.

The following year Linde and colleagues published a systematic review of 23 randomised double-blind trials including 1757 patients (Linde *et al.* 1996). A meta-analysis of 13 placebo-controlled trials ($n = 925$) indicated that there were significantly more treatment responders among patients taking *Hypericum* monopreparations than placebo (relative benefit: 2.67, 95% confidence interval [CI]: 1.78–4.01). The pooled result from three trials ($n = 317$) comparing *Hypericum* monopreparations with low-dose tricyclic antidepressants suggested that there was no significant difference in treatment effect (1.10, 95% CI: 0.93–1.31).

Some of the main criticisms of the studies that had been published up to this point centred on the vague classifications of depression resulting in highly heterogeneous patient groups and the short treatment periods (often 4 weeks or less) that provide no indication of relapse rates (De Smet and Nolen 1996). The authors of two subsequent systematic reviews applied stricter criteria for selecting studies in the attempt to provide more precise estimates of the treatment effect.

Kim and colleagues included only trials of monopreparations in which patients were diagnosed using well-defined criteria (Diagnostic and Statistical Manual of Mental Disorders, fourth edition [DSM-IV] or third edition revised [DSM-III-R] or International Classification of Diseases, tenth revision [ICD-10]) and assessed with the Hamilton Depression (HAM-D) scale (Kim *et al.* 1999). This resulted in two placebo-controlled trials ($n = 169$) and four comparative trials ($n = 482$) in the meta-analysis. Although smaller than that reported by Linde *et al.*, the effect of *Hypericum* was significantly different to placebo (relative benefit: 1.5, 95% CI: 1.0–1.9) and no different from the low-dose tricyclics (1.1, 95% CI: 0.9–1.3).

Williams *et al.* included only trials lasting at least 6 weeks in their review (Williams *et al.* 2000). Fourteen such studies were located with six placebo-controlled trials ($n = 419$) and six comparative trials ($n = 851$) in the meta-analysis. The results also indicated that *Hypericum* was superior to placebo (relative benefit: 1.9, 95% CI: 1.2–2.8) and similar to low-dose tricyclics (1.2, 95% CI:1.0–1.4). However, tests for publication bias suggested such bias existed hence treatment effects may be overestimated.

Nonetheless, these authors along with those of other reviews on the subject (Volz 1997, Wheatley 1998, Stevinson and Ernst 1999, Gastor and Holroyd 2000, Linde and Mulrow 2000, Nangia *et al.* 2000), concluded that on the basis of the overall body of evidence, *Hypericum* is an effective treatment for mild to moderate depression and may be as effective as low-dose tricyclic antidepressants.

Methodological issues

In addition to the imprecise diagnostic criteria and short time frames mentioned above, many trials had other methodological limitations that weakened the reliability of their results. Most did not provide an adequate description of randomisation methods, sample size calculation or blinding procedures. Many did not account for withdrawals or conduct intention-to-treat analyses. Some did not use accepted, validated outcome measures and some neglected to monitor compliance and adverse events. Trials comparing *Hypericum* with conventional medication have been criticised for using low doses of outdated antidepressants and for lacking sufficient statistical power to detect any difference that might exist between the treatments. More recent trials have been designed with greater rigor in order to provide more convincing data.

Subsequent equivalence trials

Philipp and colleagues conducted an 8-week multicentre trial testing *Hypericum* (STEI 300; 1050 mg daily) against imipramine (100 mg daily) and placebo (Philipp *et al.* 1999). Patients ($n = 263$) had a diagnosis of moderate depression according to ICD-10. The reduction from baseline of HAM-D scores for *Hypericum* (mean ± standard deviation [SD]: -15.4 ± 8.1) was significantly greater than for placebo (-12.1 ± 7.4) and was not significantly different to imipramine (-14.2 ± 7.3). Responder rates (defined as the number of patients with at least a 50% improvement from baseline) were 76% for *Hypericum*, 67% for imipramine and 63% for placebo.

Another comparison of *Hypericum* (ZE 117; 500 mg daily) with imipramine (150 mg daily) was performed by Woelk *et al.* (2000) in a 6-week multicentre trial (Woelk 2000). Patients ($n = 324$) were diagnosed as mild to moderately depressed according to ICD-10. The reduction in HAM-D scores for *Hypericum* (mean: 12.00) and imipramine (12.75) was not significantly different between the groups (-0.75, 95% CI: -1.90 to 0.40). Responder rates ($\geq 50\%$ improvement from baseline) were 43% for hypericum and 40% for imipramine.

Harrer and colleagues reported the first trial to test *Hypericum* against a modern antidepressant (Harrer *et al.* 1999). Patients ($n = 149$) diagnosed with mild to moderate depression according to ICD-10 criteria received either *Hypericum* (LoHyp-57; 400 mg daily) or fluoxetine (10 mg daily) for 6 weeks. The reduction in HAM-D scores was not significantly different between *Hypericum* (mean: -8.7) and fluoxetine (-9.07). The responder rate was 71% with *Hypericum* and 72% with fluoxetine.

A further comparison of *Hypericum* (ZE 117; 500 mg daily) and fluoxetine (20 mg daily) was conducted by Schrader *et al.* in a 6-week multicentre trial with mild to moderately depressed patients ($n = 238$) diagnosed according to ICD-10 (Schrader 2000). The reduction in HAM-D scores was not significantly different between *Hypericum* (mean: 8.11, 95% CI: 9.0–7.3) and fluoxetine (7.25, 95% CI: 8.1–6.4). The responder rate ($\geq 50\%$ improvement from baseline) was 60% for *Hypericum* and 40% for fluoxetine.

The methodological quality of these recent trials is considerably higher than the majority of the earlier ones. Of particular importance are the aspects of trial design that enable the therapeutic equivalence of treatments to be evaluated (Jones *et al.* 1996). The evidence behind suggestions that *Hypericum* is as effective as conventional antidepressants for treating mild to moderate depression is now more convincing, although not yet conclusive.

Sertraline has also been compared with *Hypericum* in a study by Brenner and colleagues although this was just a small pilot study, not an equivalence trial (Brenner *et al.* 2000). Mild to moderately depressed patients ($n = 30$) diagnosed according to DSM-IV were randomised to *Hypericum* (LI 160; 600 mg daily for 1 week then 900 mg daily for 6 weeks) or sertraline (50 mg daily for 1 week then 75 mg daily for 6 weeks). Reductions in HAM-D scores were not significantly different between *Hypericum* (mean ± SD: −8.4 ± 6.5) and sertraline (−9.1 ± 5.2). Responder rates (≥50% improvement from baseline) were 47% with *Hypericum* and 40% with sertraline.

Subsequent placebo-controlled trials

Shelton and colleagues conducted a trial involving patients ($n = 200$) diagnosed with major depression according to DSM-IV and scoring at least 20 on the HAM-D scale (Shelton *et al.* 2001). They were randomised to *Hypericum* (LI 160; 900 mg daily) or placebo for 8 weeks, with the *Hypericum* dose increased to 1200 mg after 4 weeks in the absence of an adequate therapeutic response. Changes in HAM-D scores did not significantly differ between *Hypericum* (estimated endpoint mean: 14.2) and placebo (14.9). Response rates defined as a HAM-D score of 12 or less (representing at least a 50% improvement from baseline) were 27% for *Hypericum* and 19% for placebo.

The results of this trial along with those from earlier studies involving patients with moderate to severe depression (Osterheider *et al.* 1992, Vorbach *et al.* 1997), suggest that the antidepressant efficacy of *Hypericum* may be limited to mild to moderate depressive disorders. There is currently little evidence that it is effective for severe or other forms of depression, although a preliminary study suggested it may have some potential for treating seasonal affective disorder (Kasper 1997).

Two other placebo-controlled trials have been published as conference abstracts (Montgomery *et al.* 2000, Randløv *et al.* 2001) and although both report results in favour of *Hypericum* for mild to moderately depressed patients, the studies were most notable for the dramatic placebo responses observed. Considerable interest will surround the results of the ongoing trial in the United States testing *Hypericum* against placebo and sertraline in mild to moderately depressed patients (Holden 1998). The 6-month trial funded by the National Institutes of Health is probably the first to be independent of commercial sponsorship.

Safety profile

When evaluating the value of a treatment, efficacy data must be assessed in relation to evidence of risks. A systematic review of the evidence relating to the safety of *Hypericum* from 1998 (Ernst *et al.* 1998) included information about adverse events from clinical trials, drug monitoring studies, case reports and surveillance schemes. Collectively, the data indicated that *Hypericum* was well tolerated. Clinical trial data suggested that the incidence and nature of adverse events was similar to that of placebo. In an open study in which 3250 patients taking *Hypericum* (900 mg daily Jarsin® 300) were monitored over 4 weeks (Woelk *et al.* 1994), adverse events were reported in 79 individuals (2.4%). Gastrointestinal symptoms (nausea, abdominal pain, appetite loss, diarrhoea) were the most frequent complaints, followed by allergic reactions, fatigue, anxiety and dizziness. Forty eight patients (1.45%) withdrew from treatment. Three other drug monitoring studies produced similar results (Albrecht *et al.* 1994, Grube *et al.* 1997, Meier *et al.* 1997).

There were however no reliable data on long-term safety. Since then, two further open surveillance studies published as conference abstracts have provided data from a larger sample ($n = 11,296$) (Zeller 2000) and over a longer treatment period (12 months) (Woelk *et al.* 2000) confirming that *Hypericum* extracts are generally well tolerated.

In another systematic review, the safety profile of *Hypericum* was compared with those of several conventional antidepressants (Stevinson and Ernst 1999). On the basis of the available evidence, *Hypericum* appeared to be better tolerated than synthetic antidepressants, but the most notable finding of the review was the relative lack of reliable information about the safety of *Hypericum*.

The subsequent publication of various case reports suggest that more serious adverse events of *Hypericum* are possible including subacute toxic neuropathy (Bove 1998), psychotic relapse in schizophrenia (Lal and Iskandar 2000), delirium (Khawaja *et al.* 1999), serotonin syndrome (Parker 2001) and hair loss (Parker 2001). Several cases of hypomania have also been published (Stevinson and Ernst submitted). These are all recognised as adverse effects of conventional antidepressant drugs suggesting that, unsurprisingly, *Hypericum* possesses some of the adverse pharmacological effects of synthetic antidepressants along with the therapeutic properties.

Similarly, some of the drug interactions associated with modern antidepressants (Nemeroff *et al.* 1996) also apply to *Hypericum*. Preliminary evidence first emerged from case reports, and *in vivo* and pharmacokinetic studies indicating that *Hypericum* affects the metabolism of several concomitant medications resulting in reduced plasma concentrations of the drug (Ernst 1999). Further experimental studies have confirmed that *Hypericum* is a potent inducer of several cytochrome P450 enzymes (Roby *et al.* 2000) and the transport protein P-glycoprotein (Durr *et al.* 2000). This has implications for patients taking a wide range of prescribed medications and pre-clinical studies have demonstrated the potential for interactions with digoxin (Johne *et al.* 1999), indinavir (Piscitelli *et al.* 2000) and amitriptyline (Roots *et al.* 2000). Cases of acute heart transplant rejections of patients on ciclosporin, (Barone *et al.* 2000, Breidenbach *et al.* 2000, Karliova *et al.* 2000, Rushitzka *et al.* 2000), reduced anticoagulant effects of warfarin and inter-menstrual bleeding with oral contraceptives (Yue *et al.* 2000) have all been reported, illustrating the potentially serious consequences of these herb–drug interactions.

Other such interactions that are particularly relevant to *Hypericum* involve serotonin reuptake inhibitors and triptans for treating migraine. Cases have been reported of elderly patients experiencing symptoms characteristic of serotonin syndrome after taking *Hypericum* and sertraline concurrently (Lantz *et al.* 1999). No cases of harmful interactions with alcohol have been identified and randomized trials have suggested no detrimental effects at normal therapeutic doses in healthy volunteers (Schmidt *et al.* 1993, Herberg *et al.*, 1994, Friede *et al.* 1998).

Concerns about photosensitivity were raised by reports of toxic reactions in light-skinned grazing animals exposed to bright sunlight after consuming large quantities of *Hypericum* flowers (Bombardelli and Morazzoni 1995). At the therapeutic doses used for treating depression, risks to humans appear low. A few cases of reversible photosensitive reactions have been reported (Golsch *et al.* 1997, Lane-Brown 2000) but a pharmacokinetic study indicated increased sensitivity to ultraviolet light only with very high doses (3600 mg containing 11.3 mg total hypericin) (Brockmöller *et al.* 1997).

As with other herbs, the safety of *Hypericum* during pregnancy and lactation has not been systematically investigated and it should therefore be avoided at these times (Grush *et al.* 1998).

Conclusion

Weighing up the existing evidence of the benefits and risks of *Hypericum*, it is possible to conclude that when taken without concomitant medication, the herb is an effective and well tolerated treatment for mild to moderate depression. It appears that it may be similarly effective as conventional antidepressants with the possible advantage of superior tolerability. However, *Hypericum* does not represent a risk-free therapy and its efficacy in the long-term has not been established. Further evidence is required in order to define more precisely the potential role of *Hypericum* in the treatment of depression.

References

Albrecht, M., Hübner, W.D., Podzuweit, H. and Schmidt U. (1994) St. John's wort extract in the treatment of depression: observations on the use of Jarsin capsules. *Der Kassenarzt* 41, 45–54.

Barone, G.W., Gurley, B.J., Ketel, B.L., Lightfoot, M.L. and Abul-Ezz, S.R. (2000) Drug interaction between St. John's wort and cyclosporine. *Ann. Pharmacol.* 34, 1013–16.

Bombardelli, E. and Morazzoni, P. (1995) *Hypericum perforatum. Fitoterapia* 66, 43–68.

Bove, G.M. (1998) Acute neuropathy after exposure to sun in a patient treated with St. John's wort. *Lancet* 352, 1121–2.

Breidenbach, Th., Hoffmann, M.W., Becker, Th., Schlitt, H. and Klempnauer, J. (2000) Drug interaction of St. John's wort with cyclosporin. *Lancet* 355, 1912.

Brenner, R., Azbel, V., Madhusoodanan, S. and Pawlowska, M. (2000) Comparison of an extract of hypericum (LI 160) and sertraline in the treatment of depression: a double-blind randomized pilot study. *Clin. Therapeut.* 22, 411–19

Brockmöller, J., Reum, T., Bauer, S., Kerb, R., Hübner, W. and Roots, I. (1997) Hypericin and pseudohypericin: pharmacokinetics and effects on photosensitivity in humans. *Pharmacopsychiatry* 30(Suppl), 94–101.

De Smet PAGM and Nolen, W.A. (1996) St. John's wort as an antidepressant. *Br. Med. J.* 313, 241–2.

Durr, D., Steiger, B., Kullak-Ublick, G.A., Rentsch, K.M., Steinert, H.C., Meier, P.J. and Fattinger, K. (2000) St John's wort induces intestinal P-glycoprotein/MDR1 and intestinal and hepatic CYP3A4. *Clin. Pharmacol. Ther.* 68, 598–604.

Ernst, E. (1995) St. John's wort, an anti-depressant? A systematic, criteria-based review. *Phytomed.* 2, 67–71.

Ernst, E. (1999) Second thoughts about safety of St. John's wort. *Lancet* 354, 2014–16.

Ernst, E., Rand, J.I., Barnes, J. and Stevinson, C. (1998) Adverse effects profile of the herbal antidepressant St. John's wort (*Hypericum perforatum* L.) *Eur. J. Clin. Pharmacol.* 54, 589–94.

Friede, M., Hasenfuss, I. and Wüsternberg, P. (1998) Alltagssicherheit eines pflanzlichen Antidepressivums aus Johanniskraut. *Fortschr. Med.* 116, 131–5.

Gastor, B. and Holroyd, J. (2000) St. John's wort for depression: a systematic review. *Arch. Intern. Med.* 160, 152–6.

Golsch, S., Vocks, E., Rakoski, J., Brockow, K. and Ring, J. (1997) Reversible Erhöhung der Photosensitivität im UV-B-Bereich durch Johanniskrautextrakt-Präparate. *Hautarzt* 48, 249–52.

Grube, B., Schermuck, S., Hopfenmuller, W. and Michel, F. (1997) Use of a *Hypericum* extract in mild, transient depressive disorders. *Eur. J. Clin. Res.* 9, 293–302.

Grush, L.R., Nierenberg, A., Keefe, B. and Cohen, L.S. (1998) St. John's wort during pregnancy. *JAMA* 280, 1566.

Harrer, G., Schmidt, Kuhn U. and Biller, A. (1999) Comparison of equivalence between the St. John's wort extract LoHyp-57 and fluoxetine. *Arzneim Forsch/Drug Res.* 49, 289–96.

Herberg, K.W. (1994) Testing of psychotropic herbaceous agents: alternative to psychopharmacotherapy? *Therapiewoche* 44, 704–13.

Holden, C. (1998) NIH to explore St. John's wort. *Science* 278, 91.

Johne, A., Brockmöller, J., Bauer, S., Maurer, A., Langheinrich, M. and Roots, I. (1999) Pharmacokinetic interaction of digoxin with an herbal extract from St. John's wort (*Hypericum perforatum*). *Clin. Pharmacol. Ther.* 66, 338–45.

Jones, B., Jarvis, P., Lewis, J.A. and Ebbut, A.F. (1996) Trials to assess equivalence: the importance of rigorous methods. *Br. Med. J.* 313, 36–9.

Karliova, M., Treichel, U., Malagò, M., Frilling, A., Gerken, G. and Broelsch, C.E. (2000) Interaction of *Hypericum perforatum* (St. John's wort) with cyclosporin A metabolism in a patient after liver transplantation. *J. Hepatol.* 33, 853–5.

Kasper, S. (1997) Treatment of seasonal affective disorder (SAD) with hypericum extract. *Pharmacopsychiatry* 30(Suppl): 89–93.

Khawaja, I.S., Marotta, R.F. and Lippermann, S. (1999) Herbal medicines as a factor in delirium. *Psychiatric Services* 50, 969–70.

Kim, H.L., Streltzer, J. and Goebert, D. (1999) St. John's wort for depression: a meta-analysis of well-defined clinical trials. *J. Nerv. Ment. Dis.* 187, 532–9.

Lal, S. and Iskandar, H. (2000) St. John's wort and schizophrenia. *Can. Med. Assoc. J.* 163, 262–3.

Lane-Brown, M.M. (2000) Photosensitivity associated with herbal preparations of St. John's wort (*Hypericum perforatum*). *Med. J. Aust.* 172, 302.

Lantz, M.S., Buchalter, E. and Giambanco, V. (1999) St. John's wort and antidepressant drug interactions in the elderly. *J. Geriatr. Psychiatr. Neurol.* 12, 7–10.

Linde, K. and Mulrow, C.D. (2000) St. John's wort for depression (Cochrane Review). In The Cochrane Library, Issue 1, 2000. Oxford: Update Software.

Linde, K., Ramirez, G., Mulrow, C.D., Pauls, A., Weidenhammer, W. and Melchart, D. (1996) St. John's wort for depression – an overview and meta-analysis of randomised clinical trials. *Br. Med. J.* 313, 253–8.

Meier, B., Liske, E. and Rosinus, V. (1997) Wirksamkeit und Verträglichkeit eines standardisierten Johanniskraut-Vollextrakts (ZE 117) bei Patienten mit depressiver Symptomatik unterschiedlicher Schweregrade-eine Anwendebeobachtung. *Forsch Komplementärmed* 4, 87–93.

Montgomery, S.A., Hübner, W.D. and Grigoleit H.G. (2000) Efficacy and tolerablility of St. John's wort extract compared with placebo in patients with a mild to moderate depressive disorder. *Phytomed.* Suppl II:107.

Nangia, M., Syed, W. and Doraiswamy, P.M. (2000) Efficacy and safety of St. John's wort for the treatment of major depression. *Pub. Health Nutr.* 3, 487–94.

Nemeroff, C.B., De Vane, C.L. and Pollock, B.G. (1996) Newer antidepressants and the cytochrome P450 system. *Am. J. Psychiatry* 153, 311–20.

Osterheider, M., Schmidtke, A. and Beckmann, H. (1992) Behandlung depressiver Syndrome mit Hypericum (Johanniskraut) -eine placebokontrollierte Doppelblindstudie. *Fortschr Neurologie/Psychiatrie* 60(Suppl 2), 210–11.

Parker, V., Wong, A.H.C., Boon, H.S. and Seeman, M.V. (2001) Adverse reactions to St. John's wort. *Can. J. Psychiatry* 46, 77–9.

Philipp, M., Kohnen, R. and Hiller, K-O. (1999) Hypericum extract versus imipramine or placebo in patients with moderate depression: randomised multicentre study of treatment for eight weeks. *Br. Med. J.* 319, 1535–9.

Piscitelli, S.C., Burstein, A.H., Chaitt, D., Alfaro, R.M. and Falloon, J. (2000) Indinavir concentrations and St. John's wort. *Lancet* 355, 547–8.

Randløv, C., Thomsen, C., Winther, K. and Mehlsen, J. (2001) Effects of hypericum in mild to moderately depressed outpatients: a placebo-controlled clinical trial. *Altern. Ther. Health Med.* 7, 108–9.

Roby, C.A., Anderson, G.D., Kantor, E., Dryer, D.A. and Burstein, A.H. (2000) St. John's wort: effect on CYP3A4 activity. *Clin. Pharmacol. Ther.* 67, 451–7.

Roots, I., Johne, A., Schmider, J., Brockmöller, J., Maurer, A., Störmer, E. and Donath, F. (2000) Interaction of a herbal extract from St. John's wort with amitriptyline and its metabolites. *Clin. Pharmacol. Ther.* 67, 159.

Rushitzka, F., Meier, P.J., Turina, M., Lüscher, T.F. and Noll, G. (2000) Acute heart transplant rejection due to St. John's wort. *Lancet* 355, 348–9.

Schrader, E. (2000) On behalf of the Study Group. Equivalence of St. John's wort extract (Ze 117) and fluoxetine: a randomised controlled study in mild-moderate depression. *Int. Clin. Psychopharmacol.* 15, 61–8.

Schmidt, U., Harrer, G., Kuhn, U., Berger-Deinert, W. and Luther, D. (1993) Wechselwirkungen von Hypericum-Extrakt mit Alkohol. *Nervenheilkunde* 6, 314–19.

Shelton, R.C., Keller, M.B., Gelenberg, A., Dunner, D.L., Hirschfeld, R., Thase, M.E., Russell, J., Lydiard, R.B., Crits-Cristoph, P., Gallop, R., Todd, L., Hellerstein, D., Goodnick, P., Keitner, G., Stahl, S.M. and Halbreich, U. (2001) Effectiveness of St. John's wort in major depression: a randomised controlled trial. *JAMA* **285**, 1978–86.

Stevinson, C. and Ernst, E. (1999) *Hypericum* for depression: an update of the clinical evidence. *Eur. Neuropsychopharmacol.* **9**, 501–5

Stevinson, C. and Ernst, E. (1999) Safety of hypericum in patients with depression: a comparison with conventional antidepressants. *CNS Drugs* 11, 125–32.

Stevinson, C. and Ernst, E. Mania induced by St. John's wort? (Submitted).

Volz, H.P. (1997) Controlled clinical trials of *Hypericum* extracts in depressed patients: an overview. *Pharmacopsychiatry* **30**(Suppl), 72–6.

Vorbach, E., Arnoldt, K. and Hübner, W. (1997) Efficacy and tolerability of St. John's Wort extract LI 160 versus imipramine in patients with severe depressive episodes according to ICD-10. *Pharmacopsychiatry* **30**(Suppl), 81–5.

Wheatley, D. (1998) *Hypericum* extract: potential in the treatment of depression. *CNS Drugs* 9, 431–40.

Williams, J.W., Mulrow, C.D., Chiquette, E., Hitchcock, Noël. P., Aguilar, C. and Cornell, J. (2000) A systematic review of newer pharmacotherapies for depression in adults; evidence report summary. *Ann. Intern. Med.* **132**, 743–56.

Woelk, H. (2000) For the Remotiv/imipramine Study Group. Comparison of St. John's wort and imipramine for treating depression: randomised controlled trial. *Br. Med. J.* **32**, 536–9.

Woelk, H., Beneke, M., Gebert, I., Rappard, F. and Rechziegler, H. (2000) *Hypericum* extract ZE 117: open long-term study in patients with mild-moderate depression. *Phytomed.* Suppl II,109.

Woelk, H., Burkard, G. and Grünwald J. (1994) Benefits and risks of the *Hypericum* extract LI 160. Drug monitoring study with 3250 patients. *J. Geriatr. Psychiatry Neurol.* 7(Suppl I), 34–8.

Yue, Q-Y., Bergquist, C. and Gerdén, B. (2000) Safety of St. John's wort (*Hypericum perforatum*). *Lancet* **355**, 576–7.

Zeller, K. (2000) A convincing safety profile of St. John's wort extract (Laif®600) showed in a large post marketing surveillance. *Phytomed.* Suppl II, 107.

Index

Index